Algebras in Analysis

Algebras in Analysis

PROCEEDINGS OF AN INSTRUCTIONAL CONFERENCE
ORGANIZED BY THE LONDON MATHEMATICAL SOCIETY
AND THE UNIVERSITY OF BIRMINGHAM
(A NATO ADVANCED STUDY INSTITUTE)

Edited by
J. H. WILLIAMSON
Department of Mathematics, University of York,
York, England

1975
ACADEMIC PRESS
London · New York · San Francisco
A Subsidiary of Harcourt Brace Jovanovich, Publishers

ACADEMIC PRESS INC. (LONDON) LTD.
24/28 Oval Road,
London NW1

United States Edition published by
ACADEMIC PRESS INC.
111 Fifth Avenue
New York, New York 10003

Library of Congress Catalog Card Number: 74 18509
ISBN: 0 12 757150 7

Printed in Great Britain by
PAGE BROS (NORWICH) LTD
Mile Cross Lane
Norwich

Contributors

I. G. Craw *Department of Mathematics, University of Aberdeen, Aberdeen, Scotland.*

R. E. Harte *Department of Mathematics, University College, Cork, Ireland.*

H. Helson *Department of Mathematics, University of California, Berkeley, California, U.S.A.*

B. E. Johnson *Department of Mathematics, University of Newcastle, England.*

R. V. Kadison *Department of Mathematics, University of Pennsylvania, Philadelphia, U.S.A.*

C. F. Skau *Department of Mathematics, University of Oslo, Norway.*

J. L. Taylor *Department of Mathematics, University of Utah, Salt Lake City, Utah, U.S.A.*

L. Waelbroeck *Department of Mathematics, Free University of Brussels, Belgium.*

Preface

The London Mathematical Society organized, in conjunction with the Scientific Affairs Committee of NATO (as part of the Advanced Study Institutes programme) and the University of Birmingham, an Instructional Conference on some parts of functional analysis, held in Birmingham from 3 to 22 September 1973.

The aim of the meeting was to provide an opportunity for both students of mathematics and for established mathematicians to learn something of recent developments in functional and harmonic analysis that are likely to be of importance during the next few years. There were some formal courses of lectures, and also informal discussion and seminar sessions, to which various members of the Conference contributed.

This volume contains the invited lectures and one or two of the seminar talks (in some cases in expanded form). The lectures vary in the background assumed and in the level of sophistication required of the reader; it is hoped that all may be found useful as introductions to current developments in analysis.

Thanks are due to all those who contributed to the mathematical programme of the Conference, notably the five invited lecturers and those members who took part in the seminar sessions. Thanks are also due to many who helped in the organization and running of the Conference, and in the preparation of manuscripts for the press. In particular mention should be made of Mr Peter Ludvik, for invaluable editorial assistance, of Mrs Judy Jackson, for her efficient and patient typing and re-typing of material; above all of Dr H. C. Wilkie, the Conference Secretary, for his unobtrusive but masterly organisation of the Conference itself.

February 1975 J. H. WILLIAMSON

Contents

CONTRIBUTORS ... ... *v*
PREFACE ... ... *vii*

PART I. INVITED LECTURES

1. Analyticity on Compact Abelian Groups

HENRY HELSON

Introduction 2
1. Groups with Ordered Duals... 3
2. Groups with Archimedean Duals 17
3. Analyticity 28
4. The Structure of Cocycles 37
5. Special Results 50
6. Cocycles and Ergodic Theory 57
References 60

2. Banach Algebras: Introductory Course

B. E. JOHNSON

1. The Definition 63
2. Examples 64
3. Commutative Banach Algebras—Spectrum and Resolvent ... 65
4. The One-variable Functional Calculus 70
5. Regularity and the Hull-kernel Topology 72
6. Non-commutative Banach Algebras 74
7. Modules and Irreducible Representations 75
8. Uniqueness of Norm for Banach Algebras 78
9. Approximate Units and Factorisation 80

3. Introduction to Cohomology in Banach Algebras

B. E. JOHNSON

1. Extensions of Banach Algebras 84
2. Derivations and Automorphisms of Banach Algebras 87

3. The Hochschild Cohomology Groups 89
4. Cohomology and Fixed Points 90
5. Reflexive Modules and Modules with a Reflexive Auxiliary Norm 91
6. Amenable Groups and Algebras 94
7. Modules which are not Dual Modules 95
8. Helemskiĭ's Cohomology Theory 96
References 99

4. Operator Algebras

R. V. KADISON

Introduction 101
1. Some C^*-algebra Basics 102
2. Von Neumann Algebra Basics 107
3. Algebraic Structure in von Neumann Algebras 112
4. Action of von Neumann Algebras on Spaces 116

5. Banach Algebras and Topology

J. L. TAYLOR

Introduction 118
1. Axioms for Cohomology 120
2. Consequences of the Axioms 126
3. Čech Cohomology 131
4. Idempotents and Logarithms 138
5. The Functor K_1 144
6. The Functor K_0 148
7. Further Properties of K_0 154
8. Relations between K_0 and K_1 159
9. The Bott Periodicity Theorem 163
10. K-theory 169
11. Applications of Cohomology to Harmonic Analysis 175
12. K-theory and Harmonic Analysis 181
References 185

6. The Holomorphic Functional Calculus and Non-Banach Algebras

L. WAELBROECK

Introduction 187

Part I

1. Holomorphic Functional Calculus in Banach Algebras 189

Part II

2. Spaces and Algebras with Bounded Structures 197

3. Examples 200
4. Commuative b-algebras with Idempotent Boundedness ... 207
5. Regular Elements and Equiregular Sets 209
6. Continuous Inverse Algebras 215
7. Locally Pseudo-convex Spaces and Algebras 217
Part III
8. Heuristic Considerations 221
9. Indentities Involving Differential Forms 225
10. The Spectrum 234
11. The Holomorphic Functional Calculus and Non-compact Spectra 241
12. Applications 244
References 249

PART II. ABSTRACTS AND SEMINAR TALKS

Abstracts 255

Seminar Talks 265
The Gleason Parts of A(D^) *by* I. G. CRAW 265
A Silov Boundary for Systems? *by* R. E. HARTE 268
Orthogonal Measures on the State Space of a C^***-algebra** *by* C. F. SKAU 272

INDEX 305

PART I

INVITED LECTURES

1. Analyticity on Compact Abelian Groups

H. HELSON

Department of Mathematics
University of California
Berkeley, California, USA

Introduction ... 2
1. Groups with Ordered Duals ... 3
   1.1. The groups ... 3
   1.2. Beurling's first theorem ... 4
   1.3. Szegö's theorem ... 6
   1.4. Consequences of Szegö's theorem ... 8
   1.5. Generalization to L^p, $p \neq 2$ ... 10
   1.6. Invariant subspaces in L^p, $p \geqslant 1$ ... 12
   1.7. Analytic measures ... 13

2. Groups with Archimedean Duals ... 17
   2.1. The flow on K ... 17
   2.2. Almost periodic functions ... 19
   2.3. Spectral resolution of an invariant subspace ... 20
   2.4. Spectral resolution (continued) ... 25
   2.5. An existence theorem ... 26

3. Analyticity ... 28
   3.1. The spaces L^2_v, H^2_v ... 28
   3.2. Cocycles and subspaces ... 31
   3.3. The condition for $\mathcal{M}_f$ to be simply invariant ... 34
   3.4. The modulus of an analytic function ... 36

4. The Structure of Cocycles ... 37
   4.1. The problem ... 37
   4.2. A criterion for triviality ... 38
   4.3. The counterexample ... 39
   4.4. Refinements ... 41
   4.5. The differential equation ... 42
   4.6. An equivalence theorem ... 44
   4.7. A representation theorem ... 46

4.8. Versions of cocycles 47
4.9. Gamelin's representation theorem 48

5. Special Results 50
5.1. Continuous cocycles 50
5.2. The theorem of Wagner and Milaszewicz 53
5.3. Generators of invariant subspaces 54
5.4. Unsolved problems 56

6. Cocycles on a Flow 57
6.1. Analyticity on a flow 57
6.2. Cocycles and ergodic theory 59
References 60

INTRODUCTION

This is an expanded version of lectures given at the Conference. The subject has grown too large for this author to write a comprehensive account of it. The necessary sacrifice has been made at both ends: there is little historical perspective, and the difficult task of surveying the present state of affairs has been carried out less thoroughly than intended.

For the connections with older work it is fortunate that three excellent books [**15**, **26**, **38**] cover parts of the subject and show where it came from. At the other end, at least references will be made to much recent work. In the area covered the exposition is intended to be detailed enough to follow. Most of the results are from a series of papers by this author and collaborators. Many proofs have been simplified, and this perhaps justifies writing a connected account of the theory.

The first chapter treats Szegö's theorem and its many consequences, all valid on a compact abelian group whose dual is linearly ordered.

Section 2 specializes to groups whose duals are archimedean-ordered. A good deal of machinery is developed: almost periodic functions, cross sections of a flow, the unitary group and the cocycle associated with a simply invariant subspace. A proof is given for the recent result that every simply invariant subspace contains non-null functions of constant modulus.

Section 3 deals with analyticity. The condition for a function in L^2 to generate a simply invariant subspace is obtained. Inclusion of invariant subspaces is related to analyticity of the quotient of their cocycles. The modulus of functions in H^2 is characterized, and the theorem of Arens on the modulus of a continuous analytic function is proved.

Section 4 treats the structure of cocycles. Non-trivial cocycles are constructed and the spectrum of the associated spectral resolutions studied. Two representation theorems (one due to Gamelin) are obtained. Gamelin's theorem that every cocycle has a strict version is given a simpler proof.

In Section 5 we collect several theorems that seem interesting and promising, but that do not yet fit into a larger pattern.

The last Section shows how these results make sense on a flow of general type. Some technical difficulties arise, but probably the main theorems are true very generally. References are made to work already published in this direction. On the other hand there is a possibility of fruitful application of techniques here to ergodic theory, but these applications are just beginning.

This paper is mainly expository; however the following theorems are being published here for the first time: 26, 30, 31, 32, 33, 34 with its corollary. References to papers written by this author (alone or with a coauthor) have systematically been omitted. Classical results about Fourier series and transforms have been used freely. An excellent source for this information is [**28**].

The author expresses his gratitude to Paul Muhly for friendly advice in connection with this writing.

1. GROUPS WITH ORDERED DUALS

1.1. The groups

K is a compact abelian group dual to a discrete group Γ. Elements of K are called x, y; those of Γ are denoted by λ, τ. The Haar measure σ on K is finite, and normalized so that $\sigma(K) = 1$. Each x in K is a function on Γ with values $x(\lambda)$; for λ in Γ let χ_λ be the character on K defined by $\chi_\lambda(x) = x(\lambda)$. The duality of compact and discrete abelian groups is supposed to be known.

Lebesgue spaces are always on K and based on $d\sigma$, unless otherwise mentioned. A function f in L^1 has Fourier coefficients and Fourier series

$$a_\lambda(f) = \int f(x)\,\overline{\chi_\lambda(x)}\,d\sigma(x), \qquad f(x) \sim \sum a_\lambda(f)\,\chi_\lambda(x). \tag{1}$$

The range of integration or of summation will be omitted when, as here, it is all of K or of Γ. The elementary machinery of Fourier analysis is assumed known: Fejér's theorem, the existence of approximate identities, properties of convolution, the Parseval relation. This information can be found in [**28**] or [**38**].

We assume that Γ is *ordered.* That is, a semigroup Γ_+ is given in Γ such that Γ is the disjoint union $\Gamma_+ \cup \{0\} \cup \{-\Gamma_+\}$. The elements of Γ_+ are *positive*; those of $-\Gamma_+$ are *negative.* A function f in L^1 is called *analytic* if $a_\lambda(f) = 0$ for all $\lambda < 0$. H^p $(1 \leqslant p \leqslant \infty)$ is the subspace of L^p consisting of all analytic functions in L^p. For technical reasons it is important also to define H^p_0, the set of functions in H^p with $a_0 = 0$.

The first example of such a pair of groups is naturally the circle T and the integer group Z in its natural order. In this chapter we shall be extending results known in this case. However we shall mainly be interested in cases where Γ has no smallest positive element, and beginning in the next chapter we shall assume that Γ is a subgroup of the real line, not isomorphic to Z.

The second example is the two-dimensional torus T^2, dual to the lattice group Z^2. The character pairing is given by

$$((e^{ix}, e^{iy}), (m, n)) = e^{i(mx+ny)}, \tag{2}$$

and Fourier series have the familiar form

$$f(e^{ix}, e^{iy}) \sim \sum a_{mn} e^{i(mx+ny)}. \tag{3}$$

An order relation can be introduced into Z^2 in two essentially different ways, and these two kinds of ordering illustrate the possibilities that have to be dealt with in generalizing from the circle group. Let L be a line with irrational slope in the plane, containing the origin. We take for Z^2_+ all lattice points on one side of L. Analytically, Z^2_+ is the set of (m, n) such that $m\alpha + n\beta > 0$ for a certain pair of real numbers α, β having irrational ratio. The mapping that carries (m, n) to $\lambda = m\alpha + n\beta$ is an order-preserving isomorphism of Z^2 with a subgroup of R (the real number system). This order is archimedean, and Z^2_+ has no smallest element.

The other kind of order is obtained if L has rational slope. Now Z^2_+ is to contain the lattice points on one side of L, together with those on one ray of L from the origin. This order is non-archimedean, and Z^2 has a smallest positive element.

For each order relation we obtain a class of analytic functions. The theorems of this chapter are valid for all order relations, and those of succeeding chapters at least for all archimedean orders. Put another way, these theorems are not fine enough to distinguish among groups or among different orders in the same group. Now the additive group of rational numbers is a group Γ quite different from Z^2, and it would be interesting to find differences in their function theory.

An *invariant subspace* is a closed subspace of L^2 that contains with each of its elements f also $\chi_\lambda f$ for all *positive* λ in Γ. Such subspaces were first studied on the circle by Beurling [**5**]. The description of invariant subspaces is the main problem we shall study.

1.2. Beurling's first theorem

Let $\mathscr{M}$ be an invariant subspace on K. Set $\mathscr{M}_\lambda = \chi_\lambda \cdot \mathscr{M}$ (the set of all products $\chi_\lambda f$ with f in $\mathscr{M}$). The family of subspaces $\mathscr{M}_\lambda$ decreases as λ increases in Γ. We define

$$\mathscr{M}_+ = \bigcap_{\lambda<0} \mathscr{M}_\lambda, \qquad \mathscr{M}_- = \text{closure of} \bigcup_{\lambda>0} \mathscr{M}_\lambda. \tag{4}$$

Then $\mathscr{M}_{-} \subset \mathscr{M} \subset \mathscr{M}_{+}$, and $\mathscr{M}$ may coincide with either or both of its versions $\mathscr{M}_{+}$, $\mathscr{M}_{-}$.

THEOREM 1. *Let $\mathscr{M}$ be an invariant subspace larger than $\mathscr{M}_{-}$. Then $\mathscr{M} = q \,.\, H^2$, where q is measurable on K and $|q(x)| = 1$ almost everywhere.*

Beurling proved this theorem for the case that $\mathscr{M}$ is contained in $H^2(T)$, as a corollary of a deeper result that we shall call Beurling's second theorem [**5**]. In $H^2(T)$ the hypothesis that $\mathscr{M} \neq \mathscr{M}_{-}$ is redundant. The discovery that the first theorem is easy to prove using the geometry of Hilbert space has been important for progress in several directions. The earliest relevant papers are [**18, 25**].

To prove the theorem, find a function q of norm one in $\mathscr{M}$ orthogonal to $\mathscr{M}_{-}$. Then q is orthogonal to $\chi_\lambda q$ for positive λ in Γ:

$$\int |q|^2 \chi_\lambda \, d\sigma = 0 \qquad (\lambda > 0). \tag{5}$$

By taking the complex conjugate of this relation we find the same formula for $\lambda < 0$. That is, $a_\lambda(q) = 0$ except for $\lambda = 0$. This implies that $|q|$ is constant, necessarily $|q| = 1$ a.e. Now q is in $\mathscr{M}$, and multiplication by q preserves norm in L^2. Hence $q \,.\, H^2$ is a closed subspace of L^2, contained in $\mathscr{M}$.

Since q is orthogonal to $\chi_\lambda \cdot \mathscr{M}$ ($\lambda > 0$), we have $\chi_\lambda q$ orthogonal to $\mathscr{M}$ ($\lambda < 0$). That is, $\chi_\lambda q$ is in $\mathscr{M}$ for non-negative λ, and orthogonal to $\mathscr{M}$ for negative λ. The family $\{\chi_\lambda q\}$ is an orthonormal basis for L^2; hence $q \cdot H^2$ fills $\mathscr{M}$, and the proof is finished. (The final simple form of this proof is due to T. P. Srinivasan.)

The converse of the theorem is trivial: if q is a unitary function on K, then $(q \cdot H^2)_{-} = q \cdot H_0^2 \neq q \cdot H^2$.

For any f in L^2 let $\mathscr{M}_f$ be the smallest invariant subspace containing f. Thus $\mathscr{M}_f$ is the closed span of the functions $\chi_\lambda f$, $\lambda \geqslant 0$. On applying the theorem just proved to $\mathscr{M}_f$ we find the following generalization of a classical theorem.

COROLLARY. *Let f belong to H^2 and have a first non-vanishing Fourier coefficient. Then $f(x) \neq 0$ a.e.*

The hypothesis that f has a first non-vanishing Fourier coefficient implies that $\mathscr{M}_f \neq (\mathscr{M}_f)_{-}$. Hence $\mathscr{M}_f = q \cdot H^2$ for some unitary function q. But all the functions in $\mathscr{M}_f$ vanish almost everywhere on the set where f does; at the same time q is in $\mathscr{M}_f$ and vanishes on no set of positive measure. The result follows.

On the circle every non-null function in H^2 has a first non-vanishing Fourier coefficient, and so the conclusion of the theorem holds. Usually this fact is expressed by saying that the radial limits of a function in $H^2(T)$ vanish at most on a null set, unless the function is null. Our formulation shows that the existence of radial limits (a relatively deep fact) is not really involved in the question of where the boundary function can vanish.

An analytic unitary function is called *inner*. A function f in H^2 such that $\mathscr{M}_f = H^2$ is called *outer*. (The terminology is Beurling's.) Theorem 1 yields a second

COROLLARY. *Let f belong to H^2 and have a first non-vanishing Fourier coefficient. Then $f = qh$ with q inner and h outer. The factoring is unique, up to multiplication of q and h by constant factors of modulus one.*

$\mathscr{M}_f = q \cdot H^2$ for some unitary q. Since f is analytic, $\mathscr{M}_f$ is contained in H^2 so q is inner. Then $f = qh$ for some h in H^2. Now $\mathscr{M}_h = \mathscr{M}_{\bar{q}f} = \bar{q} \cdot \mathscr{M}_f = H^2$, and h is outer.

If $f = pq$ with p inner and g outer, then $\mathscr{M}_f = q \cdot H^2$ is equal to $p \cdot H^2$. Hence $\bar{p}q \cdot H^2 = H^2$, so $\bar{p}q$ is analytic. Similarly $p\bar{q}$ is analytic. Hence the Fourier series of $\bar{p}q$ contains only the constant term. It follows that $\bar{p}q$ is constant, as was to be proved.

These theorems leave open the question whether a non-null analytic function can vanish on a set of positive measure if it has no first non-vanishing Fourier coefficient. In this generality the problem is not interesting: when Z^2 is given a non-archimedean order it is easy to construct functions in $H^2(T^2)$ vanishing on large open sets but not everywhere. However we shall find a positive result (more difficult than the one just proved) when we specialize to archimedean order relations. In the meanwhile we shall have to attach a hypothesis about first non-vanishing Fourier coefficient to certain theorems.

1.3. Szegö's theorem

This beautiful theorem of Szegö will be the basis of much that follows [**40**].

THEOREM 2. *Let w be a non-negative summable function on K. Then*

$$\exp\left(\int \log w \, d\sigma\right) = \inf_P \int |1 + P|^2 w \, d\sigma, \tag{6}$$

where P ranges over all finite sums $\sum_{\lambda > 0} a_\lambda \chi_\lambda$. The infimum is zero if and only if $\int \log w \, d\sigma = -\infty$.

The formula is easy to prove if w is suitably restricted; the difficulty lies in approximating general functions. We shall need the following facts. If f is bounded on K, then it is the pointwise limit almost everywhere of trigonometric polynomials f_n having the same bound as f, the same mean value as f, and analytic if f is analytic. If f is real, square-summable and bounded below, then the f_n can be chosen real with the same lower bound. Indeed we convolute f with a suitable approximate identity and choose a subsequence that converges almost everywhere.

A first useful consequence is that the infimum in (6) is unchanged if P ranges over the larger set of continuous analytic functions with mean value zero.

The proof involves several steps. First suppose w is bounded from 0, so $\log w$ is in L^2 and bounded below. Call its Fourier coefficients b_λ. Assume for the moment that $b_0 = 0$. Define g in H^2 by

$$g(x) \sim \sum_{\lambda > 0} b_\lambda \chi_\lambda(x). \tag{7}$$

Then $\log w = 2 \operatorname{Re} g$, $w = |e^g|^2$.

Now find real trigonometric polynomials u_n with mean value 0, bounded below, convergent to $\log w$ almost everywhere. Let g_n be the analytic part of u_n, so that $u_n = 2 \operatorname{Re} g_n$. The dominated convergence theorem implies that

$$\lim \int |e^{-g_n}|^2 \, w \, d\sigma = \lim \int |e^{g-g_n}|^2 \, d\sigma = 1. \tag{8}$$

The exponential on the left is $1 + P$ where P is continuous, analytic, with mean value 0 (this is obvious from the power series expansion). Hence it is one of the functions envisaged in the infimum of (6), and we have

$$\exp\left(\int \log w \, d\sigma\right) = 1 \geqslant \inf_P \int |1 + P|^2 \, w \, d\sigma. \tag{9}$$

Both sides of (6) are homogeneous; hence the restriction that $b_0 = 0$ is inessential, and inequality holds under the assumption that w is bounded from 0.

For the next step we suppose w bounded above as well as bounded from 0, and let $b_0 = 0$ as before. Then e^g (defined above) is $1 + h$, with h bounded, analytic, with mean value 0. For any P the Parseval relation gives

$$\int |1 + P|^2 \, w \, d\sigma = \int |(1 + P)(1 + h)|^2 \, d\sigma \geqslant 1. \tag{10}$$

This inequality is opposite to (9). Now (6) is proved if $\log w$ is bounded, and half proved if it is merely bounded below.

We complete the proof assuming that w is bounded from 0. Set

$w_n = \min(w, n)$ for each positive integer n. For each n and P we have by (6)

$$\int |1 + P|^2 w \, d\sigma \geqslant \int |1 + P|^2 w_n \, d\sigma \geqslant \exp\left(\int \log w_n \, d\sigma\right). \tag{11}$$

The monotonic convergence theorem gives

$$\int |1 + P|^2 w \, d\sigma \geqslant \exp\left(\int \log w \, d\sigma\right). \tag{12}$$

If we take the infimum over P we get the inequality opposite to (9).

Without hypothesis on w let $w_n = \max(w, 1/n)$. For each n we have

$$\exp\left(\int \log w_n \, d\sigma\right) = \inf_P \int |1 + P|^2 w_n \, d\sigma \geqslant \inf_P \int |1 + P|^2 w \, d\sigma. \tag{13}$$

The monotonic convergence theorem once more gives

$$\exp\left(\int \log w \, d\sigma\right) \geqslant \inf_P \int |1 + P|^2 w \, d\sigma. \tag{14}$$

In particular the infimum vanishes if $\int \log w \, d\sigma = -\infty$.

Finally, for each n and P

$$\exp\left(\int \log w \, d\sigma\right) \leqslant \exp\left(\int \log w_n \, d\sigma\right) \leqslant \int |1 + P|^2 w_n \, d\sigma. \tag{15}$$

Taking the limit in n, and then the infimum over P, gives the inequality opposite to (14), and shows that $\int \log w \, d\sigma = -\infty$ if the infimum is 0. This completes the proof of Szegö's theorem.

1.4. Consequences of Szegö's theorem

Szegö's theorem is a kind of quantitative refinement of Beurling's first theorem, and as we shall see is easily equivalent to Beurling's second theorem. First we improve the results of Section 1.2.

COROLLARY. *Let f belong to L^2. $\mathcal{M}_f$ has the form $q \cdot H^2$, q a unitary function, if and only if $\log |f|$ is summable.*

$\mathcal{M}_f$ is larger than $(\mathcal{M}_f)_-$ if and only if f cannot be approximated in L^2 by linear combinations of $\chi_\lambda f$, $\lambda > 0$; that is, if and only if

$$\inf_P \int |(1 + P) f|^2 \, d\sigma > 0. \tag{16}$$

By Theorem 2, this is equivalent to the summability of $\log |f|$. Theorem 1 and its converse assert that this condition is necessary and sufficient for $\mathcal{M}_f$ to have the form $q \cdot H^2$.

COROLLARY. *If f is in H^2 and $\log|f|$ is summable, then $f = qh$ with q inner and h outer.*

For then $\mathcal{M}_f \neq (\mathcal{M}_f)_-$, and this is all that was needed in the proof of factoring.

JENSEN'S INEQUALITY. *For f in H^1,*

$$\int \log|f|\,d\sigma \geqslant \log|a_0(f)|. \tag{17}$$

For each trigonometric polynomial P of the sort appearing in (6), $(1 + P)^2 f$ is in H^1 with mean value $a_0(f)$. Hence

$$\int |(1 + P)^2 f|\,d\sigma \geqslant |a_0(f)|. \tag{18}$$

The infimum over P of the left side is evaluated by (6) and we obtain (17).

COROLLARY. *If f is in H^1 and has a first non-vanishing Fourier coefficient, then $\log|f|$ is summable.*

After multiplying f by a character we may assume $a_0(f) \neq 0$. Then (17) gives the result.

COROLLARY. *A non-negative summable function w is $|g|^2$ for some g in H^2 with $a_0(g) \neq 0$ if and only if $\log w$ is summable. In that case g can be chosen to be outer, and then is unique up to a constant factor.*

If $w = |g|^2$ for g in H^2, $a_0(g) \neq 0$, then $\log w$ is summable by the last corollary.

In the other direction, suppose $\log w$ is summable. Set $f = w^{\frac{1}{2}}$. We have shown that $\mathcal{M}_f = q \cdot H^2$ for some unitary function q. Thus $f = qg$ for some g in H^2, which is outer by an argument we have used before. Hence $a_0(g) \neq 0$ and we have $w = |g|^2$ as required.

Finally we must show that an outer function is determined by its modulus up to a constant factor. Suppose g and h are outer functions with the same modulus. Then $g = qh$ for some unitary q. We have $H^2 = \mathcal{M}_g = q \cdot \mathcal{M}_h = q \cdot H^2$. Thus q is analytic; similarly $\bar{q}$ is analytic, so q is constant as asserted.

BEURLING'S MAIN THEOREM. *g in H^2 is outer if and only if*

$$\int \log|g|\,d\sigma = \log|a_0(g)| > -\infty. \tag{19}$$

Suppose that g is outer in H^2. Then $a_0(g) \neq 0$. From the definition of an outer function it follows that functions Pg, with P as in (6), can approximate any function in H^2 with mean value 0. Choose P so Pg approximates $-(g - a_0(g))$. Then $(1 + P)g$ approximates the constant $a_0(g)$ in the norm of L^2. Hence

$$\exp\left(\int \log |g|^2 \, d\sigma\right) = \inf_P \int |(1 + P)g|^2 \, d\sigma \leqslant |a_0(g)|^2. \tag{20}$$

Jensen's inequality is in the opposite sense, so (19) holds if g is outer.

If (19) holds, then

$$\inf_P \int |(1 + P)g|^2 \, d\sigma = |a_0(g)|^2 > 0. \tag{21}$$

By the Parseval relation, the only way for $(1 + P)g$ to have norm close to $|a_0(g)|$ is for $(1 + P)g$ to be close to the constant function $a_0(g)$ in L^2. Thus $\mathcal{M}_g$ contains a non-null constant function, and consequently fills H^2. This completes the proof.

1.5. Generalization to L^p, $p \neq 2$

The proof of Theorem 1 depended on the geometry of Hilbert space. The exponent 2 is not essential in Szegö's theorem, however; the theorem has been proved with arbitrary positive exponent on groups with ordered duals [**15**]. In this section we shall derive the theorem with arbitrary exponent from the case already studied. The general theorem provides a convenient way to define and study H^p for $0 < p < 1$ on these groups. (The definition by means of Fourier coefficients makes sense only for $p \geqslant 1$.)

THEOREM 2′. *With the notation of Theorem 2, we have for any positive* p

$$\exp\left(\int \log w \, d\sigma\right) = \inf_P \int |1 + P|^p \, w \, d\sigma. \tag{22}$$

The result will follow easily from Theorem 2 and this

LEMMA. *With the same notation,*

$$\inf_P \int |1 + P|^p \, w \, d\sigma = \inf_P \int |e^P|^p \, w \, d\sigma. \tag{23}$$

Each e^P is uniformly the limit of functions $1 + P$; hence the left side of (23) is not greater than the right side. We have to prove, roughly, that the infimum on the left is approached with functions $1 + P$ of the special form e^P.

Fix a P and write $1 + P = qg$, with q inner and g outer. Then g is a bounded analytic function, and $a_0(q)\, a_0(g) = 1$. Therefore $|a_0(g)| \geqslant 1$. If we set $f = g/a_0(g)$ we have

$$\int |1 + P|^p\, w\, d\sigma = \int |g|^p\, w\, d\sigma \geqslant \int |f|^p\, w\, d\sigma. \tag{24}$$

Now f is outer with mean value 1; hence $\log |f|$ is summable with mean value 0. Choose real trigonometric polynomials u_n with mean value 0, uniformly bounded above, tending to $\log|f|$ almost everywhere. By the dominated convergence theorem

$$\int |f|^p\, w\, d\sigma = \lim \int e^{u_n p}\, w\, d\sigma. \tag{25}$$

But $e^{u_n} = |e^Q|$ where Q is an analytic trigonometric polynomial with mean value 0; hence

$$\int |f|^p\, w\, d\sigma \geqslant \inf_P \int |e^P|^p\, w\, d\sigma. \tag{26}$$

Combining (24) and (26) completes the proof of (23).

The right side of (23) is obviously independent of p, $p > 0$. Hence the left side is independent of p, and Theorem 2′ follows from Theorem 2.

The definition of H^p cannot be given in terms of Fourier coefficients when $p < 1$. Instead H^p may be defined as the subspace of L^p obtained by closing the set of analytic trigonometric polynomials in the metric of L^p. We find the same spaces as before if $p \geqslant 1$; by Theorem 2′ we get a proper subspace of L^p, at any rate, if $0 < p < 1$. The next two theorems provide other desirable properties.

THEOREM 3. *If f is in H^p $(0 < p < 1)$ and in L^1, then it is in H^1.*

The intersection of H^p with L^1 is closed in L^1, and invariant under group translations. Hence the convolution $k * f$ is in this intersection, for any k in L^1. If $a_\lambda(f) \neq 0$ for some negative λ, then $\chi_\lambda * f = a_\lambda(f)\, \chi_\lambda$ belongs to H^p. This contradicts Theorem 2′, so f is in H^1 as asserted.

THEOREM 4. *H^p $(0 < p < 1)$ consists exactly of those f in L^p such that fg is in H^2 for some bounded outer function g.*

Suppose f is a non-null function in H^p. Then $\log |f|$ is summable above; hence we can find a bounded outer function g such that $|g| = \min(1, |f|^{-1})$. If $\{Q_n\}$ is a sequence of analytic trigonometric polynomials converging to f in L^p, then $Q_n g$ tends to fg in L^p. But each $Q_n g$ is in H^p, so fg is in H^p. Since fg is bounded it is in H^2, by the last theorem.

Conversely, suppose f is in L^p and fg in H^2, where g is a bounded outer function, say $|g| \leqslant 1$. Find h_n outer with $|h_n| = \min(n, |g|^{-1})$. Then gh_n is outer (Beurling's theorem), $|gh_n| \leqslant 1$, and $|gh_n|$ increases to 1 almost everywhere. Hence

$$\log |a_0(g)\, a_0(h_n)| = \int \log |gh_n| \, d\sigma \to 0. \tag{27}$$

If h_n is normalized so $a_0(g)\, a_0(h_n) > 0$ this means that $a_0(g)\, a_0(h_n)$ tends to 1. Since $|gh_n| \leqslant 1$, the Parseval relation implies that gh_n tends to the function constantly equal to 1 in the norm of L^2. A subsequence converges pointwise. On this subsequence, fgh_n tends to f almost everywhere, and in the metric of L^p by the dominated convergence theorem. Since fgh_n is in H^2 (and so in H^p) for each n, we conclude that f is in H^p as we wished to prove.

This is as far as we shall pursue this subject.

1.6. Invariant subspaces in L^p, $p \geqslant 1$

The definitions of *invariant subspace* and *outer function* make sense in L^p for any p. We shall generalize Beurling's two theorems to L^p, $p \geqslant 1$. Such results are due to several people; see **[15, 39]**.

LEMMA. *The bounded functions in any invariant subspace of L^p, $1 \leqslant p < \infty$, are dense in the subspace.*

Let f belong to an invariant subspace $\mathscr{M}$ of L^p. For each n let h_n be the outer function with $|h_n| = \min(1, n|f|^{-1})$ and $a_0(h_n) > 0$. As in the proof of the last theorem, $|h_n| \leqslant 1$, $|h_n|$ tends to 1 monotonically, so h_n tends to the constant function 1 in norm, and on a subsequence almost everywhere. Hence on the subsequence fh_n tends to f in L^p. But fh_n is in $\mathscr{M}$ and is bounded. Thus each element of $\mathscr{M}$ is a limit of bounded functions in $\mathscr{M}$.

The next assertion is dual to what has just been proved.

LEMMA. *Let $\mathscr{M}$ be an invariant subspace of L^p, $1 \leqslant p < \infty$, and $\bar{\mathscr{M}}$ its closure in L^1. Then $\bar{\mathscr{M}} \cap L^p = \mathscr{M}$.*

There is nothing to prove if $p = 1$. Otherwise let r be the exponent conjugate to p, and $\tilde{\mathscr{M}}$ the set of all g in L^r such that

$$\int fg \, d\sigma = 0 \tag{28}$$

for all f in $\mathscr{M}$. Then $\tilde{\mathscr{M}}$ is an invariant subspace of L^r. If f is in $\bar{\mathscr{M}}$ then (28) holds for all bounded g in $\tilde{\mathscr{M}}$. If f is also in L^p, then (28) holds for all g in $\tilde{\mathscr{M}}$ by the previous lemma. Hence f is in $\mathscr{M}$ as was to be proved.

THEOREM 1′. *Every invariant subspace of* L^p $(1 \leqslant p < \infty)$ *such that* $\mathscr{M} \neq \mathscr{M}_-$ *is* $q \cdot H^p$ *for some unitary function* q.

Suppose $1 \leqslant p < 2$. Then $\mathscr{N} = \mathscr{M} \cap L^2$ is an invariant subspace of L^2 (in particular it is closed), and $\mathscr{N} \neq \mathscr{N}_-$ by the first lemma above. Hence $\mathscr{N} = q \cdot H^2$ for some unitary function q. By the same lemma $\mathscr{M}$ is the closure in L^p of $q \cdot H^2$, which is $q \cdot H^p$ as we wished to show.

If $p > 2$ the result is proved by a similar application of the second lemma, or by duality from the case $p < 2$.

The same result holds in L^∞ for subspaces $\mathscr{M}$ closed in the weak star-topology of L^∞ as the dual of L^1. We assume that $\mathscr{M}$ is not equal to the star-closure of $\mathscr{M}_-$. The proof is immediate by duality from L^1.

BEURLING'S MAIN THEOREM FOR $1 \leqslant p < \infty$. *A function* g *in* H^p *is outer in* H^p *if and only if it satisfies* (19).

The proof needs no new idea, and is based on Theorem 2′ as the original statement was based on Theorem 2. We see in the same way that for g in H^∞, (19) is necessary and sufficient for g to generate a star-dense invariant subspace.

The theorem removes an ambiguity in the definition of outer function. If g is in H^p, then it belongs to H^r for each $r < p$. By Beurling's criterion, or by the preceding lemmas, g is outer in one space H^r only if it is outer in every one to which it belongs.

1.7. Analytic measures

According to the classical definition, H^p $(1 \leqslant p < \infty)$ is the set of functions f analytic for $|z| < 1$ and satisfying

$$\int_0^{2\pi} |f(re^{ix})|^p \, dx \leqslant M < \infty, \qquad 0 < r < 1. \tag{29}$$

If $p > 1$ this condition implies that $f_r(e^{ix}) = f(re^{ix})$ has a limit as r increases to 1, in the weak star-topology of L^p as the dual of L^q $(p^{-1} + q^{-1} = 1)$. The limit function f reproduces f_r by convolution with the Poisson kernel, from which it follows that f_r tends to f in the norm of L^p. This argument shows that the space H^p of functions defined on the boundary of the circle with Fourier coefficients null on the left coincides with the boundary functions of elements of the space defined by (29).

For $p = 1$ the argument fails, because L^1 is not the dual of L^∞. Instead we get star-convergence of the measures $f_r(e^{ix})\,dx$ as linear functionals on the space of continuous functions on T, and the limiting measure μ is not obviously absolutely continuous. A surprising theorem of F. and M. Riesz states that μ is *mutually absolutely continuous* with Lebesgue measure (unless

$f = 0$). This shows the two definitions of H^1 are equivalent, and brings into the same statement the result about non-vanishing of boundary values of analytic functions.

The Riesz theorem turns out to have more than one generalization. This section gives an extension, to the extent possible, to the context of groups with ordered duals. The ideas used in the proof will also give a proof of Szegö's theorem that is free of function theory [**24**].

The development will be based on the following generalization of Szegö's theorem, proved by Kolmogoroff and Krein on the circle group.

THEOREM 5. *Let μ be a positive, bounded Borel measure on K with absolutely continuous part $w d\sigma$ (w non-negative in L^1). For any positive p,*

$$\inf_P \int |1 + P|^p \, d\mu = \exp\left(\int \log w \, d\sigma\right), \tag{30}$$

where P ranges over analytic trigonometric polynomials with mean value 0.

In other words, the infimum of Szegö's theorem is not increased if a singular measure is added to $w \, d\sigma$.

We shall use the theorem only for the case $p = 2$, and we prove it first for that case. Form L^2_μ, the Lebesgue space based on μ. Denote by S the closure of the set of all $1 + P$, with P as in the statement of the theorem. This is a closed convex set; denote by $1 + H$ its unique element of smallest norm. For any complex z and positive λ in Γ the functions $(1 + H)(1 + z\chi_\lambda)$ and $1 + H + z\chi_\lambda$ belong to S and have smallest norm for $z = 0$. This leads to two relations:

$$\int |1 + H|^2 \bar{\chi}_\lambda \, d\mu = 0, \qquad \int (1 + H)\, \bar{\chi}_\lambda \, d\mu = 0 \qquad (\lambda > 0). \tag{31}$$

By taking the complex conjugate of the first equation we get the same fact for $\lambda < 0$. That is, the Fourier–Stieltjes coefficients $a_\lambda(|1 + H|^2 \, d\mu) = 0$ for all $\lambda \neq 0$. Hence

$$|1 + H|^2 \, d\mu = \kappa \, d\sigma \tag{32}$$

for some constant κ. The conclusion we need is that $1 + H = 0$ almost everywhere for the singular part of μ.

In the same way define S_0 to be the closure of all $1 + P$ in L^2_w, the Lebesgue space based on $w \, d\sigma$. If $1 + H_0$ is the minimal element of S_0 it satisfies relations analogous to (31). Only the second is of interest:

$$\int (1 + H_0)\, \bar{\chi}_\lambda \, w \, d\sigma = 0 \qquad (\lambda > 0). \tag{33}$$

Now (33) *characterizes* the minimal element of S_0. Since $(1 + H)\, d\mu =$

$(1 + H)\, w\, d\sigma$, it appears that $1 + H$ (considered as an element of L_w^2) satisfies this minimal condition as well as $1 + H_0$. Hence $1 + H = 1 = H_0$ a.e. $(d\sigma)$, and we have

$$\inf_P \int |1 + P|^2\, d\mu = \int |1 + H|^2\, d\mu = \int |1 + H|^2\, w\, d\sigma$$

$$= \inf_P \int |1 + P|^2\, w\, d\sigma. \tag{34}$$

This proves the theorem when $p = 2$.

We shall show that

$$\inf_P \int |1 + P|^p\, d\mu = \inf_P \int |e^P|^p\, d\mu. \tag{35}$$

As in the proof of Theorem 2′ the right side is obviously independent of p, so the theorem for arbitrary p will follow from the special case just proved.

The factoring $1 + P = qg$ into inner and outer parts that served before will not work now, because the factors may not be continuous and integration against $d\mu$ is not defined. The idea can be rescued, however. Fix $1 + P$ and write $v = \log |1 + P|$, a function that is continuous on K except for negative infinities. The Jensen inequality gives $a_0(v) \geqslant 0$. Define $v_n = \max(v, -n)$; then v_n is continuous and $a_0(v_n) \geqslant 0$. Thus v_n is the uniform limit of real trigonometric polynomials with non-negative mean value. Hence

$$\int e^{v_n p}\, d\mu \geqslant \inf_u \int e^{up}\, d\mu, \tag{36}$$

where u ranges over real trigonometric polynomials with mean value 0. But each such e^u is $|e^Q|$ where Q is an analytic trigonometric polynomial with mean value 0. We get

$$\int |1 + P|^p\, d\mu = \lim \int e^{v_n p}\, d\mu \geqslant \inf_Q \int |e^Q|^p\, d\mu, \tag{37}$$

where Q ranges over the same set as P above. This gives one inequality in (35), and the other is trivial.

Since the proof for $p = 2$ depended on the geometry of Hilbert space, the reduction at the end could not easily be avoided.

Actually we have met the minimal function $1 + H_0$ of the first part of the proof before. The next theorem gives the details.

THEOREM 6. *Let w be non-negative on K, with w and* $\log w$ *summable. Set*

$g = (1 + H_0)^{-1}$. Then g is in H^2, outer, $a_0(g) = 1$, and $w = \kappa |g|^2$ where $\kappa = \exp \int \log w \, d\sigma$.

From (32), with $d\mu = w\,d\sigma$, we have $|1 + H_0|^2 w = \kappa$, and this constant is evaluated by Szegö's theorem. Thus $w = \kappa |g|^2$ and g is in L^2. Now (33) becomes

$$\kappa \int (1 + H_0)|g|^2 \bar{\chi}_\lambda \, d\sigma = \kappa \int \bar{g}\bar{\chi}_\lambda \, d\sigma = 0 \qquad (\lambda > 0), \tag{38}$$

which says that g is analytic. Let $1 + P_n$ (trigonometric polynomials of the usual type) tend to $1 + H_0$ in L^2_w. Then $(1 + P_n)g$ tends to the constant function 1 in L^2. Hence g is outer and $a_0(g) = 1$. This completes the proof.

Here is the first part of a generalized Riesz theorem.

THEOREM 7. *Let ν be a complex measure on K such that $a_\lambda(\nu) = 0$ for all negative λ in Γ. Then the absolutely continuous and the singular parts of ν have the same property.*

Write $d\nu = f d\sigma + d\nu_s$, where f is in L^1 and ν_s is singular with respect to σ. We are to prove that f is in H^1. Set $\mu = |\nu| = w\,d\sigma + d\mu_s$. Denote by $\mathcal{N}$ the closure in L^2_μ of the set of analytic trigonometric polynomials. The theorem will be an easy consequence of this

LEMMA. *$\mathcal{N}$ contains the element φ of L^2_μ equal to 1 almost everywhere ($d\sigma$) and to 0 almost everywhere ($d\mu_s$).*

For if the lemma is true, let Q_n be analytic trigonometric polynomials tending to φ in L^2_μ. For $\lambda > 0$ we have

$$0 = \int \chi_\lambda Q_n \, d\nu \to \int \chi_\lambda \varphi \, d\nu = \int \chi_\lambda f \, d\sigma; \tag{39}$$

thus f is in H^1 as required.

In proving the lemma there is no loss of generality in assuming $\log w$ summable. For otherwise treat the measure $d\mu' = (w + 1)\,d\sigma + d\mu_s$. Convergence of trigonometric polynomials in $L^2_{\mu'}$ implies convergence in L^2_μ, so the lemma will hold in L^2_μ if it holds in $L^2_{\mu'}$.

Let $1 + H$ and $1 + H_0$ be the functions defined in the proof of Theorem 5. We know by Theorem 6 that $(1 + H_0)^{-1}$ is in H^2; let Q_n be analytic trigonometric polynomials tending to it in H^2. Then $Q_n(1 + H_0)$ converges to the constant function 1 in L^2_w. Now $Q_n(1 + H)$ equals $Q_n(1 + H_0)$ almost everywhere ($d\sigma$), but vanishes almost everywhere ($d\mu_s$). Hence $Q_n(1 + H)$ tends to φ in L^2_μ. The approximating functions are in $\mathcal{N}$, so the lemma is proved and the theorem as well.

The second part of the generalized Riesz theorem is

THEOREM 8. *If* ν *is a singular measure with* $a_\lambda(\nu) = 0$ *for all* $\lambda < 0$, *then also* $a_0(\nu) = 0$.

Set $\mu = |\nu|$. By Theorem 5, trigonometric polynomials P (analytic with mean value 0) approximate the constant function 1 in L^2_μ. Write $d\nu = \psi\, d\mu$ where ψ is a bounded function. Then ψ is a limit in L^2_μ of functions $P\psi$. For each P we have $a_0(P\psi\, d\mu) = 0$; it follows that $a_0(\psi\, d\mu) = 0$, which is what we wanted to prove.

Theorems 7 and 8 imply the theorem of F. and M. Riesz on the circle: if ν is a measure such that $a_n(\nu) = 0$ for $n < 0$, then ν is absolutely continuous. By the first theorem it suffices to show that any singular measure with that property must vanish. The second theorem implies that $a_0(\nu) = 0$, and by induction all the Fourier–Stieltjes coefficients of ν are 0.

More generally, if ν is a measure such that $a_\lambda(\nu)$ is different from 0 only for a sequence $\lambda_1 < \lambda_2 < \ldots$, then ν is absolutely continuous. This implies a theorem on T^2 proved by Bochner [**6**].

Generalizations of the Riesz theorem have gone far [**8, 9, 10, 12, 16, 30**]. The version given above is related to Szegö's theorem and has extensions to function algebras. On the other hand the beautiful proof of Mandrekar and Nadkarni is based on cocycles and makes sense on a flow. The importance of the F. and M. Riesz theorem is partly due to the fact that it is one of those results that are more general than any of their proofs.

2. GROUPS WITH ARCHIMEDEAN DUALS

2.1. The flow on K

R. Arens [**1, 2**] first recognized that it is natural to assume the order relation in Γ is *archimedean*. This means effectively [**38**] that Γ is a subgroup of the real line, and we shall simply assume that this is the case. Among such groups only those isomorphic to the integers possess a first positive element. This fact makes harmonic analysis on the circle rather special; therefore we assume without further mention that Γ is not isomorphic to the integers, and K is not a circle.

Γ is now a subgroup of the line R, dense in R but endowed with the discrete topology. For each real number t let e_t be the element of K defined by $e_t(\lambda) = \exp it\lambda$. It is easy to verify that the mapping from t to e_t is a one–one continuous homomorphism of R into K. The image K_0 is a dense subgroup of K, and $\sigma(K_0) = 0$.

The distinguished one-parameter subgroup K_0 defines a flow on K: x is carried in time t to $x + e_t$. We shall be studying this flow from a particular point of view, and later the same ideas will be carried over to flows of general type. In this section we establish some basic properties of this flow on K.

For any function f defined on K, and real t, let $T_t f(x) = f(x + e_t)$. Since Haar measure is invariant under translation, each T_t is a unitary operator in L^2 and $\{T_t\}$ is a continuous unitary group.

THEOREM 9. *If f is in L^1 and $T_t f = f$ for each t, then f is a constant function. For any continuous function h on K,*

$$\lim_{T\to\infty} \frac{1}{2T} \int_{-T}^{T} h(e_t)\, dt = \int h\, d\sigma. \tag{1}$$

Both statements (which say in different ways that the flow is ergodic) are proved by computations with Fourier series. By the definition of Fourier coefficients, $T_t f$ has coefficients $e^{it\lambda} a_\lambda(f)$. If $T_t f = f$ (equality in L^1, that is almost everywhere), then $e^{it\lambda} a_\lambda(f) = a_\lambda(f)$ for all λ and for every t. Hence only $a_0(f)$ is different from 0.

The relation (1) is verified by direct (and familiar) calculation if h is a trigonometric polynomial. Equality persists under uniform convergence, so it holds for all continuous functions. The theorem is proved.

Choose and fix a positive element γ of Γ. Let K_γ be the closed subgroup of K consisting of all x such that $x(\gamma) = 1$. Every y in K has a unique representation in the form $x_1 + e_t$ with x_1 in K_γ and $0 \leqslant t < 2\pi/\gamma$. Indeed, choose t in this interval so $y(\gamma) = \exp it\gamma$. Then $y(\gamma) = e_t(\gamma)$, so $x_1 = y - e_t$ is in K_γ.

The mapping from (x_1, t) to $x_1 + e_t$ carries $K_\gamma \times [0, 2\pi/\gamma)$ continuously onto K. The inverse mapping is continuous except at points of K_γ. It follows that Borel sets are taken to Borel sets in both directions. Furthermore, if σ_1 is normalized Haar measure on K_γ, then $(\gamma/2\pi) d\sigma_1 \times dt$ is carried by the mapping to $d\sigma$ on K. Indeed let $d\mu$ be the measure obtained by carrying the product measure to K. Its Fourier–Stieltjes coefficients are

$$a_\lambda(\mu) = \int \overline{\chi_\lambda(x)}\, d\mu(x) = \frac{\gamma}{2\pi} \int_{K_\gamma} \int_0^{2\pi/\gamma} \overline{\chi_\lambda(x_1)}\, e^{-it\lambda}\, dt\, d\sigma_1(x_1). \tag{2}$$

If λ is a non-zero multiple of $\gamma/2\pi$ the integral with respect to t is 0. If λ is not a multiple of $\gamma/2\pi$, then χ_λ is a non-zero character of K_γ and the integral over K_γ vanishes. Hence only $a_0(\mu)$ is different from zero, so μ is a multiple of σ. Both measures are normalized, so they are equal.

Thus K is represented measure-theoretically, and almost topologically, as a product space. The usefulness of this decomposition was shown by deLeeuw and Glicksberg [**8**].

A *cross-section* for the flow associated with K_0 is a set having exactly one point in common with each coset of K_0. K_γ is not a cross-section: it intersects each coset of K_0 in an arithmetic sequence (for example its intersection with K_0 consists of all e_t where t is an integral multiple of $2\pi/\gamma$). But K_γ is the best substitute we can find for a cross-section, for we have

THEOREM 10. *No Baire set intersects each coset of K_0 in exactly one point.*

Suppose first that Γ is countable. Then K is a complete metric space, and Baire sets are the same as Borel sets. Suppose C is a Borel cross-section for the flow. If C is contained in K_γ, the sets $C + e_{t_n}$ ($t_n = 2\pi n/\gamma$, $n = 0, \pm 1, \ldots$) are Borel sets forming a disjoint covering of K_γ. The Haar measure σ_1 of K_γ is invariant under translation in K_γ, so these sets all have the same measure. Since σ_1 is a finite measure this is impossible.

If C is not contained in K_γ, we view it as a Borel subset of $K_\gamma \times [0, 2\pi/\gamma)$ as explained above. Let P be the projection that carries (x_1, t) in the product space to x_1 in K_γ. It is true, although not obvious, that PC is a Borel subset of K_γ [**3**, 3.3.2]. Now PC is a Borel cross-section contained in K_γ, and the impossibility of this has just been proved.

I am indebted to Dr. W. Moran for the rest of the proof. We drop the assumption that Γ is countable and suppose that C is a Baire cross-section for the cosets of K_0 in K. Translate C if necessary so that $C \cap K_0 = \{0\}$. Then [**17**, p. 287] there is a closed subgroup H of K such that K/H is separable, C is a union of cosets of H, and πC (where π is the canonical homomorphism of K on K/H) is a Borel set in K/H. (The last assertion holds because π carries H-invariant Borel sets in K to Borel sets in K/H.)

Since H is part of C we have $H \cap K_0 = \{0\}$. Hence π is one-one on K_0. Thus πK_0 is a one-parameter subgroup of K/H that is not periodic. We shall show that πC is a cross-section for the cosets of πK_0. It is obvious that πC intersects each coset of πK_0 in at least one point. Suppose πC intersects a coset $\pi K_0 + (x + H)$ in points πc_1, πc_2 (c_1, c_2 in C). Then $\pi(c_1 - c_2)$ is in πK_0; that is, $c_1 - c_2 = e_t + y$ for some real t and y in H. But $C + H = C$, so c_1 and $c_2 + y$ are two elements of C differing by e_t. This implies $t = 0$, and hence $\pi c_1 = \pi c_2$. Thus πc is a cross-section in the separable group K/H, which is impossible. The theorem is proved.

2.2. Almost periodic functions

Let f be continuous on K. Then $F(t) = f(e_t)$ is continuous on the line. If trigonometric polynomials f_n converge uniformly to f on K, then $F_n(t)$ tends uniformly to $F(t)$ on R. These functions have the form

$$F_n(t) = \sum_k a_k e^{it\lambda_k} \tag{3}$$

where the exponents λ_k are in Γ. Thus F is uniformly almost periodic (almost periodic in the sense of Bohr) with exponents in Γ. Conversely, every such almost periodic function is obtained in this way, for if a sequence of exponential polynomials (3) converges uniformly on the line, then the corresponding trigonometric polynomials converge uniformly on K (because K_0 is dense in K).

If f is merely in a Lebesgue class, $F(t) = f(e_t)$ is not defined, because K_0 is a null set in K. However the Fubini theorem shows that $F_x(t) = F(x + e_t)$ is determined for almost every t, for almost every x. If we change f on a null set of K, the new functions F_x are equal to the old ones almost everywhere on R, except in a null set of x. Furthermore, f is a null function on K if and only if F_x is a null function on R for almost all x. In particular, a measurable set E in K is a null set if and only if its intersection with almost every coset of K_0 is a linear null set.

Let f be continuous on K with Fourier coefficients a_λ. Then we have the formula of Bohr:

$$a_\lambda = \lim_{T\to\infty} \frac{1}{2T}\int_{-T}^{T} F(t)\, e^{-it\lambda}\, dt, \tag{4}$$

which is verified easily for trigonometric polynomials, and then for continuous functions (compare Theorem 9). If f is, say, in L^1 the formula does not make sense, but we might expect the analogous statistical result:

$$a_\lambda = \lim_{T\to\infty} \frac{1}{2T}\int_{-T}^{T} f(x + e_t)\, \bar{\chi}_\lambda(x + e_t)\, dt \qquad \text{(a.e. } x\text{)}. \tag{5}$$

This is in fact true by the individual ergodic theorem. Indeed more is true: for almost every x

$$\lim_{T\to\infty} \frac{1}{2T}\int_{-T}^{T} |f(x + e_t)|\, dt = \int |f|\, d\sigma; \tag{6}$$

hence if a sequence f_n converges to f in L_1, then $F_{n,x}$ converges to F_x for almost every x, in the Besicovitch norm defined by the left side of (6). Hence F_x is a Besicovitch function (that is, the limit of exponential polynomials in Besicovitch norm) for almost every x. This result is due to D. Lowdenslager.

We shall not need the notion of Besicovitch almost-periodicity, but we shall study functions f on K systematically by means of the associated functions F_x defined on the line.

2.3. Spectral resolution of an invariant subspace

Let $\mathcal{M}$ be an invariant subspace of L^2. Define $\mathcal{M}_\lambda = \chi_\lambda \cdot \mathcal{M}$ for each λ in Γ. The dichotomy of the next theorem is a fundamental result.

THEOREM. 11. *Either* $\mathcal{M}_\lambda = \mathcal{M}$ *for all* λ, *or else*

$$\bigcap \mathcal{M}_\lambda = \{0\}, \text{ closure of } \bigcup \mathcal{M}_\lambda = L^2. \tag{7}$$

In the first case $\mathcal{M}$ *consists of all the functions supported on some fixed measurable subset of* K.

Suppose $\mathcal{M}_\lambda = \mathcal{M}$ for all λ. Let q be the projection of the constant function 1 on $\mathcal{M}$. Then $1 - q$ is orthogonal to $\mathcal{M}$, and thus to $\chi_\lambda q$ for all λ:

$$\int (1 - q)\bar{q}\bar{\chi}_\lambda \, d\sigma = 0, \quad \text{all } \lambda \text{ in } \Gamma. \tag{8}$$

Thus $(1 - q)\bar{q}$ is a function in L^1 whose Fourier coefficients all vanish. Hence it is the null function; that is, q is the characteristic function of some set E. Since $\chi_\lambda q$ is in $\mathcal{M}$ for all λ, $\mathcal{M}$ contains all the functions of L^2 supported on E.

The orthogonal complement of $\mathcal{M}$ contains $1 - q$, and therefore also $\chi_\lambda(1 - q)$ for all λ. Hence the complement contains all functions vanishing on E. We conclude that $\mathcal{M}$ is exactly the set of functions supported on E, as we wished to prove.

A subspace of this type, invariant under multiplication by χ_λ for all λ, has for Fourier transform a subspace invariant under translations. Such subspaces, in L^1 or L^2 on the real line, were first studied by N. Wiener. For this reason we call $\mathcal{M}$ a *Wiener subspace* if $\mathcal{M} = \mathcal{M}_\lambda$ for all λ.

The proof that an invariant subspace $\mathcal{M}$ not of Wiener type satisfies (7) is more difficult. First we shall derive the result from this fact, which will be proved later: *if* f *is in* H^1 *and vanishes on a set of positive measure, then* $f = 0$. A second proof will emerge below (p. 32).

The intersection in (7) is a Wiener subspace, consisting of all functions in L^2 supported on some set E. Let f and g be any members of $\mathcal{M}$ and $\mathcal{M}^\perp$, respectively. The invariance of $\mathcal{M}$ implies that $f\bar{g}$ is in H^1. Since $\mathcal{M}$ contains all functions supported on E, g must vanish on E. Thus $f\bar{g} = 0$ on E, and if E has positive measure must vanish everywhere. But trivially $\mathcal{M}$ contains every f such that $f\bar{g} = 0$ for all g in $\mathcal{M}^\perp$. The set of such f is obviously closed and invariant under multiplication by χ_λ for all λ. Hence $\mathcal{M}$ is a Wiener subspace unless E is a null set. That proves the first relation in (7).

The second relation is a consequence of the first one. Let $\mathcal{N}$ be the set of all f in L^2 such that $\bar{f}$ is orthogonal to $\mathcal{M}$. Then $\mathcal{N}$ is an invariant subspace, and not of Wiener type if $\mathcal{M}$ is not. The intersection of all $\mathcal{N}_\lambda$ is trivial; it follows that the union of all $\mathcal{M}_\lambda$ is dense.

An invariant subspace not of Wiener type is called *simply invariant*. We fix a simply invariant subspace $\mathcal{M}$ for study, normalized by the condition $\mathcal{M} = \mathcal{M}_+$. Denote by P_λ the orthogonal projection of L^2 on $\mathcal{M}_\lambda$. As λ increases through Γ, P_λ decreases from the identity I to 0, on account of (7).

For real numbers λ not in Γ we define P_λ so the family $\{P_\lambda\}$ is continuous from the left. Then $(I - P_\lambda)$ is a resolution of the identity in L^2, to which Stone's theorem associates the unitary group

$$V_t = -\int_{-\infty}^{\infty} e^{it\lambda}\, dP_\lambda. \tag{9}$$

We want to describe those unitary groups in L^2 that arise from simply invariant subspaces.

For τ in Γ define $S_\tau f = \chi_\tau f$. Then $\{S_\tau\}$ is a unitary group acting in L^2.

THEOREM 12. *The families* $\{P_\lambda\}$ *and* $\{V_t\}$ *associated with a normalized invariant subspace* $\mathcal{M}$ *satisfy*

$$P_{\lambda+\tau} = S_\tau P_\lambda S_{-\tau}; \qquad V_t S_\tau = e^{it\tau} S_\tau V_t \qquad (t, \lambda \in R \text{ and } \tau \in \Gamma). \tag{10}$$

Every left-continuous family of projections $\{P_\lambda\}$ *and every continuous unitary group* $\{V_t\}$ *satisfying* (10) *are obtained from a unique simply invariant subspace.*

In the first relation of (10), the right side is a self-adjoint projection whose range is the same as that of the projection on the left, so the projections are equal. It is obvious, conversely, that such a family of projections is obtained from a unique normalized simply invariant subspace.

The second part of (10), called the *Weyl commutation relation* for the pair of unitary groups, is proved by this calculation:

$$S_\tau^* V_t S_\tau = -\int_{-\infty}^{\infty} e^{it\lambda}\, d(S_{-\tau} P_\lambda S_\tau) = -\int_{-\infty}^{\infty} e^{it\lambda}\, dP_{\lambda-\tau} = -e^{it\tau}\int_{-\infty}^{\infty} e^{it\lambda}\, dP_\lambda$$
$$= e^{it\tau} V_t. \tag{11}$$

Conversely, if $\{V_t\}$ is a unitary group satisfying (10), then the spectral family in (9) produced by Stone's theorem in the other direction (with the projections decreasing, and continuous from the left) satisfy (10), by suitably rearranging the terms of (11). Thus $\{V_t\}$ is associated with a normalized simply invariant subspace, whose uniqueness is obvious.

THEOREM 13. *Each continuous unitary group* $\{V_t\}$ *in* L^2 *satisfying* (10) *has this form:* $V_t = A_t T_t$, *where for each* t *the operator* A_t *is multiplication by a measurable function* $A_t(x)$ *satisfying* (i) $|A_t| = 1$ a.e. *on* K; (ii) A_t *moves continuously in* L^2 *as a function of* t; (iii) $A_{t+u} = A_t T_t A_u$ *for all real* t, u.

Conversely, if $\{A_t\}$ *is any such family of functions on* K, *then* $V_t = A_t T_t$ *defines a continuous unitary group in* L^2 *that stisfies* (10).

Given a unitary group $\{V_t\}$ satisfying (10), define $A_t = V_t 1$. (Then (ii) holds by hypothesis.) Using the commutation relation we find for each τ in Γ

$$V_t\chi_\tau = V_tS_\tau 1 = e^{it\tau}S_\tau V_t 1 = e^{it\tau}\chi_\tau A_t = A_t T_t \chi_\tau . \tag{12}$$

By linearity, the formula $V_t f = A_t T_t f$ holds for all trigonometric polynomials f. V_t znc T_t both preserve norm; hence A_t must have modulus 1 almost everywhere. Finally (iii) holds because $V_{t+u} = V_t V_u$.

In the other direction, such a family $\{A_t\}$ defines operators V_t that are unitary by (i), form a group by (iii), and a strongly continuous group by (ii).

A family $\{A_t\}$ satisfying the conditions of the theorem will be called a *cocycle*. We have established a one-one correspondence between normalized simply invariant subspaces of L^2 and cocycles on K. We are going to obtain information about subspaces by studying their cocycles.

For H^2, the simplest simply invariant subspace, the projections are given by

$$P_\lambda \sum a_\tau \chi_\tau = \sum_{\tau \geqslant \lambda} a_\tau \chi_\tau . \tag{13}$$

Hence $V_t\chi_\tau = e^{it\tau}\chi_\tau = T_t\chi_\tau$. It follows that $V_t = T_t$, $A_t = 1$ for all t.

The projections associated with $q \cdot H^2$, where q is a unitary function, are $qP_\lambda\bar{q}$, where P_λ is given by (13). From (9) we see that $V_t = qT_t\bar{q}$, meaning that $V_t f = qT_t(\bar{q}f)$ for each f in L^2. That is, $A_t = qT_t\bar{q}$. A cocycle $A_t(x) = q(x)\bar{q}(x+e_t)$ is called a *coboundary*.

The same calculation shows that if $\mathscr{M}$ has cocycle A_t, then $q \cdot \mathscr{M}$ has cocycle $A_t(qT_t\bar{q})$. Two cocycles are called *cohomologous* if one is a coboundary times the other. We have proved now that two subspaces are related by a unitary function if and only if their cocycles are cohomologous.

By analogy with the circle group we ask whether every simply invariant subspace is $q \cdot H^2$ for some unitary function q. There are two kinds of subspace not of this form for trivial reasons. When we exclude them we are left with a problem of some difficulty.

First it is easy to see by Fourier argument that H_0^2, the functions in H^2 with mean value 0, is a simply invariant subspace not equal to $q \cdot H^2$ for any q. More generally, the same is true for $p \cdot H_0^2$ where p is a unitary function. Now H_0^2 is $(H^2)_-$, and this kind of subspace is excluded if we restrict attention to subspaces normalized by the condition $\mathscr{M} = \mathscr{M}_+$. The problem is not very serious, for by Theorem 1 the subspaces $\mathscr{M}_+$ and $\mathscr{M}_-$ are at most one dimension apart.

The second kind of exceptional subspace exists if Γ is not the group of all real numbers. Let λ be a real number not in Γ. Define $\mathscr{M}$ to be the set of functions in L^2 whose coefficients a_τ all vanish for $\tau < \lambda$. The cocycle of $\mathscr{M}$ is $A_t(x) = e^{-it\lambda}$, constant in x for each t. A Fourier argument shows that this is not a coboundary .We see also that $\mathscr{M} = \mathscr{M}_+ = \mathscr{M}_-$.

A cocycle will be called *trivial* if it is the product of a coboundary and a cocycle that is constant in x for each t. The interesting question is to decide whether every cocycle is trivial. We shall show in Section 4 that the answer is negative.

We return to the proof of Theorem 11. Let us review the development without assuming (7). Denote by $\mathcal{M}_\infty$ the intersection of all $\mathcal{M}_\lambda$, and by $\mathcal{M}_{-\infty}$ the closure of their union. These are Wiener subspaces, consisting of all functions in L^2 supported on sets E and F, respectively (so that $E \subset F$). Let $\mathcal{N}$ be the set of all functions of $\mathcal{M}$ supported on $G = F - E$. Then $\mathcal{N}$ is an invariant subspace and $\mathcal{M} = \mathcal{N} \oplus \mathcal{M}_\infty$. It is easy to see that $\mathcal{N}_\infty = \{0\}$.

Let P_λ be orthogonal projection from $\mathcal{N}_{-\infty}$ (the set of all functions supported on G) to $\mathcal{N}_\lambda$. We write (9) once more in the Hilbert space $\mathcal{N}_{-\infty}$. The commutation relations (10) still hold. We want to prove a version of Theorem 13, and that requires some care.

The problem is to interpret the calculation (12) in $\mathcal{N}_{-\infty}$. We set $A_t = V_t 1_G$, where 1_G is the characteristic function of G. Then the analogue of (12) says

$$V_t(1_G \chi_\tau) = V_t S_\tau 1_G = e^{it\tau} S_\tau V_t 1_G = A_t T_t \chi_\tau, \tag{14}$$

since A_t vanishes outside G. By linearity, $V_t(1_G f) = A_t T_t f$ for all trigonometric polynomials f. This implies that

$$\int_G |A_t(x) f(x + e_t)|^2 \, d\sigma(x) \tag{15}$$

is independent of t, if f is a trigonometric polynomial. But then the same fact holds for all f in L^2.

Now A_t is different from 0 almost everywhere on G. Indeed, $V_t(1_G f)$ vanishes wherever A_t does if f is a trigonometric polynomial, but such functions are dense in the Hilbert space. Choose $f = 1 - 1_G$ in (15). For $t = 0$ the integral vanishes; hence it is 0 for all t. But the ergodicity of the translations T_t implies that (15) is positive for some t, unless G is a null set or all of K. In the first case $E = F$, which means that $\mathcal{M}$ was a Wiener subspace. The other possibility implies that E is a null set and F is K, which is what we wanted to prove.

(This proof is taken from [**19**].)

When Theorem 11 has been established it implies the fact about analytic functions used in the first proof. Let h be in H^2 and vanish on a set of positive measure. Then $(\mathcal{M}_h)_{-\infty}$ consists of functions vanishing where h does, and so is not L^2. Hence $\mathcal{M}_h$ is a Wiener subspace. But the only Wiener subspace contained in H^2 is $\{0\}$, so h is the null function. If h is merely in H^1 and vanishes on a set of positive measure, its product with an appropriate bounded outer function is in H^2, and the result follows. (An independent proof of this fact will be given later.) More generally, we obtain from Theorem 11 this

COROLLARY. *If f belongs to a simply invariant subspace and vanishes on a set of positive measure, then $f = 0$.*

2.4. Spectral resolution (continued)

This section carries the analysis of the spectral resolution a little further, before we change the point of view. We study a simply invariant subspace $\mathcal{M}$ with associated objects $\{P_\lambda\}$, $\{V_t\}$, $\{A_t\}$.

For each f in L^2, $-d(P_\lambda f, f)$ is a finite positive measure on the line. Let L_a be the set of all f such that this measure is absolutely continuous with respect to Lebesgue measure, L_s the set of f such that it is continuous but singular, and L_d the set of f for which it is a discrete measure. Then [27, p. 516] these are mutually orthogonal closed subspaces of L^2, with sum equal to L^2. This is a general fact about spectral resolutions; in the present special context the subspaces are invariant under multiplication by χ_τ for all τ, and thus are Wiener subspaces. Indeed if f belongs to one of the subspaces, we have $-d(P_\lambda S_\tau f, S_\tau f) = -d(P_{\lambda-\tau} f, f)$ by (10), so that $S_\tau f$ is in the same subspace. This fact implies that the spectral resolution is of pure type:

THEOREM 14. *One of L_a, L_s, L_d is all of L^2.*

Let $\mathcal{N}$ be one of the subspaces different from $\{0\}$. $\mathcal{N}$ is invariant under all P_λ. Hence $P_0\mathcal{N}$ is an invariant subspace of L^2 contained in $\mathcal{N} \cap \mathcal{M}$. It is easy to see that it contains non-null functions (otherwise $P_\lambda \mathcal{N}$ would be trivial for every λ, and $\mathcal{N}$ would be trivial). Hence $P_0\mathcal{N}$ is simply invariant. By Theorem 11 the Wiener subspace $\mathcal{N}$ must be all of L^2, and the theorem is proved.

We shall say that any of $\mathcal{N}$, $\{P_\lambda\}$, $\{V_t\}$, $\{A_t\}$ are of absolutely continuous, or singular continuous, or discrete type if the corresponding subspace is non-zero.

THEOREM 15. *A cocycle A is trivial if and only if it is of discrete type. This is the case if and only if the mapping from t to A_t is almost periodic as a function from R to L^2.*

Suppose that A is of discrete type. Find λ so that $P_{\lambda-0} \neq P_{\lambda+0}$, and let q be an element of norm one in the range of $P_{\lambda-0} - P_{\lambda+0}$. Then

$$V_t q = A_t T_t q = e^{it\lambda} q, \tag{16}$$

from which it follows that $|q|$ is constant (because invariant under all T_t), and $A_t = (\exp it\lambda) q T_t \bar{q}$. Thus A is trivial.

Suppose A is a trivial cocycle, given by the same formula. For y in K set

$$A_y(x) = \chi_\lambda(y) q(x) \bar{q}(x + y). \tag{17}$$

Then the mapping from y to A_y is continuous from K to L^2; its restriction to K_0 is simply $A_{e_t} = A_t$. The existence of such an extension is what is meant by "almost periodic" in the statement of the theorem.

Finally suppose the cocycle is almost periodic. Then the inner product $(A_t, 1)$ is almost periodic in the scalar sense. This function has the representation

$$(A_t, 1) = (V_t 1, 1) = -\int_{-\infty}^{\infty} e^{it\lambda}\, d(P_\lambda 1, 1). \tag{18}$$

This Fourier-Stieltjes transform is almost periodic only if the measure $-d(P_\lambda 1, 1)$ is discrete. Hence A is a discrete cocycle.

The three implications that have been proved establish the theorem.

Let f be a non-null element of L^2. Then $\chi_\lambda f$ necessarily moves *discontinuously* in L^2 as λ varies in Γ, if Γ is given the ordinary topology of real numbers. (For a square-summable sequence is not close to its near translates.) Thus it is reasonable to expect that the projection on $\mathcal{M}_\lambda = \chi_\lambda \cdot \mathcal{M}$ should be discontinuous, if $\mathcal{M}$ is an invariant subspace. We have shown that this is true if and only if the associated cocycle is trivial. In constructing a non-trivial cocycle we shall be finding indirectly an invariant subspace $\mathcal{M}$ such that $\mathcal{M}_\lambda$ contracts continuously as λ increases through R. There seems to be no other proof that such subspaces exist.

It is an interesting problem, largely unsolved, to characterize the measures that can be obtained as $-d(P_\lambda 1, 1)$ from invariant subspaces.

Theorem 14 has no precise origin, but has been generally known for some time. Perhaps the same should be said generally for results flowing from the spectral theorem.

2.5. An existence theorem

On the circle group every simply invariant subspace is $q \cdot H^2$ for some unitary function q. This is false on every other group K, but the following weaker result is still very useful.

THEOREM 16. *Every simply invariant subspace contains a unitary function.*

Let $\mathcal{M}$ be a simply invariant subspace with unitary group $\{V_t\}$ and cocycle A. We shall find a function g in $\mathcal{M}$ such that $\log|g|$ is summable. We have shown that $\mathcal{M}_g = q \cdot H^2$ for some unitary q, and we have $q \in \mathcal{M}_g \subset \mathcal{M}$, so the existence of g proves the theorem.

The proof depends on two lemmas.

LEMMA 1. *Let μ be normalized Haar measure on T^∞, the infinite-dimensional torus, and let $\{a_1, a_2, \ldots\}$ be any square-summable sequence of numbers. Then*

$$\int_{T^\infty} \log\left|\sum_1^\infty a_n e^{i\theta_n}\right| d\mu(e^{i\theta_1},\ldots) \geqslant \max \log|a_n|. \tag{19}$$

We may assume that a_1 has greatest modulus among the numbers a_j. Then (19) is a case of Jensen's inequality. To see this we shall identify the dual of T^∞, and introduce a suitable order relation.

A character of T^∞ is given by the formula

$$\chi(e^{i\theta_1}, e^{i\theta_2},\ldots) = e^{i\Sigma n_j\theta_j}, \tag{20}$$

where $(n_1, n_2, \ldots)$ is a sequence of integers equal to 0 from some point on. The group operation is addition in each coordinate. This lattice group is isomorphic to a subgroup of the line: choose a rationally independent set of real numbers $\{\alpha_j\}$ and map $(n_1, n_2, \ldots)$ to $\lambda = \sum n_j\alpha_j$. Thus T^∞ and its dual are groups of the type we are studying. A Fourier series on T^∞ has the form

$$f(e^{i\theta_1},\ldots) \sim \sum a(n_1, n_2, \ldots)\, e^{in_1\theta_1} e^{in_2\theta_2}\ldots, \tag{21}$$

where each term contains only finitely many exponential factors. Let f be the function with Fourier series

$$f(e^{i\theta_1},\ldots) \sim \sum_1^\infty a_n\, e^{-i\theta}\, e^{i\theta_n}. \tag{22}$$

The constant term is a_1. The other terms correspond to lattice points $(-1,\ldots,1,\ldots)$ where all the missing entries are 0. For such a lattice point, $\lambda = -\alpha_1 + \alpha_n$. If $\alpha_1 < \alpha_n$ for each $n > 1$, then (22) is a Fourier series of analytic type. Thus (19) (with a_1 on the right side) is true by Jensen's inequality, and the lemma is proved.

LEMMA 2. *Let F and G be disjoint measurable subsets of K, and γ any positive number less than $\sigma(F)\sigma(G)$. Then we can find arbitrarily large positive t such that*

$$\sigma(F \cap T_tG) = \gamma, \qquad T_tG = G - e_t. \tag{23}$$

The ergodicity of the translation group has this consequence: for any f, g in L^2

$$\lim_{T\to\infty} \frac{1}{T}\int_0^T \int f(x)g(x + e_t)\, d\sigma(x)\, dt = \int f d\sigma \int g\, d\sigma. \tag{24}$$

(The formula is easy to prove by writing down the Fourier series of f and g.) Take for f and g the characteristic functions of F and of G, respectively. The inner integral in (24) is $\sigma(F \cap T_tG)$. This quantity is continuous, vanishes for $t = 0$, and by (24) is close to $\sigma(F)\sigma(G)$ on the average for large t. The result follows.

Now let f be any non-null function in $\mathscr{M}$. We are going to define two sequences of subsets of K, and a sequence of real numbers, inductively. Find ε positive so that $|f| \geqslant \varepsilon$ on a set G_1 of positive measure δ. Let F_1 be the complement of G_1.

Suppose $n > 1$, and F_k, G_k have been defined for $k < n$ so that the G_k are disjoint, and each F_k is the complement of $G_1 \cup \ldots \cup G_k$. Let t_n be the smallest positive number such that $\sigma(F_{n-1} \cap T_{t_n} G_1) = \frac{1}{2}\delta\sigma(F_{n-1})$. (There is such a number by Lemma 2.) Define $G_n = F_{n-1} \cap T_{t_n} G_1$. Then G_n is disjoint from each G_k $(k < n)$. Take for F_n the complement of $G_1 \cup \ldots \cup G_n$.

Thus we obtain sequences of sets F_k, G_k such that the G_k are mutually disjoint, and each F_k is the complement of $G_1 \cup \ldots \cup G_k$. For $n > 1$ their measures are

$$\sigma(F_n) = (1 - \delta)\left(1 - \frac{\delta}{2}\right)^{n-1}, \qquad \sigma(G_n) = \frac{\delta}{2}(1 - \delta)\left(1 - \frac{\delta}{2}\right)^{n-2}. \tag{25}$$

Since $\sigma(F_n)$ tends to 0 the union of the G_n covers K. And $\sigma(G_n)$ tends to 0 with exponential rapidity. These are the facts we shall need.

Take $f_1 = f$, and set $f_n = V_{t_n} f$ for $n > 1$. The spectral representation (9) shows that these functions are in $\mathcal{M}$. By Theorem 13 we have $|f_n| = |T_{t_n} f|$. Hence $|f_n(x)| \geqslant \varepsilon$ for x in $T_{t_n} G_1$, and this set contains G_n.

Let θ stand for the point $(e^{i\theta_1}, e^{i\theta_2}, \ldots)$ in T^∞. For each θ the function

$$g_\theta = \sum_1^\infty n^{-2} e^{i\theta_n} f_n \tag{26}$$

is in $\mathcal{M}$. We shall show that $\log |g_\theta|$ is summable for almost all θ.

By the Fubini theorem and Lemma 1 we have

$$\int_{T^\infty} \int_K \log |g_\theta| \, d\sigma(x) \, d\mu(\theta) \geqslant \int_K \sup_n \log |n^{-2} f_n(x)| \, d\sigma(x)$$

$$\geqslant \sum_1^\infty \int_{G_n} \log (n^{-2}\varepsilon) \, d\sigma = \sum_1^\infty \log (n^{-2}\varepsilon) \, \sigma(G_n) > -\infty. \tag{27}$$

Thus the inner integral in the first term is finite for almost all θ, as was to be proved.

This is as far as the spectral theorem will take us by itself. The next step is to develop the notion of analyticity, and afterwards we try to combine the two kinds of result.

3. ANALYTICITY

3.1. The spaces L^2_ν, H^2_ν

ν is the measure on the line defined by $d\nu(t) = (1 + t^2)^{-1} \, dt$. L^2_ν is the

Lebesgue space of functions on the line based on this measure. If F is in L^2_ν, then $F(t)(1-it)^{-1}$ is in $L^2(-\infty, \infty)$ and this correspondence is a linear isometry. We are led to introduce such a space by the fact that the functions $F_x(t) = f(x + e_t)$ are not in $L^2(-\infty, \infty)$ for f in L^2, even if f is for example a constant function, but these restrictions are in L^2_ν for almost every x by the Fubini theorem.

H^2_ν is the subspace of L^2_ν consisting, as the notation suggests, of its analytic functions, but the definition is complicated by the fact that functions in L^2_ν do not have Fourier transforms. Here are four equivalent characterizations.

(a) *H^2_ν is the space of functions F on the line that are boundary functions of $F(t + iu)$ analytic in the upper half-plane, such that*

$$\int_{-\infty}^{\infty} |F(t + iu)|^2 \, d\nu(t) \tag{1}$$

is bounded by a constant independent of u, for $u > 0$.

(b) *It is the space of all F in L^2_ν such that the Fourier transform of $F(t)(1 - it)^{-1}$, defined by*

$$\int_{-\infty}^{\infty} F(t)(1 - it)^{-1} e^{-itv} \, dt, \tag{2}$$

vanishes almost everywhere for negative v. (The integrand is merely square-summable, and the transform is in the sense of Plancherel.)

(c) *It is the space of functions on the line obtained from $H^2(T)$ by mapping the disc conformally on the upper half-plane.*

(d) *It is the closure in L^2_ν of the span of exponentials $\chi_v(t) = \exp itv$ with $v \geqslant 0$.*

The methods needed for the equivalence proofs, and indeed most of the results, can be found in [**26**].

The following criterion, which is a consequence of the Plancherel theorem, will be useful later.

LEMMA. *A function F in L^2_ν belongs to H^2_ν if and only if*

$$\int_{-\infty}^{\infty} F(t)\, G(t)(1 - it)^{-2} \, dt = 0 \quad \textit{for all } G \textit{ in } H^2_\nu. \tag{3}$$

Let $F(t, x)$ be a Borel function on $R \times K$ and satisfy

$$\int\int_{-\infty}^{\infty} |F(t, x)|^2 \, d\nu(t)\, d\sigma(x) < \infty. \tag{4}$$

If we set $F_x(t) = f_t(x) = F(t, x)$, then the Fubini theorem implies that F_x is in L^2_ν for almost every x, and f_t in L^2 for almost every t. Furthermore the inner product

$$(f_t, g) = \int F(t, x)\,\bar{g}(x)\,d\sigma(x) \tag{5}$$

is in L^2_ν, for any g in L^2.

We say that F is *weakly analytic* if (5) is a function in H^2_ν for each g in L^2; F is *pointwise analytic* if F_x is in H^2_ν for almost every x. The following result is a fundamental tool.

LEMMA. *For Borel functions on $R \times K$ satisfying* (4) *the two definitions of analyticity are equivalent.*

For G in H^2_ν and g in L^2 the integral

$$\int_{-\infty}^{\infty}\int F(t, x)\,\bar{g}(x)\,G(t)\,(1 - it)^{-2}\,d\sigma(x)\,dt \tag{6}$$

exists absolutely. Suppose first F is pointwise analytic. Then for each G, the integral in t vanishes, so (6) is 0. Changing the order of integration we see that (5) is a function in L^2_ν whose integral against $G(t)\,(1 - it)^{-2}$ vanishes for every G. By the first lemma (5) is in H^2_ν. Since g is arbitrary, the definition says that F is weakly analytic.

If F is weakly analytic, then reversing the steps we find

$$\int_{-\infty}^{\infty} F(t, x)\,G(t)\,(1 - it)^{-2}\,dt = 0 \quad \textit{almost everywhere} \tag{7}$$

for each G in H^2_ν. However the exceptional null set of x depends on G. H^2_ν is a separable space; therefore (7) is true, except on a single null set, for all G in a dense subset of the space. Now F_x is in L^2_ν for almost every x; at such points the integral depends continuously on G. Thus indeed (7) holds for all G at once outside a fixed null set of x. Hence F is pointwise analytic, as we wished to prove.

Note that both kinds of analyticity refer to the variable t.

A cocycle $A_t(x)$ is only defined up to null sets of x for each t. We want to study it as a function of t for fixed x. The technical difficulty is met by the following well known

LEMMA. *Let f_t belong to L^2 for t real, and suppose the mapping from t to f_t is continuous. There is a Borel function $F(t, x)$ on $R \times K$ such that $F(t, x) = f_t(x)$ almost everywhere in x for each t.*

It is enough to prove the lemma when $f_t = 0$ outside some interval $[a, b]$. Let $\{t_n\}_1^\infty$ be an enumeration of the rationals in the interval, with $t_1 = a$. For each n choose a Borel function $\tilde{f}_{t_n}$ in the Lebesgue class of f_{t_n}. Now for positive integers k and t in the interval define $f_{t,k} = \tilde{f}_{t_n}$, where t_n is the largest one of $\{t_1, \ldots, t_k\}$ not exceeding t. Set $f_{t,k} = 0$ for t outside the interval. Thus $f_{t,k}(x)$ is a kind of step function; it is obviously a Borel function on $R \times K$. By the assumed continuity, $f_{t,k}$ tends to f_t in L^2 uniformly in t.

On passing to a subsequence (without changing notation) we can make the approximating sequence satisfy

$$\sum_1^{\infty} \| f_{t,k+1} - f_{t,k} \| < \infty \quad \text{for all } t. \tag{8}$$

Then for every t

$$\sum_1^{\infty} | f_{t,k+1}(x) - f_{t,k}(x) | < \infty \quad \text{a.e.} \tag{9}$$

so that $f_{t,k}(x)$ converges almost everywhere. Set the limit function equal to 0 where the sequence diverges. The function $F(t, x)$ so obtained is a Borel function equal to $f_t(x)$ almost everywhere, for each t.

We call F a *Borel version* of $\{f_t\}$. Two Borel versions are equal almost everywhere in x for every t; hence by the Fubini theorem they are equal for almost all t, except in a null set of x.

A cocycle A is *analytic* if any, and thus every Borel version $A(t, x)$ is analytic in the sense of the equivalent definitions given above.

3.2. Cocycles and subspaces

The notion of analyticity just introduced for cocycles will be used in this section to develop the connection between subspaces and their cocycles. The results are generalizations of facts that are easy to prove on the circle group.

THEOREM 17. *Let $\mathcal{M} = \mathcal{M}_+$ be a simply invariant subspace with cocycle A. A Borel function f in L^2 belongs to $\mathcal{M}$ if and only if $A(t, x)\, f(x + e_t)$ is analytic.*

For any g in L^2 we have

$$(A_t T_t f, g) = (V_t f, g) = -\int_{-\infty}^{\infty} e^{it\lambda}\, d(P_\lambda f, g). \tag{10}$$

If f is in $\mathcal{M}$ the measure on the right is carried on the positive real axis, and this function of t is in H^2_ν by criterion (a). If f is not in $\mathcal{M}$, then for some g (for example f) the measure has some mass on the left. Then (10) is not in H^2_ν. If it were, then the integral (10) extended over the negative axis would be a non-constant function belonging to H^2_ν together with its complex conjugate. Criterion (c) shows this is impossible. Thus f is in $\mathcal{M}$ if and only if $A(t, x)\, f(x + e_t)$ is weakly analytic.

In particular, *f in L^2 belongs to H^2 if and only if $f(x+e_t)$ is in H^2_ν for almost every* x. Now this has been proved without using Theorem 11, because we know the spectral resolution attached to H^2 directly. If f is a non-null element of H^2, then $f(x+e_t)$ is non-null in H^2_ν for almost every x; this function vanishes at most on a null set of the line by criterion (c); hence $f \neq 0$ a.e. If f is merely in H^1 we get the same result by multiplying f with a suitably small function in H^∞. Thus we have the independent proof that was promised in connection with Theorem 11 that a non-null function in H^1 is different from 0 almost everywhere.

For any simply invariant subspace $\mathcal{M}$, let $\tilde{\mathcal{M}}$ be the set of all g in L^2 such that fg is in H^1 for all f in $\mathcal{M}$. For example, $\tilde{H}^2 = H^2$. $\tilde{\mathcal{M}}$ is a simply invariant subspace, and is always normalized: $\tilde{\mathcal{M}} = (\tilde{\mathcal{M}})_+$. We also have $(\mathcal{M}_+)^\sim = (\mathcal{M}_-)^\sim$ for all $\mathcal{M}$. Evidently $\tilde{\mathcal{M}}$ is closely related to the orthogonal complement of $\mathcal{M}$: we have precisely $\tilde{\mathcal{M}} = (\bar{\mathcal{M}}^\perp)_+$, where the bar denotes complex conjugation.

THEOREM 18. *If A is the cocycle of $\mathcal{M}$, then $\bar{A}$ is the cocycle of $\tilde{\mathcal{M}}$.*

For the proof it is convenient to introduce the space K^2_ν analogous to H^2_ν, consisting of functions in L^2_ν that are analytic in the lower half-plane. Complex conjugation carries H^2_ν onto K^2_ν. Therefore their intersection consists of constant functions.

Theorem 17 has this complement: *f in L^2 is orthogonal to $\mathcal{M}_-$*if and only if $A(t,x)\,f(x+e_t)$ is in K^2_ν for almost every x.* (The proof is exactly like that of the theorem.) Equivalently, f is orthogonal to $\bar{\mathcal{M}}_-$ if and only if $A(t,x)\bar{f}(x+e_t)$ is conjugate-analytic; that is $\bar{A}(t,x)\,f(x+e_t)$ is analytic. Thus $\bar{A}$ is the cocycle of $(\bar{\mathcal{M}}_-)^\perp = (\bar{\mathcal{M}}^\perp)_+ = \tilde{\mathcal{M}}$, as was to be proved. (We use the fact, which is easy to prove, that a cocycle cannot be attached as in Theorem 17 to two distinct subspaces.)

The following lemma, needed for the next theorems, is interesting itself.

LEMMA. *Let $\mathcal{M}$ be a simply invariant subspace, and h a bounded function such that $h\cdot\mathcal{M}$ is contained in $\mathcal{M}$. Then h is in H^∞.*

There is an obvious, but tedious way to prove the lemma. Let A be the cocycle of $\mathcal{M}$. Then for each positive integer n, and any bounded function f in $\mathcal{M}$, the function $h^n(x+e_t)\,f(x+e_t)\,A(t,x)$ has a bounded analytic extension to the upper half-plane of t, for almost every x. On function-theoretic grounds this is possible only if $h(x+e_t)$ itself has such an extension.

Here is a prettier proof, based on the fact that if a function in $\mathcal{M}$ vanishes on a set of positive measure, then it vanishes identically. The idea is closely related to the proof of a related fact by Muhly [**32**].

Let $\mathcal{N}$ be the set of all bounded functions with the property of h. Then $\mathcal{N}$

is a closed subalgebra of L^∞ containing H^∞. Adding a constant to h if necessary, we may assume that h is invertible in $\mathcal{N}$. Let g be the outer function with the same modulus as h: $h = qg$ with q unitary. Now g^{-1} is analytic (the proof is easy); hence both $q = g^{-1}h$ and $\bar{q} = gh^{-1}$ are in $\mathcal{N}$.

Since $\mathcal{N}$ is uniformly closed each sum

$$\sum_{-\infty}^{\infty} a_n q^n \qquad \text{with} \qquad \sum_{-\infty}^{\infty} |a_n| < \infty \tag{11}$$

belongs to $\mathcal{N}$. If q is not a constant function we can choose the coefficients so the function k so obtained vanishes on a set of positive measure in K, but not identically. If f is any non-null function in $\mathcal{M}$ then $f \neq 0$ a.e.; thus kf vanishes on a set of positive measure but not identically. However kf is in $\mathcal{M}$, so this is impossible. The contradiction shows that q was constant, so $h = qg$ is analytic as was to be proved.

The next theorem does not mention cocycles but is closely related to the other results of this section. It is interpolated here because it is a corollary of the preceding lemma. For references see **[15]**.

THEOREM 19. *H^∞ is maximal among weakly star-closed subalgebras of L^∞.*

Let $\mathcal{N}$ be a proper star-closed subalgebra of L^∞ containing H^∞. Denote by $\mathcal{M}$ the closure of $\mathcal{N}$ in L^2. The star-closure of $\mathcal{N}$ implies (by lemmas in Section 1) that $\mathcal{M}$ is a proper, and therefore simply invariant subspace of L^2. Each element h of $\mathcal{N}$ multiplies $\mathcal{M}$ into itself; by the lemma it is in H^∞, and the theorem is proved.

We return to our subject. For any cocycle A let $\mathcal{M}_A$ be the associated normalized simply invariant subspace: $\mathcal{M}_A = (\mathcal{M}_A)_+$. Given two invariant subspaces $\mathcal{M}$ and $\mathcal{N}$, denote by $\mathcal{M} \cdot \mathcal{N}$ the smallest closed subspace of L^2 containing all products fg, where f and g are bounded members of $\mathcal{M}$, $\mathcal{N}$ respectively. (Recall that bounded functions are dense in each invariant subspace.) Such products are simply invariant, and we have

THEOREM 20. *For any cocycles A and B, $\mathcal{M}_A \cdot \mathcal{M}_B = \mathcal{M}_{AB}$ or $(\mathcal{M}_{AB})_-$.*

One inclusion is easy. For f and g as described, $A(t, x) f(x + e_t)$ and $B(t, x) g(x + e_t)$ are bounded functions in H_v^2 for almost every x, by Theorem 17. Hence their product is in H_v^2; by the same theorem, fg is in $(\mathcal{M}_{AB})_+ = \mathcal{M}_{AB}$.

In the other direction, we prove the theorem first in the special case $AB = 1$. If $\mathcal{M}_A \cdot \tilde{\mathcal{M}}_A$ is not H^2 or H_0^2, then $(\mathcal{M}_A \cdot \tilde{\mathcal{M}}_A)^\sim$ is an invariant subspace containing H^2 properly. Hence it contains a bounded function h that is not analytic: $h \cdot (\mathcal{M}_A \cdot \tilde{\mathcal{M}}_A) \subset H^2$. This implies that the functions of $h \cdot \tilde{\mathcal{M}}_A$ all multiply $\mathcal{M}_A$ into H^1, or in other words $h \cdot \tilde{\mathcal{M}}_A \subset \tilde{\mathcal{M}}_A$. By the last lemma this is impossible, and the special case is proved.

Now suppose in general that $\mathcal{M}_A \cdot \mathcal{M}_B$, which is contained in $\mathcal{M}_{AB}$, has a different cocycle. Then $(\mathcal{M}_A \cdot \mathcal{M}_B)^{\sim}$ contains $\tilde{\mathcal{M}}_{AB}$ strictly. Hence there is a bounded function h such that $h \cdot (\mathcal{M}_A \cdot \mathcal{M}_B) \subset H^2$, but $h \cdot \mathcal{M}_{AB}$ is not contained in H^2. We draw the conclusion that $h \cdot \mathcal{M}_A \subset \tilde{\mathcal{M}}_B$, and then

$$h \cdot \mathcal{M}_A \cdot \tilde{\mathcal{M}}_A \subset \tilde{\mathcal{M}}_A \cdot \tilde{\mathcal{M}}_B \subset \tilde{\mathcal{M}}_{AB}. \tag{12}$$

(The last inclusion uses Theorem 18 and the easy half of this theorem.) The special case proved above gives $h \cdot H^2$ or $h \cdot H_0^2$ for the left side. If we multiply (12) by $\mathcal{M}_{AB}$ we get

$$h \cdot (\mathcal{M}_{AB})_- \subset \tilde{\mathcal{M}}_{AB} \cdot \mathcal{M}_{AB} \subset H^2. \tag{13}$$

The minus sign is of no importance here, and this contradicts our assumption about h. The contradiction proves the theorem.

The fussiness in the statement and proof is necessary: if $\mathcal{M}$ is a subspace of continuous type $\mathcal{M} \cdot \tilde{\mathcal{M}} = H_0^2$, if $\mathcal{M}$ is discrete the product is H^2.

THEOREM 21. *$\mathcal{M}_A$ is contained in $\mathcal{M}_B$ if and only if $A^{-1}B$ is an analytic cocycle.*

Let $C = A^{-1}B$. If C is analytic, then $B(t, x) f(x + e_t)$ is in H_v^2 for almost every x if $A(t, x) f(x + e_t)$ is. Hence $\mathcal{M}_A \subset \mathcal{M}_B$ by Theorem 17.

To prove the converse we first note this special case of the theorem: $\mathcal{M}_A$ contains H^2 if and only if A is analytic. Equivalently (by Theorem 18), $\mathcal{M}_A$ is contained in H^2 if and only if $\bar{A}$ is analytic. The first of these assertions is proved by taking $f = 1$ in the statement of Theorem 17, and observing that $\mathcal{M}$ contains H^2 if and only if it contains 1.

Now suppose $\mathcal{M}_A \subset \mathcal{M}_B$. Then $\tilde{\mathcal{M}}_B \subset \tilde{\mathcal{M}}_A$, and so $\mathcal{M}_{\bar{C}} = (\mathcal{M}_A \cdot \tilde{\mathcal{M}}_B)_+ \subset H^2$. The cocycle of the subspace on the left is conjugate-analytic by the remark just made, so C is analytic as required.

The theorems of this section are function-theoretic in character. Since they translate function-theoretic information into algebraic statements, they may allow us to avoid function-theoretic reasoning in some proofs to come.

3.3. The condition for $\mathcal{M}_f$ to be simply invariant

Szegö's theorem on the circle says that $\mathcal{M}_f$, the invariant subspace generated by f in L^2, is simply invariant if and only if $\log |f|$ is summable. On other groups K (always with archimedean dual) the condition is sufficient (because then $\mathcal{M}_f = q \cdot H^2$, q unitary, and this subspace is obviously simply invariant) but it is not clear whether it is necessary. In fact it is not.

For f in L^2 define the function

$$\rho(f) = \int_{-\infty}^{\infty} \log |f(x + e_t)| \, d\nu(t). \tag{14}$$

The set on which $\rho = -\infty$ is invariant under translations of K_0, and thus has measure 0 or 1. We shall write $\rho > -\infty$ or $\rho = -\infty$ to describe the two cases.

THEOREM 22. *For f in L^2, $\mathcal{M}_f$ is simply invariant if and only if $\rho(f) > -\infty$.*

The original proof [**24**] was based on Szegö's theorem. The one that follows is function-theoretic; it uses Szegö's theorem only in the circle.

Suppose $\mathcal{M}_f$ is simply invariant. Find a bounded function g in $\tilde{\mathcal{M}}_f$ such that fg is not the null function. Then fg is in H^2, and if $\rho(fg) > -\infty$ then also $\rho(f) > -\infty$. Thus it will suffice to show that $\rho(f) > -\infty$ for every non-null f in H^2.

For any such f, Theorem 17 says that $f(x + e_t)$ is in H_v^2 for almost all x. Thus $f(x + e_t)$ is the image of a function in $H^2(T)$ under a conformal map that carries the disc to the upper half-plane. If 0 is carried to i, this map carries Lebesgue measure on the circle to v on the line. By Szegö's theorem on the circle, $\log|(F(e^{iu})|$ is summable for any non-null function F in $H^2(T)$. Since $f(x + e_t)$ is not the null function (almost all x) we have $\rho(f) > -\infty$.

In the other direction the proof depends on this fact: if f_n converges to f in L^2, then $f_n(x + e_t)$ tends to $f(x + e_t)$ in L_v^2 for almost every fixed x, at least on a subsequence of n. The proof depends merely on Fubini's theorem.

Now suppose f is in L^2 and $\rho(f) > -\infty$. We may suppose f non-negative. There is a real function $k(t, x)$ such that $f(x + e_t) \exp ik(t, x)$ is in H_v^2 for almost every x. Indeed carry f by conformal map to the unit circle. Its logarithm is summable; hence it is the modulus of an outer function in $H^2(T)$. This outer function, mapped back to the half-plane, is $f(x + e_t) \exp ik(t, x)$. ($k(t, x)$ is determined for each x up to an additive constant; there is no reason to think it can be defined so as to be measurable in (t, x).)

If $\mathcal{M}_f$ is a Wiener subspace it is all of L^2. Thus any g in L^2 is the limit of functions $P_n f$ where the P_n are analytic trigonometric polynomials. Then $(P_n f)(x + e_t)$ converges to $g(x + e_t)$ in L_v^2 for almost every x on a subsequence of n; hence also $(P_n f)(x + e_t) \exp ik(t, x)$ tends to $g(x + e_t) \exp ik(t, x)$. The approximating functions are in H_v^2 for almost every x, so the limit function is too. But this conclusion is not true for all g, for example if g vanishes on a set of positive measure but not identically. The contraction shows that $\mathcal{M}_f$ is simply invariant, and the theorem is proved.

COROLLARY. *If f is a non-null element of a simply invariant subspace, then $\rho(f) > -\infty$.*

Let f belong to L^2. If $\log|f|$ is in L^1 then $\rho(f) > -\infty$ by the Fubini theorem. The following example shows that the converse implication is false.

Let F be a closed set in K having positive measure and disjoint from K_0.

Denote the characteristic function of F by k. Then

$$\int_{-\infty}^{\infty} k(x - e_t)\, d\nu(t) \tag{15}$$

is positive almost everywhere, but is not essentially bounded from 0; for the integral is continuous and vanishes for $x = 0$ (the proof is easy). Therefore we can find a positive function h that is not summable, but satisfies

$$\iint_{-\infty}^{\infty} k(x - e_t)\, h(x)\, d\nu(t)\, d\sigma(x) < \infty. \tag{16}$$

This implies

$$\iint_{-\infty}^{\infty} h(x + e_t)\, d\nu(t)\, k(x)\, d\sigma(x) < \infty. \tag{17}$$

Hence the inner integral is finite for almost all x in F. But the set where it is finite is invariant under translations from K_0, so it is finite for almost all x. Thus $\rho(e^{-h}) > -\infty$, but h is not summable.

3.4. The modulus of an analytic function

We have shown, as a corollary of Szegö's theorem, that a non-negative summable function w is $|g|^2$ for some g in H^2 with $a_0(g) \neq 0$ if and only if $\log w$ is summable. On the circle, this result characterizes the modulus of functions in H^2. On K the question remains whether $\log|g|$ can fail to be summable if $a_0(g) = 0$.

THEOREM 23. *Let w be non-negative in L^2. There is a non-null function g in H^2 such that $w = |g|$ if and only if $\rho(w) > -\infty$.*

If g is non-null in H^2 then $\mathscr{M}_g$ is simply invariant. Hence $\rho(g) > -\infty$ by the last theorem.

Suppose that $\rho(w) > -\infty$. It has been known for a long time that some g in H^2 satisfies $0 < |g| \leqslant w$ almost everywhere. Indeed let h be a function in $\bar{\mathscr{M}}_w$ bounded by 1; then $g = hw$ is such a function.

Theorem 16, which is a more recent result, asserts that h can be chosen to be a unitary function, and that proves the theorem.

From what we have shown it follows that there are functions g in H^2, and even in H^∞, such that $\log|g|$ is not summable. Conventional wisdom suggests that such a g can even be continuous. A curious theorem of Arens **[2]**, which is older than all these other results, asserts that this is impossible.

THEOREM 24. *If g is non-null, analytic and continuous, then* $\log|g|$ *is summable.*

Define

$$f(x) = \int_{-\infty}^{\infty} g(x - e_t)\, dv(t)/\pi. \tag{18}$$

Then f is non-null and continuous. Let G be a non-empty open set and δ a positive number such that $|f| \geqslant \delta > 0$ on G. Let h be continuous, non-negative, null outside G, and strictly positive somewhere in G. We have

$$-\infty < (\log \delta) \int h\, d\sigma \leqslant \int h \log |f|\, d\sigma. \tag{19}$$

Now $\pi^{-1}\, dv(t)$ is a Poisson kernel. Therefore on function-theoretic grounds

$$\log |f(x)| = \log \pi^{-1} \left| \int_{-\infty}^{\infty} g(x - e_t)\, dv(t) \right| \leqslant \pi^{-1} \int_{-\infty}^{\infty} \log |g(x - e_t)|\, dv(t). \tag{20}$$

Thus (19) can be continued:

$$\leqslant \int \int_{-\infty}^{\infty} h(x) \log |g(x - e_t)|\, dv(t)\, d\sigma(x)/\pi. \tag{21}$$

Since the integrand is bounded above we can change variables to obtain

$$-\infty < \int \int_{-\infty}^{\infty} h(x + e_t)\, dv(t) \log |g(x)|\, d\sigma(x). \tag{22}$$

The inside integral is continuous, and positive because every coset $x + K_0$ intersects the open set where h is positive. Hence this integral has a positive lower bound. Therefore $\log |g|$ is summable.

4. THE STRUCTURE OF COCYCLES

4.1. The problem

A non-null function f in $H^2(T)$ is a product qg with q inner, g outer. The outer factor is determined by the modulus of f; it is fair to say that the function-theoretic character of f is determined by the inner factor. Thus the invariant subspace $q \cdot H^2$ is an object that, from one point of view, carries the same information as f.

On groups K with archimedean duals there is no factoring theorem (except under additional hypothesis), but f in H^2 generates an invariant subspace $\mathcal{M}_f$ that represents f in a sense. Then there is no reason to restrict attention to subspaces generated by functions in H^2. We form $\mathcal{M}_f$ for any f in L^2, and we study invariant subspaces in general.

Simply invariant subspaces give rise to cocycles. Inclusion of subspaces is related to analyticity of cocycles. The classification of simply invariant subspaces ordered by inclusion is thus reduced to the classification of cocycles ordered by analyticity (see Theorem 21). Two general problems emerge: to study the structure of analytic cocycles, and to find properties of the analytic cocycles as a subset of the class of all cocycles.

We make a start on these problems in this chapter.

4.2. A criterion for triviality

Theorem 15 said that a cocycle is trivial if and only if it is almost periodic as a mapping from R to L^2. The following interesting variant was proved by C. G. R. Carlson [7].

THEOREM 25. *A cocycle A is trivial if and only if any Borel version $A(t, x)$ is a Besicovitch function of t for almost every x, with exponents belonging to a countable set that is independent of x.*

A trivial cocycle is one of the form

$$e^{it\gamma}q(x)\,\bar{q}(x + e_t) \tag{1}$$

where γ is a real number and q a unitary function. For almost every x, $\bar{q}(x + e_t)$ is a Besicovitch function of t (this is Lowdenslager's theorem, proved in Chapter 2); its product with exp $it\gamma$ is still a Besicovitch function. Moreover q has its Fourier frequencies in a countable subgroup Γ_0 of Γ. Hence the function (1) of t can be approximated in Besicovitch norm by exponential polynomials with frequencies in $\Gamma_0 + \gamma$. This proves half the theorem.

In the other direction suppose the condition is satisfied, with the frequencies of $A(t, x)$ in a fixed countable set Γ_0 for almost every x. Then we can find γ in Γ_0 such that the coefficient $a_\gamma(x)$ of $A(t, x)$ is different from 0 for all x in a set of positive measure. (Otherwise since Γ_0 is countable $A(t, x)$ would have no non-zero coefficients, except in a null set of x, whereas its norm in B^2 is 1 for each x.) Hence the quantity

$$\frac{1}{2T}\int_{-T}^{T} e^{-it\gamma}(A_t, a_\gamma)\,dt = \int_{-\infty}^{\infty} \frac{\sin T(\lambda - \gamma)}{T(\lambda - \gamma)}\,d(P_\lambda 1, a_\gamma) \tag{2}$$

has limit different from 0 as T tends to ∞. But the integrand on the right side tends to 0 except at $\lambda = \gamma$. Hence the measure carries mass at this point. This means the spectral resolution $\{P_\lambda\}$ associated with the cocycle is not continuous. By Theorem 15 the cocycle is trivial.

It is not known whether the hypothesis about the exponents can be omitted from the statement of the theorem.

A technical problem can be dealt with conveniently at this point. Let A be a cocycle on K. For each t write the Fourier series of A_t:

$$A_t(x) \sim \sum a_\lambda(t)\,\chi_\lambda(x). \tag{3}$$

The sum really extends merely over a countable set Γ_0, because the cocycle takes its values in a separable subspace of L^2. If Γ_1 is any dense subgroup of R containing Γ_0, then (3) can be interpreted as a Fourier series on the group K_1 dual to Γ_1. The functions A_t on K_1 thus determined are a cocycle on K_1 (because the defining properties of a cocycle can be expressed in terms of the coefficients $a_\lambda(t)$). The fact we want is this: A is trivial on K if and only if it is trivial on K_1. For the proof, observe that $a_0(t)$ is the Fourier–Stieltjes transform of $-d(P_\lambda 1, 1)$ on either group; hence the spectral resolution associated with the cocycle on K is of the same type as the spectral resolution of the cocycle on K_1. By Theorem 15, A is trivial on K if and only if it is trivial on K_1. This remark enables us to assume that Γ is countable in certain proofs.

4.3. The counterexample

Much of what has preceded would be empty if every cocycle were trivial, and now we shall construct non-trivial cocycles on each group K. Cocycles more or less related to the ones we have defined abound in analysis, often disguised. So far as the author is aware, the construction below is the only one known that succeeds on all K. However the claim, even if true, is not of great importance, for the literature is large and rich. The following references, at least, are directly relevant to our subject: [**14**, **15**, **23**, **25**, **35**, **46**].

By the remark of the last section we may assume that Γ is countable. Let $\{\lambda_1, \lambda_2, \ldots\}$ be positive elements of Γ decreasing to 0 and satisfying

$$\sum_1^\infty \lambda_j < \infty. \tag{4}$$

Define $\lambda_0 = 0$, $\lambda_{-n} = -\lambda_n$, and set

$$A(t, x) = \exp i\left[\sum_{j=-\infty}^{\infty} \chi_{\lambda_j}(x)\,(1 - e^{it\lambda_j})\right]. \tag{5}$$

The series converges absolutely on account of (4), and defines a cocycle. We shall choose the λ_j so as to make the cocycle non-trivial.

Find a sequence of positive numbers t_j tending to ∞ such that $e_{t_j} \to 0$ in K. This means exactly that $\exp it_j\lambda \to 1$ for every λ in Γ, and is possible because Γ is countable. Choose ε_j positive so that

$$\sum_1^\infty j\varepsilon_j < \infty. \tag{6}$$

We are going to define the sequence $\{\lambda_j\}$ and a subsequence $\{u_j\}$ of $\{t_j\}$ inductively. Set $u_1 = t_1$. Take for λ_1 any element of Γ such that

$$|\exp iu_1\lambda_1 + 1| < \varepsilon_1 \quad \text{and} \quad 0 < \lambda_1 < \pi u_1^{-1}. \tag{7}$$

(The first inequality evidently holds if λ_1 is close to πu_1^{-1}.) When u_j, λ_j have been chosen for $j < n$, take u_n to be the smallest of the numbers t_j such that

$$|\exp iu_n\lambda_j - 1| < \varepsilon_n \quad (1 \leqslant j < n) \quad \text{and} \quad u_n > \varepsilon_n^{-1}u_{n-1}. \tag{8}$$

Then find λ_n in Γ to satisfy

$$|\exp iu_n\lambda_n + 1| < \varepsilon_n \quad \text{and} \quad 0 < \lambda_n < \pi u_n^{-1}. \tag{9}$$

Then we have

$$u_n\lambda_j < \pi\varepsilon_j \qquad (n < j); \tag{10}$$

for $n = j - 1$ this follows from the second inequalities in (8) and (9), and it holds *a fortiori* for smaller n. From (10) we obtain a complement to (8):

$$|\exp iu_n\lambda_j - 1| < \pi\varepsilon_j \qquad (n < j). \tag{11}$$

The construction ensures that (4) holds, so the cocycle is defined, and we must show that it is non-trivial. The crucial fact is that at $t = u_n$ it is almost equal to $\exp 4i \operatorname{Re} \chi_{\lambda_n}$. From (5) we have

$$A(u_n, x) = \exp i\,[2\operatorname{Re} \chi_{\lambda_n}(x)\,(1 - e^{iu_n\lambda_n}) + \textstyle\sum' \chi_{\lambda_j}(x)\,(1 - e^{iu_n\lambda_j})], \tag{12}$$

where the dashed sum omits $j = n, -n$. Call the quantity in brackets $K_n(x)$. The inequalities above show that

$$|K_n(x) - 4\operatorname{Re} \chi_{\lambda_n}(x)| < 2n\varepsilon_n + 2\pi \sum_{n+1}^{\infty} \varepsilon_j \tag{13}$$

for all x. The quantity on the right tends to 0 by (6). If the cocycle is trivial, given by (1), then $qT_{u_n}\bar{q}$ tends to the constant function 1 in L^2 (because $e_{u_n} \to 0$ in K). On a subsequence $\exp iu_n\gamma$ tends to a limit. Hence on this subsequence A_{u_n} tends to a constant function in L^2.

From (13) it follows that $\exp 4i \operatorname{Re} \chi_{\lambda_n}$ converges to the same constant function. This is impossible, because the distribution of values of $\chi_\lambda(x)$ on the circle is independent of λ ($\lambda \neq 0$). The contradiction shows the cocycle is non-trivial.

The last step, involving the distribution of values of χ_λ, will yield more information in the next section, so it is worthwhile to examine it carefully. Let φ be a trigonometric polynomial on the circle and $\lambda \neq 0$. Then obviously

$$\int \varphi(\chi_\lambda(x))\, d\sigma(x) = a_0(\varphi). \tag{14}$$

By approximation the same formula holds for all continuous functions on T. If we apply this to λ_n and

$$\varphi(e^{it}) = \psi(e^{4i\cos t}) \tag{15}$$

we get

$$\int \psi(\exp 4i \operatorname{Re} \chi_{\lambda_n})\, d\sigma = a_0(\varphi) \tag{16}$$

for each n. Assuming that $\exp 4i \operatorname{Re} \chi_{\lambda_n}$ converges to the constant function $\exp i\alpha$ we should have $a_0(\varphi) = \psi(\exp i\alpha)$ for every continuous function ψ on T. Of course this is absurd.

4.4. Refinements

The cocycle constructed in the last section is of continuous type, since it is non-trivial, but we can prove more: actually the cocycle is singular continuous (see Section 2.4). That is, the associated spectral resolution $\{P_\lambda\}$ is singular with respect to Lebesgue measure. To prove this it suffices to show that the Fourier–Stieltjes transform

$$-\int_{-\infty}^{\infty} e^{it\lambda}\, d(P_\lambda 1, 1) = \int A_t\, d\sigma \tag{17}$$

does not tend to 0 as $t \to \infty$.

For large n the integral of A_{u_n} differs little from

$$\int \exp(4i \operatorname{Re} \chi_{\lambda_n})\, d\sigma = \frac{1}{2\pi} \int_0^{2\pi} e^{4i \cos t}\, dt. \tag{18}$$

A direct argument shows this quantity is not 0, so the assertion is proved.

(The proof that the integral does not vanish can be avoided if we change the cocycle a little. Multiply the exponent in (5) by a positive number r. The cocycle so obtained is like the original one; in particular (13) holds with the obvious modification. For this cocycle the integrand on the right side of (18) is $\exp 4ri \cos t$, and obviously the integral is not 0 if r is small enough.)

C. C. Moore has shown that the cocycle (5) is of absolutely continuous type if the exponents are chosen in a different way, at least if Γ contains an infinite linearly independent subset. (No doubt this hypothesis can be removed.)

The proof needs a more general version of (14). Let $\varphi_1, \ldots, \varphi_n$ be continuous functions on T, and $\lambda_1, \ldots, \lambda_n$ rationally independent elements of Γ. Then

$$\int \prod_1^n \varphi_j(\chi_{\lambda_j})\, d\sigma = \prod_1^n a_0(\varphi_j). \tag{19}$$

The fact is easily verified if the φ_j are trigonometric polynomials, and persists under uniform convergence.

Choose the λ_j $(j \geqslant 1)$ to be independent and satisfy (4). For the cocycle (5) we can compute the quantity (17):

$$\int A_t\,d\sigma = \lim \int \prod_1^n \exp 2i\,\mathrm{Re}\,[\chi_{\lambda_j}(1 - e^{it\lambda_j})]\,d\sigma$$

$$= \lim \prod_1^n \frac{1}{2\pi}\int_0^{2\pi} \exp 2i\,\mathrm{Re}\,[e^{iu}(1 - e^{it\lambda_j})]\,du$$

$$= \lim \prod_1^n \frac{1}{2\pi}\int_0^{2\pi} \exp 2i\,[\cos u - \cos(u + t\lambda_j)]\,du = \prod_1^\infty \rho(\lambda_j t), \tag{20}$$

where

$$\rho(t) = \frac{1}{2\pi}\int_0^{2\pi} \exp 2i\,[\cos u - \cos(u + t)]\,du. \tag{21}$$

We shall prove for suitable frequencies that (20) tends to 0 rapidly, so that (17) is a square-summable function of t. This implies that the measure in (17) is absolutely continuous, as we want to prove.

From (21) it is easy to see that $|\rho(t)| \leqslant 1$ for all t, and the inequality is strict on some interval, say $|\rho(t)| \leqslant \delta < 1$ on (a, b). Let $n(t)$ denote the number of points $\lambda_j t$ lying in (a, b) From (20) we have

$$\left|\int A_t\,d\sigma\right| = \left|\prod_1^\infty \rho(\lambda_j t)\right| \leqslant \delta^{n(t)}. \tag{22}$$

We choose the λ_j now to make $n(t)$ large for all large t.

The choice is not delicate. Take $\lambda_j = j^{-2} + \gamma_j$, where the γ_j are real numbers that satisfy $j^2\gamma_j \to 0$, and make $\{\lambda_j\}$ $(j \geqslant 1)$ an independent subset of Γ. The crucial condition (4) holds, and we verify that $n(t)$ exceeds a constant times $t^{\frac{1}{2}}$ for large t. Hence the right side of (22) is square-summable (and much more) over the right half-line. Now $|\rho|$ is an even function, so the same is true on the left, and the result is proved.

4.5. The differential equation

A connection was made in [**21**] between invariant subspaces (in a different context) and a linear differential equation. The same idea can be applied to the objects we are studying.

Let m be a real function defined on K. We ask for a function q satisfying

$$q'(x) = i\,m(x)\,q(x). \tag{23}$$

The differentiation can be interpreted in two ways. The pointwise derivative is

$$q'(x) = \lim_{t \to 0} t^{-1}[q(x + e_t) - q(x)]. \tag{24}$$

Or we can view differentiation as an unbounded operator in L^2. Let D be the operator defined by

$$D(\sum a_\lambda \chi_\lambda) = \sum \lambda a_\lambda \chi_\lambda \tag{25}$$

on the domain consisting of all functions in L^2 whose coefficients satisfy

$$\sum |\lambda a_\lambda|^2 < \infty. \tag{26}$$

Then D is self-adjoint and $Dq(x) = -iq'(x)$ almost everywhere, for q in the domain of D. For simplicity suppose m is bounded. Then multiplication by m in L^2 is a bounded self-adjoint operator, and $D - m$ is self-adjoint on the domain of D. The differential equation is analogous to the operator equation

$$(D - m)\, q = 0. \tag{27}$$

Let this operator have spectral resolution

$$D - m = -\int_{-\infty}^{\infty} \lambda dP_\lambda, \tag{28}$$

where the projections P_λ are chosen decreasing.

The domain of D, which is the same as that of $D - m$, is invariant under each operator S_τ (multiplication by χ_τ, τ in Γ). A straightforward calculation shows

$$S_\tau^{-1}(D - m)\, S_\tau = (D - m) + \tau I \qquad (\tau \text{ in } \Gamma). \tag{29}$$

This commutation relation, with (28), implies that $P_{\lambda+\tau} = S_\tau P_\lambda S_{-\tau}$, or in other words that $\{P_\lambda\}$ is the spectral resolution of an invariant subspace of L^2. The unitary group associated with the subspace is $V_t = \exp it(D - m)$.

In the other direction, suppose A is a cocycle given by

$$A(t, x) = \exp\left(i \int_0^t m(x + e_u)\, du \right), \tag{30}$$

where $m(x)$ is real and suitably smooth. Then a formal calculation suggests that the infinitesimal generator of $V_t = A_t T_t$ is $D + m$, and this is true under suitable hypotheses.

Thus a formal connection is established between the linear differential equation (27) and simply invariant subspaces of L^2. The study of subspaces is more general, because not all self-adjoint operators satisfying (29) can be expressed in the form $D - m$.

Let m be a bounded real function on K. We shall say that $D + m$ is *equivalent* to D if $D + m = \bar{q}Dq$, where q is multiplication by a unitary function also called q. Applied to the function 1, this equation means (27); conversely, if (27) has a non-null solution q, its modulus is necessarily constant, and if $|q| = 1$ a.e. then the operator equation holds. Thus the equivalence problem is the same as the problem of solving (27).

Given a bounded real function m, set $V_t = \exp it(D - m)$, $A_t = V_t 1$. A function q is a solution of (27) if and only if $V_t q$ is independent of q, that is $A_t T_t q = q$ for all t. A non-null function q exists with this property if and only if A is a coboundary. Similarly, A is a trivial cocycle if and only if $(D - m + \alpha)\, q = 0$ for some real α and non-null function q.

Our construction produced a cocycle (5) of the form (30), in which indeed $m(x)$ is a function with absolutely convergent Fourier series. Thus the existence of solutions of (27) is a structural question, and does not merely depend on the smoothness of $m(x)$. The refinements given in the last section give further information about the spectrum of $D - m$ for smooth functions $m(x)$.

The theorem of the next section asserts that every cocycle is cohomologous to a cocycle of special type. This, and other results of the same type, can be interpreted to say that any self-adjoint operator satisfying (29) is equivalent in the sense above to $D - m$, where $m(x)$ is a real function on K that is smooth in some sense.

4.6. An equivalence theorem

For the first time we shall have to refer to the function–theoretic description of inner functions in the upper half-plane. An account congenial to our point of view is contained in [**26**]. If $q(z)$ is inner (the upper half-plane is understood, unless mention is made otherwise), its *weight at infinity* is the largest number γ such that $q(z) \exp - i\gamma z$ is inner. If this number is 0 we say that q has no weight at infinity.

Let A be an analytic cocycle. We say that A is a *Blaschke* or a *singular* cocycle if in some Borel version $A(t, x)$ is an inner function of that type for almost every x. An analytic cocycle A has a bounded analytic extension to the upper half-plane of t (for almost every x), and the functional equation of a cocycle has the complex form

$$A(t + z, x) = A(t, x)\, A(z, x + e_t) \quad \text{a.e. } \textit{for every } t, z. \tag{31}$$

The inner function $A(z, x)$ has a weight at infinity $\gamma(x)$ for each x. Then $\gamma(x + e_t) = \gamma(x)$ for all t. Furthermore $\gamma(x)$ is a Borel function (the proof is not hard), so $\gamma(x)$ is constant, and we can speak of the weight at infinity of any analytic cocycle.

THEOREM 26. *Every cocycle A is cohomologous to a Blaschke cocycle B, which has the property that the zeros of $B(z, x)$ do not accumulate on the real axis, for almost all x.*

Theorem 16 implies immediately that every cocycle is cohomologous to an analytic cocycle. Indeed let q be a unitary function in $\mathcal{M}$, the invariant subspace associated with the cocycle A. Then $\bar{q}\mathcal{M}$ contains 1; hence its cocycle B is analytic. But A and B are cohomologous, and the assertion is proved. In order to show that B can be found in the special form described by the theorem, we have to choose q in $\mathcal{M}$ in a particular way, and this will require strengthening of Theorem 16.

Here is the information we need: *$\mathcal{M}$ contains a function g such that* $\log|g|$ *is summable, g is orthogonal to $\mathcal{M}_\tau$ for some positive τ, but g is not contained in $\mathcal{M}_\lambda$ for any positive λ.* Let us assume for the moment that g exists and finish the proof of the theorem.

Without loss of generality we may assume that $\mathcal{M}$ contains H^2, so that A is an analytic cocycle, and that g is analytic (indeed outer). For g can be factored as qh, q unitary and h outer. Now $\bar{q}\cdot\mathcal{M}$ contains 1 and therefore all of H^2, and its cocycle is cohomologous to A. Furthermore h has the same properties in $\bar{q}\cdot\mathcal{M}$ as g in $\mathcal{M}$.

Returning to original notation, we shall show under these hypotheses that the cocycle A of $\mathcal{M}$ is of the type specified in the theorem.

Since g is in $\mathcal{M}$, $A(t, x)\,g(x + e_t)$ is in H_v^2 for almost every x. But also $\chi_{-\tau}g$ is orthogonal to $\mathcal{M}$; hence $A(t, x)\,g(x + e_t)\,e^{-it\tau}$ is in K_v^2 (the space of conjugate–analytic functions) for almost every x. It follows that $A(t, x)\,g(x+e_t)$ is equal almost everywhere in t to a function that is analytic on a neighborhood of the real axis, for almost every x. On function–theoretic grounds, the inner function $A(t, x)$ must be analytic on the axis. That is, $A(t, x)$ has no singular inner divisor except perhaps a weight at infinity, and its zeros do not accumulate on the real axis. There is no weight at infinity either, for if there were it could be chosen independent of x, and then $\chi_{-\lambda}g$ would belong to $\mathcal{M}$ for some positive λ.

(The function–theoretic point in the proof can be phrased as follows: if q is inner and f is analytic on the circle, and if qf is analytic on some arc, then q is analytic on the arc. If q has zeros accumulating at a point of the circle the same is true of qf, which therefore cannot be analytic there. If q has singular part the argument is only a little more complicated.)

The theorem will be proved if we show that $\mathcal{M}$ contains a function g with the properties mentioned. To find g we have to run through the proof of Theorem 16 again. We started with any non-null function f in $\mathcal{M}$, and found g in the form of a weighted sum of functions $V_{t_j}f$. If f is orthogonal to $\chi_\tau\cdot\mathcal{M}$, then g will be too, and that is one requirement that had to be met. However g may belong to $\mathcal{M}_\lambda$ for some positive λ, even if f does not. Another step is needed to defeat this possibility.

Call the g already found g_1, and suppose it is contained in $\mathcal{M}_{\lambda_1}$ for some $\lambda_1 > 0$. Repeat the procedure to find a similar function g_2 orthogonal to

$\mathcal{M}_{\lambda_1}$, but perhaps contained in $\mathcal{M}_{\lambda_2}$ where $0 < \lambda_2 < \frac{1}{2}\lambda_1$. We continue in this way indefinitely, if necessary. Take the functions g_n to have norm 1, and set

$$g_\theta = \sum_1^\infty n^{-2} e^{i\theta_n} g_n \tag{32}$$

as in the proof of Theorem 16. For every θ in T^∞, g_θ is orthogonal to $\mathcal{M}_\tau$ but not contained in $\mathcal{M}_\lambda$ for any positive λ. For almost every θ, $\log|g|$ is summable. This completes the proof.

4.7. A representation theorem

Let $B(t, x)$ be a Borel version of the Blaschke cocycle obtained in the last theorem. For almost every x, $B(t, x)$ coincides almost everywhere in t with a function that is analytic on the real axis. However if we modify $B(t, x)$ on a null set of t we may apparently lose the cocycle. Here is an argument to show, in this case, that $B(t, x)$ actually can be analytic on the real axis of t for almost all x.

Form the two integrals

$$\frac{1}{2\varepsilon}\int_{-\varepsilon}^{\varepsilon} B_{t+u}\,du, \qquad \frac{1}{2\varepsilon}\int_{-\varepsilon}^{\varepsilon} B(t+u, x)\,du. \tag{33}$$

The first is the Bochner integral of a continuous vector-valued function, the second is the ordinary integral of a Borel function. As ε tends to 0 the first tends to B_t in L^2, because a cocycle is continuous as a mapping from R to L^2. The second integral tends to the value of the analytic function that matches $B(t, x)$ for almost all t. For any positive ε the integrals (33) represent the same function in L^2; hence their limits are equal almost everywhere on K for each t. This shows that the Borel version $B(t, x)$ can be continuous in t for almost every x, as we wished to show.

THEOREM 27. *Every cocycle A can be represented in the form*

$$A(t, x) = q(x)\,q(x + e_t)^{-1} \exp i \int_0^t m(x + e_u)\,du \tag{34}$$

where q is a unitary function on K, and m a real Borel function such that $m(x + e_u)$ is continuous in u for each x.

A version of this theorem, with an extra factor on the right side, was given in [20]. The superfluous factor was removed by Carlson [7].

After multiplying A by a coboundary, which accounts for the coboundary in (34), we have a cocycle B such that $B(t, x)$ is analytic in t for almost every x. If we set $B(t, x) = 1$ for exceptional x we can proceed without null sets.

Set $m(x) = -iB'(0, x)$, where the dash means differentiation with respect to t. Then m is a real Borel function. The functional equation of a cocycle implies

$$B'(t, x) = i\, m(n)(x + e_t)\, B(t, x), \tag{35}$$

which shows that $m(x + e_t)$ is analytic in t. The solution of this differential equation with initial value $B(0, x) = 1$ is the exponential function in (34), and the theorem is proved.

The smoothness of m on cosets of K_0 does not obviously imply much about m globally on K. From the point of view of the differential equation (27) it would be interesting to show that every cocycle is cohomologous to an exponential cocycle in which m not too large. Nevertheless the representation theorem shows that all cocycles can be approximated by cocycles with special properties; no doubt more information about the structure of cocycles can be obtained from this fact. As we have seen, however, smoothness of m does not imply that the exponential is a coboundary.

4.8. Versions of cocycles

A cocycle A_t is defined as a mapping from R to L^2. We have used in an important way the fact that every cocycle has a Borel version; the proof does not need the algebraic properties of the cocycle. Now we come to two theorems of Gamelin [**15**] whose effect is to abolish null sets from the study of cocycles.

THEOREM 28. *Every cocycle A has a Borel version such that*

$$|A(t, x)| = 1, \qquad A(t + u, x) = A(t, x)\, A(u, x + e_t) \tag{36}$$

for all t, u, x without exception.

A Borel version of A has the form (34), which must be interpreted to hold for almost every x, for each t. If we choose a definite Borel function of modulus 1 for q, and similarly a real Borel function m (as we did actually in the proof of Theorem 27), then the right side of (34) defines a Borel version of A having the required properties. Such a function is called a *strict version* of A.

THEOREM 29. *Let $A(t, x)$ be a Borel function on $R \times K$ such that*

$$|A(t, x)| = 1, \quad A(t + u, x) = A(t, x)A(u, x + e_t) \tag{37}$$

for almost all (t, u, x) in $R \times R \times K$. Then there is a cocycle A such that A_t equals $A(t, x)$ almost everywhere on K, except in a null set of t.

The idea of the proof is to define operators V_t by setting $V_t f(x) = A(t, x)$ $f(x + e_t)$, and showing that $\{V_t\}$ is a continuous unitary group in spite of difficulties with null sets. This is difficult in L^2, but works in a certain subspace.

Let H be the subspace of L^2 obtained by closing the set of functions

$$\tilde{h}(x) = \int_{-\infty}^{\infty} h(u)A(u, x)\, du, \qquad h \in L^1(-\infty, \infty). \tag{38}$$

For each such function define

$$V_t\tilde{h}(x) = \int_{-\infty}^{\infty} h(u - t)A(u, x)\, du. \tag{39}$$

For any h and k summable on the line we have $(V_t\tilde{h}, V_t\tilde{k}) = (\tilde{h}, \tilde{k})$, where the inner products are in L^2. The proof consists in writing out the triple integrals and using (37). Since functions of the form (38) are dense in H, V_t is a unitary operator in H for each real t.

The operator that carries h in $L^1(-\infty, \infty)$ to $\tilde{h}$ in L^2 reduces norm. Hence $V_t\tilde{h}$ moves continuously in L^2 for each $\tilde{h}$ of the form (38), and the same follows for every element of H. Thus $\{V_t\}$ is a continuous unitary group in H.

If we replace u by $u + t$ in (39) we get formally

$$V_t\tilde{h}(x) = A(t, x)\tilde{h}(x + e_t). \tag{40}$$

This is the formula we want, but the step has to be done carefully. When the cocycle is broken apart by (37) we get equality for almost all pairs (t, x) and the exceptional set does not depend on h. Hence for t outside a fixed null set Z we have

$$|A(t, x)| = 1 \quad \text{and} \quad V_t\tilde{h}(x) = A(t, x)\tilde{h}(x + e_t) \quad \text{a.e.} \tag{41}$$

Let A_t be the element of L^2 equal to $A(t, x)$ for t not in Z; thus (41) means $V_t\tilde{h} = A_tT_t\tilde{h}$ for such t. If H contains constant functions (and it does, once the theorem is proved), then A_t is continuous as a mapping from the complement of Z to L^2. The fact can be proved, however, merely knowing that the functions of H do not all vanish on any set of positive measure. Indeed, for t and u not in Z

$$A_tT_t\tilde{h} - A_uT_u\tilde{h} = (A_t - A_u)T_u\tilde{h} + A_t(T_t\tilde{h} - T_u\tilde{h}). \tag{42}$$

As t tends to u the left side tends to 0 in norm because the unitary group is continuous. The second term on the right also tends to 0, so the first term on the right does too. Since A_t is a unitary function, this implies that A_t tends to A_u.

The continuous extension of this mapping to points of Z defines A_t, still a multiplication, for all t, and $\{A_t\}$ is a cocycle with the required properties.

4.9. Gamelin's representation theorem

Theorem 27 gives the form of an arbitrary cocycle, but the representation is not unique, and the formula does not answer all questions about the structure of cocycles. Gamelin has extended a theorem of algebraic cohomology theory to this context; his result gives a different way of generating cocycles on K. The information given by this representation theorem is not of the same kind as that provided by (34), and thus the two approaches complement each other.

Let γ be a fixed positive element of Γ and K_γ the subgroup of K that annihilates γ. For notational simplicity let e be the element $e_{2\pi/\gamma}$ of K_0. The intersection of K_0 with K_γ consists exactly of the multiples ke of e. For functions f defined on K_γ let $Tf(x) = f(x + e)$. A *cocycle* on K_γ is a function a from the integers to $L^2(K_\gamma)$ (the Lebesgue space based on normalized Haar measure σ_1 of K_γ) that satisfies

$$a_{j+k} = a_j T^j a_k, \quad |a_k| = 1 \quad \text{a.e.} \tag{43}$$

for all integers j, k. Evidently $a_0 = 1$ a.e., and the unitary function a_1 determines a_k for all k. If we choose a definite Borel function for a_1 and define the other a_k successively by (43) we obtain a strict version of the cocycle without further trouble. It will be convenient to study cocycles in the strict sense, and to identify two strict cocycles if their generating functions agree almost everywhere.

The cocycle a is a *coboundary* if $a_k = qT^kq^{-1}$ for some unitary function q on K_γ; this is the case if $a_1 = aTq^{-1}$. A cocycle is *trivial* if it is the product of a coboundary with a cocycle that is constant in x for each k; that is, a coboundary times exp $ik\alpha$ for some real α.

Let a be a cocycle on K_γ. For x in K_γ and $0 \leqslant t < 2\pi/\gamma$ define

$$A(2\pi k/\gamma + t, x) = a_k(x). \tag{44}$$

Then for such x and t

$$A(2\pi k/\gamma + t, x) = A(2\pi k/\gamma, x)A(t, x + ke), \tag{45}$$

the second factor on the right being 1. This is the beginning of the cocycle identity on K. One must check (the details are tedious) that a strict cocycle A on K is uniquely determined by (44).

It is obvious that two cocycles on K_γ lead to the same cocycle on K if and only if they are equal almost everywhere; furthermore the extension preserves products, so the mapping from a to A is an isomorphism of the group of cocycles on K_γ into the group of cocycles on K. Gamelin's theorem asserts that *the mapping induces an isomorphism of the two cohomology groups* **[15]**.

The assertion has two distinct parts. First, every cohomology class on K contains the image of some cocycle on K_γ. Second, the extension of a cocycle a is a coboundary if and only if a is a coboundary.

Given a cocycle A on K, define $a_k(x) = A(2\pi k/\gamma, x)$ for x in K_γ. Then

a is a cocycle on K_γ; call its extension B. The first point will be established by showing that A and B are cohomologous. Define

$$q(x + e_t) = A(t, x) \qquad (x \text{ in } K_\gamma, \quad 0 \leqslant t < 2\pi/\gamma). \tag{46}$$

Then q is a unitary Borel function on K, and $q = 1$ on K_γ. Thus $A(t, x)$ $q(x)q(x + e_t)^{-1}$ is a cocycle cohomologous to A, equal to $A(t, x)$ when t is $2\pi k/\gamma$ and x is in K_γ, and constant on intervals of t so as to be an extension of a. Thus it coincides with B, and the first part is finished.

Let a be a coboundary: $a_1 = qTq^{-1}$ for a unitary function q on K_γ. Define q on all of K by setting $q(x + e_t) = q(x)$ for $0 \leqslant t < 2\pi/\gamma$ (x in K_γ). Then qT_tq^{-1} is the extension of a to K. In the other direction, if qT_tq^{-1} is a coboundary on K that extends a, then q is constant on intervals $(x, x + e)$ for x in K_γ, and a is a coboundary too. This completes the proof of Gamelin's theorem.

It is also true by an obvious argument that a is trivial if and only if its extension is trivial.

Now one can find non-trivial cocycles, at least on some groups K, by exhibiting a unitary function a_1 on K_γ that is not of the form $(\exp i\alpha)q(x)$ $q(x + e)^{-1}$, where α is real and q unitary on K_γ.

5. SPECIAL RESULTS

This chapter presents some isolated theorems about cocycles and subspaces. Some unsolved problems are mentioned in the final section.

5.1. Continuous cocycles

Aren's theorem (Section 3) shows that continuous analytic functions are privileged among all bounded analytic functions, in a way that is surprising from the point of view of the circle group. It is not obvious which simply invariant subspaces contain continuous functions. We shall show that it suffices for the cocycle of the subspace to be continuous in an appropriate sense.

A cocycle A is *continuous* if for each t, A_t coincides almost everywhere with a continuous function on K. Forelli has pointed out that a stronger kind of continuity then holds automatically: *the mapping from t to A_t is continuous as a function from R to C (the space of continuous functions on K).*

For the proof, write the Fourier series of A_t:

$$A_t(x) \sim \sum a_\lambda(t)\chi_\lambda(x). \tag{1}$$

Each $a_\lambda(t)$ is continuous on the line, because a cocycle is continuous as a mapping to L^2. Only countably many λ appear in the sum (a fact we have used before). Hence we can find a sequence $\{k_n\}$ of trigonometric polynomials on K such that $k_n * A_t$ tends to A_t in the norm of C for every t. For each n, the finite sum

$$k_n * A_t(x) = \sum a_\lambda(t)\hat{k}_n(\lambda)\chi_\lambda(x) \tag{2}$$

defines a continuous mapping from R to C. Hence A_t is the limit of a sequence of continuous mappings, and so has points of continuity. The cocycle relation shows that A_t is continuous at $t = 0$, and then at every t, as we wished to show.

THEOREM 30. *If A is a continuous cocycle, then the associated invariant subspace $\mathcal{M}$ contains non-null continuous functions. For each continuous non-null f in $\mathcal{M}$, $\log |f|$ is summable.*

With the usual notation,

$$A_t = V_t 1 = -\int_{-\infty}^{\infty} e^{it\lambda}\, d(P_\lambda 1). \tag{3}$$

Choose F in $L^1(-\infty, \infty)$ so that

$$\hat{F}(\lambda) = \int_{-\infty}^{\infty} F(t)\, e^{it\lambda}\, dt \tag{4}$$

vanishes for $\lambda < 0$. Then

$$\int_{-\infty}^{\infty} F(t) A_t\, dt = -\int_{-\infty}^{\infty} \hat{F}(\lambda)\, d(P_\lambda 1); \tag{5}$$

the integral on the left is a Bochner integral in C; that on the right is in L^2, and equality holds almost everywhere. The right side represents a function of $\mathcal{M}$. If we choose F so $\hat{F}$ does not vanish identically on the spectrum of $dP_\lambda 1$, then we have found a non-null continuous function in $\mathcal{M}$. If 1 is orthogonal to $\mathcal{M}$ this is impossible, but the argument works for $V_t \chi_\tau$ if τ is large enough.

Now $\bar{A}$ is continuous with A. Let f be any non-null continuous function in $\mathcal{M}$, and g non-null and continuous in $\tilde{\mathcal{M}}$. Then fg is continuous and non-null in H^2. By Arens' theorem $\log |fg|$ is summable. Hence $\log |f|$ is summable, and the proof is finished.

In Section 3 we constructed a function f that was non-negative with $\rho(f) > -\infty$ but $\log f$ not summable. Actually f could be chosen to be continuous. Thus continuous functions in simply invariant subspaces need not generally have summable logarithms.

The special cocycles constructed in Section 4 were all continuous, so the theorem just proved gives information about interesting subspaces.

Let $\mathcal{A}$ denote the algebra of all functions on K with absolutely convergent Fourier series. If A is a cocycle such that A_t agrees almost everywhere with an element of $\mathscr{A}$ for every t, then it follows as before that A_t is continuous as a mapping from R to $\mathscr{A}$. We ask whether this implies that the associated invariant subspace $\mathscr{M}$ contains elements of $\mathscr{A}$. There is a difficulty: the norm of A_t may grow with t so that the left side of (5), a Bochner integral in $\mathscr{A}$, may not exist for any F such that $\hat{F}$ is supported on a half-line. The result we can prove is therefore more special.

THEOREM 31. *Let m be a real function in $\mathscr{A}$ with Fourier coefficients m_λ satisfying*

$$\sum_{0<\lambda<1} |m_\lambda \log \lambda| < \infty. \tag{6}$$

Then the invariant subspace $\mathscr{M}$ associated with the cocycle

$$A(t, x) = \exp i \int_0^t m(x + e_u)\, du \tag{7}$$

contains non-null elements of $\mathscr{A}$.

We may assume $m_0 = 0$. Set

$$h_t(x) = \int_0^t m(x + e_u)\, du = -\sum m_\lambda(1 - e^{it\lambda})\,(i\lambda)^{-1}\,\chi_\lambda(x), \tag{8}$$

an element of $\mathscr{A}$ with

$$\| h_t \| = \sum |m_\lambda(1 - e^{it\lambda})\,\lambda^{-1}|. \tag{9}$$

This norm is trivially $O(t)$ as $|t| \to \infty$. We shall show that (6) implies

$$\int_{-\infty}^{\infty} \| h_t \|\,(1 + t^2)^{-1}\, dt < \infty. \tag{10}$$

By (9), the integral (10) equals

$$\sum |m_\lambda \lambda^{-1}| \int_{-\infty}^{\infty} |1 - e^{it\lambda}|\,(1 + t^2)^{-1}\, dt. \tag{11}$$

On account of (6) this sum will be finite if there is a constant k such that

$$\lambda^{-1} \int_{-\infty}^{\infty} |1 - e^{it\lambda}|\,(1 + t^2)^{-1}\, dt \leqslant k\,|\log \lambda|, \qquad 0 < \lambda < 1. \tag{12}$$

A straightforward calculation proves this inequality. Hence (10) is true.

We gather some facts we shall need in this

LEMMA. *The function $\rho(t) = \| A_t \|$ has the following properties:*

$$\rho(t) \geqslant 1, \qquad \rho(t+u) \leqslant \rho(t)\,\rho(u) \quad \textit{for all } t, u, \textit{ and}$$

$$\int_{-\infty}^{\infty} \log \rho(t)\, d\nu(t) < \infty. \tag{13}$$

The first inequality is true because the norm of a function in $\mathscr{A}$ is at least as great as its norm in L^2. Since $\mathscr{A}$ is a Banach algebra,

$$\| A_{t+u} \| \leqslant \| A_t \| \cdot \| T_t A_u \| = \| A_t \| \cdot \| A_u \|, \tag{14}$$

proving the second inequality. The last statement is essentially (10):

$$A_t = \exp ih_t, \qquad \| A_t \| \leqslant \exp \| h_t \|. \tag{15}$$

Construct the Lebesgue space L^1_ρ on the real line, with norm

$$\| F \|_\rho = \int_{-\infty}^{\infty} |F(t)|\, \rho(t)\, dt. \tag{16}$$

Such spaces have been studied in detail (for example in [**45**]). The first statement of the lemma means that L^1_ρ is contained in L^1; the second that L^1_ρ is a Banach algebra under convolution. Finally and crucially, the third assertion implies that L^1_ρ contains functions whose Fourier transforms vanish outside any prescribed interval of the line.

The Bochner integral on the left in (5) is defined for any F in L^1_ρ, and F can be chosen so that $\hat{F}$ is not null on the support of $dP_\lambda 1$ (or of $dP_\lambda \chi_\tau$ for some τ), but vanishes for $\lambda < 0$. Thus the same proof gives the conclusion in $\mathscr{A}$.

5.2. The theorem of Wagner and Milaszewicz

A kind of converse to the question dealt with in the last section is this: *if f is continuous, analytic and never* 0 *on K, is the inner factor of f continuous?* An interesting theorem on the subject was proved by J. Wagner [**44**], and improved by J. P. Milaszewicz. Before stating that result we prove an easy one.

THEOREM 32. *Let f be continuous, analytic and non-vanishing on K. If f is in $\mathscr{A}$, then the inner and outer factors of f are in $\mathscr{A}$.*

Since $\mathscr{A}$ is a Banach algebra and self-adjoint, first $|f|^2$ and then $\log|f|^2$ are in $\mathscr{A}$. Denote the Fourier coefficients of $\log|f|^2$ by b_λ. Then

$$g(x) = \tfrac{1}{2} b_0 + \sum_{\lambda > 0} b_\lambda \chi_\lambda(x) \tag{17}$$

is an analytic function in $\mathscr{A}$, and $|f| = |\exp g|$. Now $h = \exp g$ is outer and belongs to $\mathscr{A}$. It follows that fh^{-1}, the inner factor of f, is in $\mathscr{A}$, and the theorem is proved.

Milaszewicz' theorem asserts that *the inner and outer factors of f are* *continuous if f is continuous, analytic and non-vanishing on K, and has Fourier coefficients* a_λ *satisfying*

$$\sum |a_\lambda| e^{-\lambda\varepsilon} < \infty \tag{18}$$

for every positive ε. The proof has not yet been published, and we cannot go further here.

5.3. Generators of invariant subspaces

THEOREM 33. *Every simply invariant subspace is generated by two of its elements, which may be taken to be unitary functions.*

On the circle every simply invariant subspace is $q \cdot H^2$ for some unitary function q; thus it has a single generator q.

Let $\mathcal{M}$ be the subspace and A its cocycle. If $\mathcal{M} = \mathcal{M}_+ \neq \mathcal{M}_-$ then $\mathcal{M} = q \cdot H^2$ for a unitary function q, and there is nothing more to prove. If $\mathcal{M} = \mathcal{M}_- \neq \mathcal{M}_+$ there are special difficulties and we leave this case aside for the moment. Otherwise $\mathcal{M} = \mathcal{M}_+ = \mathcal{M}_-$, the case we now consider. After multiplying $\mathcal{M}$ by a unitary function (which does not affect the existence of a generating pair) we have a subspace whose cocycle is of Blaschke type, with the property of Theorem 26. We shall assume that A is a cocycle of this kind to begin with. At one point below K must be separable.

Since A is analytic, 1 is in $\mathcal{M}$, so $A_t = V_t 1$ is in $\mathcal{M}$ for each t. We shall show that 1 and A_u generate $\mathcal{M}$ for suitably chosen u.

LEMMA. 1 *and* A_u *generate* $\mathcal{M}$ *if for almost all x the Blaschke products* $A(z, x)$, $A(z + u, x)$ *have no common zeros.*

Let u be a real number such that the Blaschke products have no common zeros for almost every x. Denote by $\mathcal{N}$ the smallest invariant subspace containing 1 and A_u. $\mathcal{N}$ is contained in $\mathcal{M}$ and so is simply invariant; call its cocycle B, analytic because 1 is in $\mathcal{N}$. Since A_u is in $\mathcal{N}$, $A_u(x + e_t) B(t, x)$ is an inner function of t for almost every x. By the cocycle identity this product equals $A(t + u, x) A(t, x)^{-1} B(t, x)$. By hypothesis the first factor cannot cancel any of the poles in the upper half-plane belonging to the second factor. Hence $A(t, x)^{-1} B(t, x)$ is analytic. This implies (Theorem 21) that $\mathcal{M}$ is contained in $\mathcal{N}$ as was to be proved.

To prove the theorem we have to find u with the property of the lemma. For fixed x, the two sets of zeros are disjoint except for countably many u. We want to conclude that the sets are disjoint for some fixed u and almost every x.

Using a strict version of the cocycle, we form the set E of triples (z, u, x) such that $A(z, x) = A(z + u, x) = 0$. Let F be the set of pairs (u, x) such that (z, u, x) is in E for some z. Now E is a Borel set; as the projection of a Borel set,

F is an analytic set. Therefore F is measurable for every complete Borel measure on $R \times K$ [3, Theorem 3.2.4], in particular for the completion of $du \times d\sigma$. Since F is countable on almost every line $x = x_0$, F must be a null set of K on almost every section $u = u_0$ by the Fubini theorem. Any number u outside this null set has the property of the lemma. This completes the main part of the proof.

If $\mathcal{M} = \mathcal{M}_- \neq \mathcal{M}_+$ then $\mathcal{M} = q \cdot H_0^2$ for some unitary function q, and it suffices to find a pair of generators for H_0^2. As in the proof of Theorem 26 we can find a function g in H_0^2, not contained in $\chi_\lambda \cdot H_0^2$ for any positive λ, such that $\log |g|$ is summable. Then the inner factor p of g has the same properties. For any positive τ in Γ the functions p and χ_τ generate H_0^2, because $p(x + e_t)$ and $\chi_\tau(x + e_t)$ have no common inner divisor.

The topological argument in the proof needed K to be separable. This restriction was needed to find the number u of the lemma, that is to establish a fact about the cocycle rather than the subspace. This argument can be carried out on the separable quotient group where the cocycle really lives, so separability of K is not required for the truth of the theorem.

THEOREM 34. *Let $\mathcal{M}$ be a simply invariant subspace of L^2, and $\mathcal{N}$ an invariant subspace contained in $\mathcal{M}$. There is a unitary function q in $\mathcal{M}$ such that q and $\mathcal{N}$ generate $\mathcal{M}$.*

Let A be the cocycle of $\mathcal{M}$ and B that of $\mathcal{N}$. Once more by Theorem 26 we may assume that A is a Blaschke cocycle with $A(t, x)$ analytic on the real axis for almost every x. (This last property will be important in this proof.) Then $A = BC$ where C is analytic. Choose u by Theorem 33 so that 1 and A_u generate $\mathcal{M}$. For each real θ set $g_\theta = 1 + e^{i\theta} A_u$. We shall prove that g_θ and $\mathcal{N}$ generate $\mathcal{M}$ for almost every θ. Since $\log |g_\theta|$ is summable for almost every θ, g_θ can be replaced by a unitary function q.

Let D be the set of (θ, z, x) where $A(z, x) + e^{i\theta} A(z + u, x)$ vanishes, and E the set of (θ, z, x) where $C(z, x) = 0$. These are Borel sets in the product space. Denote by F the set of (θ, x) such that (θ, z, x) is in $D \cap E$ for some z. Then (K being separable again) F is measurable. It is easy to see that for fixed x the set of θ for which the functions have common zeros is countable. Hence F is a null set. The Fubini theorem gives the following conclusion: for any θ outside a fixed null set, the functions $A(z, x) + e^{i\theta} A(z + u, x)$ and $C(z, x)$ have no common zeros for almost every x. And the first function is the complex extension of $g_\theta(x + e_t) A(t, x)$.

Fix θ outside the exceptional set and write g for g_θ. If g and $\mathcal{N}$ do not generate $\mathcal{M}$, or at least $\mathcal{M}_-$, then $g(x + e_t) A(t, x)$ and $f(x + e_t) A(t, x)$ (all f in $\mathcal{N}$) have a common, non-trivial inner divisor for almost every x. The first function is analytic on the real axis, and thus has no singular inner divisor, except perhaps a weight at infinity. The functions $f(x + e_t) B(t, x)$ (f in $\mathcal{N}$) have no common inner divisor, and therefore a divisor of $f(x + e_t) B(t, x) C(t, x)$ for all f must divide $C(t, x)$. Thus its zeros lie in E, and it cannot divide $g(x + e_t) A(t, x)$, whose zeros are in D. Finally if $g(x + e_t) A(t, x)$ is divisible in H^2_ν by $\exp it\lambda$ ($\lambda > 0$) for more than one value of θ, then A itself would have a weight at infinity, which is not the case. Eliminating one more value of θ we have shown that g and $\mathcal{N}$ generate $\mathcal{M}$ or $\mathcal{M}_-$.

If $\mathcal{M} = \mathcal{M}_+ \neq \mathcal{M}_-$ then $\mathcal{M} = q \cdot H^2$ for some unitary q, and there is nothing to prove.

The proof that F is a null set can be carried out on a separable quotient group of K, so K does not really have to be separable.

The theorem is interesting mainly for the following corollary. Let $\mathcal{M}$ be an invariant subspace of H^2. Then $K = H^2 \ominus \mathcal{M}$ is a subspace of H^2 invariant under the adjoints of the shift operators S_λ (λ positive in Γ) in H^2. If P denotes the orthogonal projection of L^2 on H^2 these adjoints are given by

$$S_\lambda^* f = P S_\lambda^{-1} f \qquad (f \text{ in } H^2). \tag{19}$$

A closed subspace of H^2 invariant under this semigroup is called *star-invariant.*

COROLLARY. *Every star-invariant subspace of H^2 is generated by one of its elements.*

Let K be star-invariant in H^2. Denote by $\bar{K}$ the set of complex conjugates of elements of K. Then $\mathcal{M} = \bar{K} + H^2_0$ is closed, simply invariant, and contains H^2_0. By the theorem there is an element q of $\mathcal{M}$ that generates $\mathcal{M}$ with H^2_0. Let g be the conjugate-analytic part of q. Then g is in $\bar{K}$, and it is easy to see that $\bar{g}$ in K has the required property.

5.4. Unsolved problems

Is every simply invariant subspace of L^2 generated by one of its elements? Has H^2_0 a single generator? These problems are old, and apparently untouched by all that precedes. Answers will be interesting and probably difficult.

Theorem 26 makes Blaschke cocycles important for the subject. Their structure has been studied in **[22]**, but interesting problems remain. If every simply invariant subspace is singly generated, then every Blaschke cocyle has exactly the same zeros as some function in H^2. No effective way of finding such a function is known, even when the cocycle is of some restricted class. In another direction, Theorem 26 says in particular that every singular analytic cocycle is cohomologous to a Blaschke cocycle. This surprising fact has not yet been exploited.

The problem of Section 5.2 is only partly solved. Here is a related question:

which Blaschke products in the upper half-plane are almost periodic (in the sense of Bohr) on the real axis?

A continuous unitary group has associated with it a multiplicity function. For groups of the type $\{V_t\}$ this multiplicity is uniform, by Mackey's theorem on systems of imprimitivity [**29**]. It is not obvious what this multiplicity can be. A first step in studying this question was taken by M. Nadkarni [**34**], and further progress has been made recently by him and S. C. Bagchi [**4**]. The paper [**31**] is related to the question. As yet, however, we do not know the extent to which a simply invariant subspace is determined by its multiplicity.

A contraction in Hilbert space is associated with an operator-valued inner function in the circle [**19, 42**]. In the same way a semigroup of contractions indexed by the positive elements of Γ leads to an operator-valued cocycle, perhaps on a flow more general than the one we have been studying. We should expect that theorems about a single-contraction can be generalized to semigroups to the same extent that facts about inner functions can be extended to cocycles. This application justifies curiosity about generalizations of the theorems that have been presented in two directions: to operator-valued cocycles defined on general flows. Significant progress has already been made [**4, 7, 10, 11, 33**].

There is, however, a particular question that interests the author. The Corollary of Theorem 34 asserts that the semigroup $\{S_\lambda^*\}$ in any star-invariant subspace K of H^2 has a cyclic vector. Of course any semigroup similar to such a semigroup has a cyclic vector. Can one describe intrinsically some interesting class of semigroups to which the theorem applies? We might hope, for example, to generalize some of the results of [**41, 43**] to semigroups.

The part of this theory depending on Szegö's theorem has been developed much further by König, Lumer, Srinivasan and Wang, and others. The theory of cocycles seems to lead in a different direction, which will be described briefly in the next section.

6. COCYCLES ON A FLOW

6.1. Analyticity on a flow

Let X be a space with a Borel structure. A *Borel flow* on X is a family of isomorphisms φ_t (t real) of X such that $\varphi_{t+u} = \varphi_t\varphi_u$ and $\varphi_t(x)$ is a Borel function on $R \times X$. We often write x_t for $\varphi_t(x)$. Let μ be a positive Borel measure in X that is invariant under the flow and ergodic. Assume the Lebesgue space L^2 based on μ is separable. With this structure most of the machinery of the preceding chapters can be reconstructed.

Translation operators T_t are defined by $T_t f(x) = f(x_t)$. Each T_t is a unitary operator in L^2, and the group $\{T_t\}$ is weakly measurable. Hence

$$T_t = -\int_{-\infty}^{\infty} e^{it\lambda}\, dP_\lambda \tag{1}$$

for some decreasing resolution of the identity $\{P_\lambda\}$ in L^2. We choose P_λ to be continuous from the left.

Define H^2 as the range of P_0. We can prove, as on a group, that a function f in L^2 belongs to H^2 if and only if $f(x_t)$ is in H^2_v as a function of t for almost every x. This pointwise criterion enables us to operate with analytic functions almost as in the circle. For example, any f in L^2 is a sum $g + h$ where g and $\bar{h}$ are in H^2, and are uniquely determined up to a constant. If μ is an infinite measure, there is no ambiguous constant. If f is in H^2, then $\exp f$ is too if it is in L^2.

Define H^∞ to be the set of functions in L^∞ that are analytic on almost every orbit of the flow. Multiplication by a function in H^∞ carries H^2 into itself. Call a closed subspace of L^2 *invariant* if it has this property. A *Wiener subspace* is invariant under multiplication by all bounded functions; otherwise an invariant subspace is *simply invariant.* A Wiener subspace consists of all functions carried on some fixed subset of X.

A *cocycle* is a mapping from R to L^∞ such that $V_t = A_t T_t$ defines a continuous unitary group in L^2. Then A_t is a unitary function for each t and satisfies

$$A_{t+u} = A_t T_t A_u \qquad \text{(all real } t, u) \tag{2}$$

where equality holds almost everywhere on X. A family of functions A_t satisfying (2) and continuous in a weak sense determines such a group. A *coboundary* is a cocycle of the form $q(x)\, q(x_t)^{-1}$, where q is a unitary function on X. A cocycle is *trivial* if it is a coboundary times a cocycle that is constant on X for each t. Definitions of analytic, Blaschke, and singular cocycles are as before.

If the flow has a rich supply of eigenfunctions one can prove the flow is isomorphic to the standard flow on a torus that we have been studying. To generalize theorems about invariant subspaces and cocycles one must find proofs that do not mention group characters. However the flow provides information of a kind one does not have in generalizing function theory of the circle to the context of function algebras.

We know a lot about generalizations of the theorem of F. and M. Riesz in both directions [**8**, **9**, **10**, **12**, **15**, **16**, **30**]. Other aspects of function theory on flows are treated in [**7**, **11**, **33**], and in a series of papers of Muhly in press. No systematic exposition in the spirit of this essay has been written.

To illustrate the state of affairs, let us examine the connection between cocycles and invariant subspaces. Given a cocycle, the corresponding unitary group determines a spectral resolution $\{P_\lambda\}$, and it is not difficult to prove that the range of P_0 is a simply invariant subspace related to the cocycle in ways we should expect. However given the subspace it is not obvious how to recapture the cocycle. The problem is that in the absence of characters, P_0 does not determine all the P_λ in a simple way. This difficulty is the subject of an important theorem of Forelli [**10**] for the case of a continuous flow. For Borel flows the difficulty can be overcome [7], but not yet in a simple and natural way.

6.2. Cocycles and ergodic theory

Let K be a group of the usual type. Then $T \times K$ is a compact group whose dual can be ordered, to which therefore these theorems apply. However there is a natural way to define a flow on $T \times K$ by means of a cocycle on K, which is not a group of translations in $T \times K$. This construction is well known in ergodic theory, in the theory of group representations, and in algebra.

We start with a strict cocycle A on K. A Borel flow is defined in $T \times K$ by setting

$$\varphi_t(e^{iu}, x) = (A(t, x)\, e^{iu}, x + e_t). \tag{3}$$

Indeed the cocycle identity is just the composition rule for the flow. Furthermore the flow is not generally continuous.

Let $d\mu$ be the normalized product measure $(2\pi)^{-1}\, du \times d\sigma(x)$. A function f in L^2_μ can be regarded as a mapping from T to $L^2(K)$ with Fourier series

$$f(e^{iu}, x) \sim \sum_{-\infty}^{\infty} f_n(x)\, e^{niu}; \tag{4}$$

the coefficients f_n are in $L^2(K)$. The obvious computation shows that $T_t f$ has Fourier coefficients $f_n(x + e_t)\, A(t, x)^n$. The Parseval relation shows that T_t preserves norm in L^2_μ; hence μ is invariant under the flow.

A function f in L^2_μ is invariant under the flow if and only if

$$f_n(x + e_t)\, A(t, x)^n = f_n(x) \quad \text{a.e. on } K \tag{5}$$

for each t and n. If f is not null then some f_n is not null; then $|f_n|$ is constant and A^n is a coboundary. Conversely, if A^n is a coboundary for some $n \neq 0$, then we find a non-constant function invariant under the flow. Hence the flow is ergodic if and only if A^n is a coboundary only for $n = 0$.

A flow is called *weakly mixing* if it has no eigenfunctions except constant functions. An *eigenfunction* is a non-null function f in L^2_μ such that $T_t f = (\exp it\alpha) f$ for all t, where α is some real number. Our flow on $T \times K$ is weakly mixing if and only if A^n is non-trivial for each $n \neq 0$.

A function f in L^2_μ belongs to H^2 if and only if $f_n(x + e_t)\, A(t, x)^n$ is in H^2_ν for almost every x and each n. This means that f_n is in the subspace $\mathscr{M}_n$ whose cocycle is A^n for each n.

Let q be a unitary Borel function on K. A Borel isomorphism of $T \times K$ is defined by

$$\psi(e^{iu}, x) = (e^{iu}q(x), x). \tag{6}$$

If $\{\varphi_t\}$ is the flow associated with a cocycle A, then $\psi^{-1}\varphi_t\psi$ is the flow derived from $A(t, x)\, q(x)\, q(x + e_t)^{-1}$. Thus conjugacy of flows is related to equivalence of their cocycles.

The rotation flow on a group K has discrete spectrum, but the spectrum of a group $\{V_t\}$ can be much more general. The new flow on $T \times K$ defined by means of a cocycle A on K has spectrum incorporating the complexity not merely of A, but of all powers of A, since as we have been the effect of T_t in a subspace of L^2_μ is given by $f(x + e_t)\, A(t, x)^n$. Methods of constructing cocycles become thereby tools in ergodic theory. Furthermore a cocycle defined on any flow can be used in the same way to define a new flow in a product space. Thus the study of cocycles is close to the study of flows in general. This connection between cocycles and ergodic theory was shown to the author by William Parry.

The construction just mentioned is only one way in which cocycles arise in connection with transformation groups. Other ways can be found in papers of Furstenberg [**13, 14**] and of Parry [**36, 37**], for example. These points of contact between harmonic analysis and ergodic theory remain to be explored.

REFERENCES

1. R. Arens, A Banach algebra generalization of conformal mappings of the disc, *Trans. Amer. Math. Soc.* **81** (1956), 501–513.
2. R. Arens, The boundary integral of log $|\phi|$ for generalized analytic functions, *Trans. Amer. Math. Soc.* **86** (1957), 57–69.
3. William Arveson, Representations of C^*-algebras, *to appear*.
4. S. C. Bagchi, Invariant subspaces of vector-valued function spaces on Bohr groups, *Thesis, Indian Statistical Institute, Calcutta*, 1973.
5. A. Beurling, On two problems concerning linear transformations in Hilbert space, *Acta Math.* **81** (1949), 239–255.
6. S. Bochner, Boundary values of analytic functions in several variables and of almost periodic functions, *Ann. of Math.* **45** (1944), 708–722.
7. Carl G. R. Carlson, Cohomology classes in harmonic analysis, *Thesis, Stanford University*, 1972.
8. K. deLeeuw and I. Glicksberg, Quasi-invariance and analyticity of measures on compact groups, *Acta Math.* **109** (1963), 179–205.
9. F. Forelli, Analytic measures, *Pacific J. Math.* **13** (1963), 571–578.
10. F. Forelli, Analytic and quasi-invariant measures, *Acta Math.* **118** (1967), 33–59.
11. F. Forelli, Conjugate functions and flows, *Quart. J. Math. Oxford Ser.* **20** (1969), 215–233.
12. F. Forelli, What makes a positive measure the total variation of an analytic measure? *J. London Math. Soc.* **2** (1970), 713–718.
13. H. Furstenberg, Strict ergodicity and transformations of the torus, *Amer. J. Math.* **83** (1961), 573–601.

14. H. Furstenberg, The structure of distal flows, *Amer. J. Math.* **85** (1963), 477–515.
15. T. W. Gamelin, *Uniform Algebras*, Prentice-Hall, 1969.
16. I. Glicksberg, The abstract F. and M. Riesz theorem, *J. Functional Analysis* **1** (1967), 109–122.
17. P. R. Halmos, *Measure Theory*, D. van Nostrand Co., 1950.
18. P. R. Halmos, Shifts on Hilbert space, *J. Reine Angew. Math.* **208** (1961), 102–112.
19. H. Helson, *Lectures on Invariant Subspaces*, Academic Press, London and New York 1964.
20. H. Helson, Compact groups with ordered duals, *Proc. London Math. Soc.* **14A** (1965), 144–156; II, *J. London Math. Soc.* **1** (1969), 237–242; IV, *Bull. London Math. Soc.* **5** (1973), 67–69.
21. H. Helson, The differential equation of an inner function, *Studia Math.* **35** (1970), 311–321.
22. H. Helson, Structure of Blaschke cocycles, *Studia Math.* **44** (1972), 493–500.
23. H. Helson and J.-P. Kahane, Compact groups with ordered duals III, *J. London Math. Soc.* **4** (1972), 573–575.
24. H. Helson and D. Lowdenslager, Prediction theory and Fourier series in several variables, *Acta Math.* **99** (1958), 165–202; II, *Acta Math.* **106** (1961), 175–213.
25. H. Helson and D. Lowdenslager, Invariant subspaces, *Proc. Int. Symp. Linear Spaces*, Jerusalem, 1960, Macmillan (Pergamon), 1961, 251–262.
26. K. Hoffman, *Banach Spaces of Analytic Functions*, Prentice-Hall, 1962.
27. T. Kato, *Perturbation Theory for Linear Operators*, Springer-Verlag, 1966.
28. Y. Katznelson, *An Introduction to Harmonic Analysis*, John Wiley and Sons, 1968.
29. G. W. Mackey, Unitary representations of group extensions I, *Acta Math.* **99** (1958), 265–311.
30. V. Mandrekar and M. Nadkarni, Quasi-invariance of analytic measures on compact groups, *Bull. Amer. Math. Soc.* **73** (1967), 915–920.
31. V. Mandrekar, M. Nadkarni and D. Patil, Singular invariant measures on the line, *Studia Math.* **35** (1970), 1–13.
32. P. S. Muhly, Maximal weak-* Dirichlet algebras, *Proc. Amer. Math. Soc.* **36** (1972), 515–518.
33. P. S. Muhly, A structure theory for isometric representations of a class of semigroups, *J. Reine Angew. Math.* **255** (1972), 135–154.
34. M. Nadkarni, A class of measures on the Bohr group, *Pacific J. Math.* **23** (1967), 321–328.
35. S. Parrott, *Thesis, University of Michigan*, 1965.
36. William Parry, Compact abelian group extensions of discrete dynamical systems, *Z. Wahrscheinlichkeitstheorie und Verw. Gebiete* **13** (1969), 95–113.
37. William Parry, Cocycles and velocity changes, *to appear*.
38. W. Rudin, *Fourier Analysis on Groups*, Interscience, 1962.
39. T. P. Srinivasan, Simply invariant subspaces, *Bull. Amer. Math. Soc.* **69** (1963), 997–998.
40. G. Szegö, Beiträge zur Theorie der Toeplitzschen Formen (erste Mitteilung), *Math. Z.* **6** (1920), 167–202.
41. B. Sz.-Nagy, Cyclic vectors and commutants, *Linear Operators and Approximation* (Proc. of a Conf. at Oberwohlfach, 1971), Birkhäuser Verlag, 1972, 62–67.
42. B. Sz.-Nagy and C. Foiaş, *Analyse harmonique des opérateurs de l'espace de Hilbert*, Akademiai Kiado, 1967.

43. B. Sz.-Nagy and C. Foiaş, Vecteurs cycliques et quasi-affinités, *Studia Math.* **31** (1968), 35–42.
44. J. Wagner, Factorisation de fonctions analytiques sur un polydisque, *C.R. Acad. Sci. Paris* **269** (1969) Sér. A, 1147–1150.
45. J. Wermer, On a class of normed rings, *Ark. för Mat.* **2** (1953), 537–551.
46. K. Yale, Invariant subspaces and projective representations, *Pacific J. Math.* **36** (1971), 557–565.

2. Banach Algebras: Introductory Course

B. E. JOHNSON

Department of Mathematics
University of Newcastle, England

1. The Definition ... 63
2. Examples ... 64
3. Commutative Banach Algebras—Spectrum and Resolvent ... 65
4. The One-variable Functional Calculus ... 70
5. Regularity and the Hull-kernel Topology ... 72
6. Non-commutative Banach Algebras ... 74
7. Modules and Irreducible Representations ... 75
8. Uniqueness of Norm for Banach Algebras ... 78
9. Approximate Units and Factorization ... 80

1. THE DEFINITION

Definition 1.1. A *Banach algebra* $\mathfrak{A}$ is a Banach space over $\mathbb{C}$ with a multiplication making it an associative algebra and satisfying

$$\|ab\| \leqslant \|a\|\,\|b\| \qquad a, b \in \mathfrak{A} \tag{1}$$

Notes (a) If $\mathfrak{A}$ is a normed linear space with multiplication satisfying the above conditions then $\mathfrak{A}$ is a *normed algebra.* The multiplication extends to the completion of $\mathfrak{A}$ making it a Banach algebra.

(b) (1) implies that the multiplication is jointly continuous. If we replace (1) by the hypothesis that the multiplication is jointly continuous (or even, since it is equivalent by the uniform boundedness theorem, separately continuous) then there is $K > 0$ with $\|ab\| \leqslant K\|a\|\,\|b\|$ $(a, b \in \mathfrak{A})$. Replacing $\|\ \|$ by $\|\ \|'$ where $\|a\|' = K\|a\|$, we obtain a norm which satisfies (1) and is equivalent to $\|\ \|$.

(c) $\mathbb{C}$ can be replaced by $\mathbb{R}$ to get the definition of a *real Banach algebra.* Some results for complex Banach algebras carry over directly to real Banach algebras, others can be extended but with greater effort and some are not true for real Banach algebras. We shall not consider the real case at all.

Definition 1.2. A Banach algebra is *unital* if it contains an *identity element*, that is an element 1 with $1a = a = a1$, $a \in \mathfrak{A}$, and $\|1\| = 1$. An algebra has at most one identity element.

Notes (a) A Banach algebra can have a unit and not be unital. In such a case it can be renormed by an equivalent norm $\|a\|' = \sup\{\|ax\|;\ x \in \mathfrak{A}, \|x\| \leqslant 1\}$ to make it unital.

(b) If $\mathfrak{A}$ is a Banach algebra then $\mathfrak{A}^1 = \mathfrak{A} \oplus \mathbb{C}$, with multiplication

$$(a, \lambda)(b, \mu) = (ab + \lambda b + \mu a, \lambda\mu) \qquad a, b \in \mathfrak{A}, \lambda, \mu \in \mathbb{C}$$

and norm

$$\|(a, \lambda)\| = \|a\| + |\lambda|,$$

is a unital Banach algebra (the identity is (0, 1)) containing an ideal $\{(a, 0): a \in \mathfrak{A}\}$ isometric with $\mathfrak{A}$ and of codimension 1. $\mathfrak{A}^1$ is *the algebra obtained by adjoining an identity to* $\mathfrak{A}$. With some notable exceptions results for algebras with a unit can be extended to algebras with no unit by considering $\mathfrak{A}^1$ (note that the construction *can* be applied even when $\mathfrak{A}$ has a unit).

2. EXAMPLES

(a) If $\mathfrak{X}$ is a Banach space the Banach space $\mathscr{L}(\mathfrak{X})$ of all bounded linear operators on $\mathfrak{X}$, with multiplication defined by composition of functions, is a unital Banach algebra.

(b) The sequence spaces c_0, c, l^p $(1 \leqslant p \leqslant \infty)$ are Banach algebras if multiplication is performed term-by-term, that is if the product of $\{a_n\}$ and $\{b_n\}$ is the sequence whose nth term is $a_n b_n$. Only c and l^∞ are unital.

(c) In a similar way if Ω is a non void topological space then $C_0(\Omega)$, the space of continuous functions on Ω which are zero at infinity (f is zero at infinity if for each $\varepsilon > 0$ there is a compact set K in Ω with $|f(x)| < \varepsilon$ for all $x \in \Omega \backslash K$) with norm

$$\|f\| = \sup\{|f(\omega)| : \omega \in \Omega\}$$

and multiplication

$$(fg)(\omega) = f(\omega)g(\omega), \qquad \omega \in \Omega$$

is a Banach algebra. This algebra is unital if and only if Ω is compact. When Ω is compact we write $C(\Omega)$ for $C_0(\Omega)$.

(d) If G is a locally compact group then $M(G)$, the space of bounded regular complex valued Borel measures on G is an algebra if we define

$$\int f\, d\mu * \nu = \int f(gh)\, d\mu \times \nu(g, h), \qquad \mu, \nu \in M(G), \qquad f \in C_0(G).$$

When G is discrete $M(G) = l^1(G)$, the space of functions on G with $\sum_{g \in G} |a_g| < \infty$ and the multiplication amounts to defining $e_g \times e_h = e_{gh}$, where $\{e_g : g \in G\}$ is the standard basis in $l^1(G)$, and extending this by linearity and continuity to $l^1(G)$.

(e) Any closed subalgebra of a Banach algebra is itself a Banach algebra. Subalgebras of the algebras $\mathscr{L}(\mathfrak{X})$ form a wide class—in fact every Banach algebra is isometrically isomorphic with a closed subalgebra of $\mathscr{L}(\mathfrak{X})$ for some $\mathfrak{X}$. The algebra $\mathscr{L}\mathscr{C}(\mathfrak{X})$ of compact operators on $\mathfrak{X}$ is an important subalgebra (actually an ideal) of $\mathscr{L}(\mathfrak{X})$. $\mathscr{L}\mathscr{C}(\mathfrak{X})$ is not unital unless $\mathfrak{X}$ is finite dimensional.

Subalgebras of $C_0(\Omega)$ are called *function* or *uniform algebras*—their study is a subject in itself.

The basic algebra in harmonic analysis is $L^1(G)$, the ideal in $M(G)$ of measures absolutely continuous with respect to Haar measure.

3. COMMUTATIVE BANACH ALGEBRAS—SPECTRUM AND RESOLVENT

The theory of Banach algebras can be developed either by proving general results and deducing the commutative case from them or by establishing the commutative theory first and then extending to the general case. We shall adopt the second course here as the easier and more generally useful results will then appear first.

Definition 3.1. The *inverse* of an element a of a commutative Banach algebra $\mathfrak{A}$ with a unit 1 is an element a^{-1} of $\mathfrak{A}$ with $aa^{-1} = 1$. a has at most one inverse. a is *regular* if it has an inverse and *singular* if it does not. The *spectrum* $\sigma(a)$ of a is

$$\sigma(a) = \{\lambda; \lambda \in \mathbb{C}, (a - \lambda 1) \text{ has no inverse}\}.$$

If $\mathfrak{A}$ has no unit then $\sigma(a)$ is the spectrum of a in $\mathfrak{A}^1$.

The importance of the function $\lambda \mapsto (a - \lambda 1)^{-1}$ was recognised in operator theory and integral equations before the advent of Banach algebras. In these subjects the study of equations $(T - \lambda I)\xi = \eta$, T a known operator and η a known vector, is central.

For the next few pages $\mathfrak{A}$ will denote a commutative Banach algebra with a unit 1 and $1 \neq 0$.

LEMMA 3.2. *If* $\|a\| < 1$ *then* $1 - a$ *is regular. If* $\|a\| \leqslant \frac{1}{2}$ *then* $\|1 - (1-a)^{-1}\| \leqslant 1$.

Proof. The series $1 + a + a^2 + \ldots$ converges to an element, y say, of $\mathfrak{A}$ because its partial sums form a Cauchy sequence (the difference between two partial sums has norm

$$\|a^{m+1} + \ldots + a^n\| \leqslant \|a\|^{m+1} + \ldots + \|a\|^n$$
$$\leqslant \|a\|^{m+1}(1 - \|a\|)^{-1} \to 0 \text{ as } m \to \infty).$$

By continuity of multiplication

$$ya = a + a^2 + \ldots = y - 1$$

so $y(1 - a) = 1$.

If $\|a\| \leqslant \frac{1}{2}$ then

$$\|a + a^2 \ldots + a^n\| \leqslant 2^{-1} + 2^{-2} + \ldots + 2^{-n} < 1$$

so

$$\|1 - (1 - a)^{-1}\| = \lim_n \|a + a^2 \ldots + a^n\| \leqslant 1. \qquad \blacksquare$$

THEOREM 3.3. *If $a \in \mathfrak{A}$ then $\sigma(a)$ is a compact non void subset of $\mathbb{C}$. The function $\lambda \mapsto (a - \lambda 1)^{-1}$ is analytic on $\mathbb{C}\backslash\sigma(a)$ and $\|(a - \lambda 1)^{-1}\| \to 0$ as $|\lambda| \to \infty$.*

Proof. If $|\lambda > \|a\|$ then $1 - \lambda^{-1}a$ is regular by 3.2 and thus so is $(a - \lambda 1) = -\lambda^{-1}(1 - \lambda^{-1}a)$. This shows that if $\lambda \in \sigma(a)$ then $|\lambda| \leqslant \|a\|$, so that $\sigma(a)$ is bounded.

If $|\lambda| > 2\|a\|$ then

$$\|(a - \lambda 1)^{-1}\| = |\lambda|^{-1}\|(1 - \lambda^{-1}a)^{-1}\| \leqslant |\lambda|^{-1}(\|1\| + 1) \to 0 \quad \text{as} \quad |\lambda| \to \infty.$$

If $\lambda \in \mathbb{C}\backslash\sigma(a)$ then

$$0 < \|1\| \leqslant \|(a - \lambda 1)\| \, \|(a - \lambda 1)^{-1}\| \quad \text{so} \quad \|(a - \lambda 1)^{-1}\| \neq 0.$$

If $\mu \in \mathbb{C}$ and $|\lambda - \mu| < \|(a - \lambda 1)^{-1}\|^{-1}$ then $1 + (\lambda - \mu)(a - \lambda 1)^{-1}$ is regular so

$$(a - \mu 1) = (a - \lambda 1)(1 + (\lambda - \mu)(a - \lambda 1)^{-1})$$

is regular as it is a product of regular elements. Thus $\mathbb{C}\backslash\sigma(a)$ is open so $\sigma(a)$ is closed and hence compact as it is also bounded.

If

$$|\lambda - \mu| < \tfrac{1}{2}\|(a - \lambda 1)^{-1}\|^{-1} \quad \text{then} \quad \|1 + (\lambda - \mu)(a - \lambda 1)^{-1}\| \leqslant 1 + \|1\|$$

so that from the equation

$$(a - \lambda 1)^{-1} - (a - \mu 1)^{-1} = (\lambda - \mu)(a - \lambda 1)^{-1}(a - \mu 1)^{-1}$$

and the expression for $(a - \mu 1)$ above we see

$$\|(a - \lambda 1)^{-1} - (a - \mu 1)^{-1}\| \leqslant |\lambda - \mu| \, \|(a - \lambda 1)^{-1}\|^2(1 + \|1\|).$$

This tends to zero as μ tends to λ, that is $\lambda \mapsto (a - \lambda 1)^{-1}$ is continuous.

As for complex valued functions, a Banach space valued function Φ of a complex variable is analytic in an open set $D \subset \mathbb{C}$ if

$$\lim_{w \to z} (z - w)^{-1}(\Phi(z) - \Phi(w))$$

exists in norm for each $z \in D$. As

$$\|(\lambda - \mu)^{-1}((a - \lambda 1)^{-1} - (a - \mu 1)^{-1}) - (a - \lambda 1)^{-2}\|$$
$$\leqslant \|(a - \lambda 1)^{-1}\| \, \|(a - \mu 1)^{-1} - (a - \lambda 1)^{-1}\| \to 0 \text{ as } \mu \to \lambda$$

we see $\lambda \mapsto (a - \lambda 1)^{-1}$ is analytic in $\mathbb{C} \backslash \sigma(a)$.

Note that this implies that if $f \in \mathfrak{A}^*$ then $F(z)\colon z \mapsto f((a - z1)^{-1})$ is analytic in the usual sense. Thus if $\sigma(a)$ were void F would be an entire (integral) function and $F(z) \to 0$ as $|z| \to \infty$ because $(a - z1)^{-1} \to 0$ as $|z| \to \infty$. This implies F is bounded on $\mathbb{C}$ so, by Liouville's theorem, F is constant. As $F(z) \to 0$ when $|z| \to \infty$ we see $F(z) = 0$ for all $z \in \mathbb{C}$, in particular for $z = 0$. Thus $f(a^{-1}) = 0$ for all $f \in \mathfrak{A}^*$. By the Hahn–Banach theorem this shows $a^{-1} = 0$ which is impossible because $0 \neq 1 = aa^{-1}$. ■

Definition 3.4. A multiplicative linear functional on a commutative Banach algebra $\mathfrak{A}$ is a non-zero linear functional ϕ with $\phi(ab) = \phi(a)\phi(b)$ $a, b \in \mathfrak{A}$. The set of multiplicative linear functionals on $\mathfrak{A}$ will be denoted by $\Phi_{\mathfrak{A}}$. Note that if $\phi \in \Phi_{\mathfrak{A}}$ then $\phi(1) = \phi(1^2)$ so $\phi(1) = 0$ or 1. If $\phi(1) = 0$ then $\phi(a) = \phi(a \,.\, 1) = \phi(a)\phi(1) = 0$ for all $a \in \mathfrak{A}$. Thus $\phi(1) = 1$.

COROLLARY 3.5. *If $\phi \in \Phi_{\mathfrak{A}}$ then ϕ is continuous and $\|\phi\| \leqslant 1$.*

Proof. Suppose $\|a\| \leqslant 1$ and $|\phi(a)| > 1$. Then $1 - \phi(a)^{-1}a$ is regular and

$$\phi(1 - \phi(a)^{-1}a) = 1 - \phi(a)^{-1}\phi(a) = 0 \quad \text{so}$$
$$1 = \phi(1 - \phi(a)^{-1}a)\,\phi((1 - \phi(a)^{-1}a)^{-1}) = 0.$$

Thus $|\phi(a)| \leqslant 1$ whenever $\|a\| \leqslant 1$. ■

COROLLARY 3.6. *$J \subset \mathfrak{A}$ is a maximal proper ideal in $\mathfrak{A}$ if and only if $J = \operatorname{Ker} \phi$ for some $\phi \in \Phi_{\mathfrak{A}}$. In particular every maximal proper ideal is closed.*

(A *maximal* proper ideal is an ideal $J \neq \mathfrak{A}$ such that if I is an ideal with $J \subseteq I \subseteq \mathfrak{A}$ then $I = J$ or $I = \mathfrak{A}$.)

Proof. Let J be a maximal proper ideal in $\mathfrak{A}$. Then $\bar{J}$ is an ideal in $\mathfrak{A}$ containing J so $\bar{J} = J$ or $\bar{J} = \mathfrak{A}$. However if an ideal K contains a regular element a then $b = ba^{-1}a$ is in K for all b in $\mathfrak{A}$, that is $K = \mathfrak{A}$. Thus J contains no regular elements and, in particular, if $\|a - 1\| < 1$ then $a \notin J$ so $1 \notin \bar{J}$. Hence $\bar{J} = J$ and J is closed. The Banach space $\mathfrak{A}/J = \mathfrak{B}$ is a Banach algebra with the quotient multiplication having exactly two ideals $\{0\}$ and $\mathfrak{B}$. $\{b; b \in \mathfrak{B}, bc = 0 \text{ for all } c \in \mathfrak{B}\}$ is an ideal which is proper as it does not contain 1 and so is $\{0\}$. If $b \in \mathfrak{B}$, $b \neq 0$ then $\{bc; c \in \mathfrak{B}\}$ is an ideal in $\mathfrak{B}$ which is not $\{0\}$ and so is $\mathfrak{B}$. In particular $bc = 1$ for some $c \in \mathfrak{B}$. Thus $\mathfrak{B}$ is a field. Now consider the map $\psi: \lambda \mapsto \lambda 1$ of $\mathbb{C}$ into $\mathfrak{B}$. It is one to one (as $1 \neq 0$ in $\mathfrak{B}$), linear and multiplicative. If $c \in \mathfrak{B}$ let $\lambda \in \sigma(c)$ (a non void set by 3.3) so $c - \lambda 1$ has no inverse. As $\mathfrak{B}$ is a field this implies $c = \lambda 1$ so ψ is surjective. Let q be the quotient map $\mathfrak{A} \to \mathfrak{B}$. Then $\psi^{-1}q$ is a multiplicative linear functional with kernel J.

Conversely suppose $\phi \in \Phi_{\mathfrak{A}}$ and I is an ideal in $\mathfrak{A}$ containing Ker ϕ. Then $\phi(I)$ is an ideal in the one-dimensional Banach algebra $\mathbb{C}$. As $\mathbb{C}$ is a field the only ideals in $\mathbb{C}$ are $\{0\}$ and $\mathbb{C}$ and as $I \supseteq \text{Ker } \phi$, $I = \phi^{-1}\phi I$ so $I = \text{Ker } \phi$ or $\mathfrak{A}$. Ker ϕ is clearly not $\mathfrak{A}$ otherwise we would have $\phi = 0$. ■

Definition 3.7. The relative weak* topology induced on $\Phi_{\mathfrak{A}}$ by its inclusion in $\mathfrak{A}^*$ is called the *Gelfand topology*. As $\Phi_{\mathfrak{A}}$ is a bounded weak* closed subset of $\mathfrak{A}^*$ it is a compact Hausdorff space. A sequence or a net $\{\phi_n\}$ in $\Phi_{\mathfrak{A}}$ converges to ϕ in this topology if $\phi_n(a) \to \phi(a)$ for each $a \in \mathfrak{A}$. Because of the one-to-one correspondence in 3.6, $\Phi_{\mathfrak{A}}$ is usually referred to as the *maximal ideal space* of $\mathfrak{A}$.

Definition 3.8. The *Gelfand transform* is the map $a \to \hat{a}$ of $\mathfrak{A}$ into $C(\Phi_{\mathfrak{A}})$ defined by

$$\hat{a}(\phi) = \phi(a) \qquad a \in \mathfrak{A}, \quad \phi \in \Phi_{\mathfrak{A}}.$$

The fact that $\hat{a}$ is continuous is a direct consequence of the definition of the Gelfand topology. All the statements in the next result are also immediate consequences of the definitions involved.

THEOREM 3.9. *The Gelfand transform is linear, multiplicative* (*that is* $(ab)\hat{} = \hat{a}\hat{b}$) *and satisfies*

(i) $\hat{1} = 1$

(ii) $\|\hat{a}\| \leqslant \|a\| \qquad a \in \mathfrak{A}$

(iii) $\sigma(a) = \hat{a}(\Phi_{\mathfrak{A}})$.

Proof of (iii). If $\lambda \in \sigma(a)$ then $(a - \lambda 1)\mathfrak{A}$ is a proper ideal (it does not contain 1) and a Zorn's Lemma argument shows that there is a maximal proper ideal J in $\mathfrak{A}$ with $(a - \lambda 1) \in (a - \lambda 1)\mathfrak{A} \subseteq J$. Let $\phi \in \Phi_{\mathfrak{A}}$ with $J = \text{Ker } \phi$ (3.6) so $\phi(a - \lambda 1) = 0$ and hence $\hat{a}(\phi) = \lambda$. Thus $\sigma(a) \subseteq \hat{a}(\Phi_{\mathfrak{A}})$. (Strictly speaking part of this argument should have preceded 3.8.—without it $\Phi_{\mathfrak{A}}$ might be void). Conversely, the fact that the Gelfand transform is multiplicative and

$\hat{1} = 1$ shows that if $a - \lambda 1$ is regular in $\mathfrak{A}$ then $(a - \lambda 1)\hat{} = \hat{a} - \lambda 1$ is regular in $C(\Phi_{\mathfrak{A}})$, so $\hat{a}$ does not take the value λ in $\Phi_{\mathfrak{A}}$. ∎

We now consider the case in which $\mathfrak{A}$ has no unit. All the above results apply in $\mathfrak{A}^1$, so by Definition 3.1 and Theorem 3.3 the spectrum of each element of $\mathfrak{A}$ is a compact non-void set. If $\phi \in \Phi_{\mathfrak{A}}$ then defining $\tilde{\phi}(a, \lambda) = \phi(a) + \lambda$ extends ϕ to an element $\tilde{\phi}$ of $\Phi_{\mathfrak{A}^1}$ which is continuous by 3.5. If the linear span of the set $\mathfrak{A}\mathfrak{A} = \{ab : a, b \in \mathfrak{A}\}$ is not all of $\mathfrak{A}$ then 3.6 only holds if we restrict attention to ideals J not containing $\mathfrak{A}\mathfrak{A}$. These are the *modular* maximal ideals which we shall study in greater depth when we consider non-commutative algebras. Definition 3.7 is still valid but we can only assert that $\Phi_{\mathfrak{A}} \cup \{0\}$ is weak*-closed in $(\mathfrak{A}^1)^*$, so $\Phi_{\mathfrak{A}}$ is only locally compact and might be void. If it is not then 3.8 and 3.9 apply except that the Gelfand transform maps into $C_0(\Phi_{\mathfrak{A}})$ and we have $\sigma(a) = \hat{a}(\Phi_{\mathfrak{A}}) \cup \{0\}$.

We shall look at some of the examples of Banach algebras which we have given in the light of the constructions in this section. In $C(\Omega)$, where Ω is a compact Hausdorff space, for each $\omega \in \Omega$, $\phi_\omega(f) = f(\omega)$ defines an element of $\Phi_{C(\Omega)}$. The map $\omega \mapsto \phi_\omega$ is a homeomorphism of Ω onto $\Phi_{C(\Omega)}$ and the Gelfand transform is merely the induced isomorphism of $C(\Omega)$ onto $C(\Phi_{C(\Omega)})$, which is clearly a trivial construction.

When $\mathfrak{A} = M(G)$, G a locally compact abelian group, $\mathfrak{A}$ is a commutative unital Banach algebra. If χ is a continuous character on G, that is a continuous map of G into $\mathbb{C}$ with

$$|\chi(g)| = 1, \chi(gh) = \chi(g)\chi(h) \qquad g, h \in G,$$

then

$$\phi_\chi(\mu) = \int \chi \, d\mu$$

defines an element ϕ_χ of $\Phi_{M(G)}$. The map $\chi \mapsto \phi_\chi$ is a homeomorphism of $\hat{G}$, the set of continuous characters on G with the topology of uniform convergence on compact subsets of G, *into* $\Phi_{M(G)}$. Restricting ϕ_χ to $L^1(G)$ the map $\chi \mapsto \phi_\chi$ is a homeomorphism of $\hat{G}$ onto $\Phi_{L^1(G)}$. Identifying $C_0(\hat{G})$ with $C_0(\Phi_{L^1(G)})$ in this way the Gelfand transform is just the inverse Fourier transform.

The value of the Gelfand transform is that it replaces a complicated multiplication by the simple one of multiplying functions in $C(\Phi_{\mathfrak{A}})$. To balance this there is often great difficulty in deciding whether a given f in $C(\Phi_{\mathfrak{A}})$ is $\hat{a}$ for some $a \in \mathfrak{A}$ and in estimating what $\|a\|$ is when such an a exists.

We conclude with a definition:

Definition 3.10. The *radical* of a commutative Banach algebra is the kernel of the Gelfand transform. The algebra is *semi-simple* if the Gelfand transform is one-to-one.

4. THE ONE-VARIABLE FUNCTIONAL CALCULUS

Throughout this section $\mathfrak{A}$ will denote a commutative Banach algebra with a unit and a an element of $\mathfrak{A}$. One approach to the problem of recognising elements of $\mathfrak{A}^\wedge$ is to see whether elements of $C(\Phi_{\mathfrak{A}})$ generated from elements of $\mathfrak{A}^\wedge$ in various ways are in $\mathfrak{A}^\wedge$. A related problem is that of interpreting $f(a)$ when f is a complex valued function of a complex variable and $a \in \mathfrak{A}$. It is clear what this shall mean if f is a polynomial—just replace the variable by a in the polynomial expression. If f is the exponential function then $f(a)$ is $1 + a + (a^2/2!) + \ldots$, the series converging in norm; and the same would apply to any other entire function. If f is a rational function then we can factorize the denominator to get

$$f(z) = P(z)(z - \lambda_1)^{-1} \ldots (z - \lambda_n)^{-1}$$

and then put

$$f(a) = P(a)(a - \lambda_1 1)^{-1} \ldots (a - \lambda_n 1)^{-1},$$

provided that the poles of f are all in $\mathbb{C}\backslash\sigma(a)$. The Cauchy integral formula gives a way of extending this to define $f(a)$ for any function f analytic in a neighbourhood of $\sigma(a)$ and this is a natural limit to the theory.

First of all $\mathcal{O}(\sigma(a))$ is the algebra of germs of functions analytic in a neighbourhood of $\sigma(a)$. It is the quotient of the algebra $\mathcal{O}_0 = \{f; f$ is a complex valued analytic function with domain a neighbourhood of $\sigma(a)\}$, where the algebraic operations are defined pointwise, by the equivalence relation $f \sim g$ if $f = g$ in some neighbourhood of $\sigma(a)$. It is necessary to check that the algebraic operations do indeed lift to the quotient but this is not difficult.

If f is analytic with domain a neighbourhood N of $\sigma(a)$ then $\sigma(a)$ and $\mathbb{C}\backslash N$ are a positive distance apart and so there is a contour, even a polygonal contour Γ, in N such that the winding number is 0 for all points of $\mathbb{C}\backslash N$ and 1 for all points of $\sigma(a)$. Such a contour will be called *admissible* for f.

Definition 4.1. If $f \in \mathcal{O}_0$ and Γ is admissible for f then

$$f(a) = -\frac{1}{2\pi i}\int_\Gamma (a - \lambda 1)^{-1} f(\lambda)\, d\lambda.$$

The integral here is a Riemann integral and is defined by taking the limit of sums corresponding to dissections of Γ in the usual way. The existence of this limit follows from the uniform continuity of $\lambda \mapsto (a - \lambda 1)^{-1}$ on Γ just as for complex valued integrands.

THEOREM 4.2. *$f(a)$ is independent of the choice of Γ. If $f \sim g$ in $\mathcal{O}_0$ then $f(a) = g(a)$ so the map $\mathcal{O}_0 \to \mathfrak{A}$ which we have constructed lifts to a map $\mathcal{O}(\sigma(a)) \to \mathfrak{A}$ which is linear and multiplicative. Moreover*

(i) $j(a) = a$ *where* $j(z) = z, \quad z \in \mathbb{C}$
$1(a) = 1.$

(ii) *If $f \in \mathcal{O}_0$ then $f(a)\hat{} = f \circ \hat{a}$ and so $f(\sigma(a)) = \sigma(f(a))$.*

(iii) *If $f \in \mathcal{O}_0$ and g is analytic in a neighbourhood of $\sigma(f(a))$ then $(g \circ f)(a) = g(f(a))$.*

Proof. If $\tilde{f}(a)$ corresponds to the curve $\tilde{\Gamma}$ then

$$f(a) - \tilde{f}(a) = -\frac{1}{2\pi i}\int_{\Gamma-\tilde{\Gamma}} (a - \lambda 1)^{-1} f(\lambda) d\lambda$$

where points outside N and points in $\sigma(a)$ have winding number zero with respect to $\Gamma - \tilde{\Gamma}$. Thus the integrand is analytic at all points with non-zero winding number (3.3) and the integral is zero by a Banach space form of Cauchy's theorem. (The appeal to such a form can be avoided by applying an element F of $\mathfrak{A}^*$ to the integral to reduce to the scalar valued case and completing the argument as at the end of 3.3.)

If $f \sim g$ then by choosing Γ so that $f = g$ on Γ we see $f(a) = g(a)$.

It is easy to see that the map $f \mapsto f(a)$ is linear. If $f, g \in \mathcal{O}_0$ then we choose admissible contours Γ, Δ for f, g such that the points of Γ have winding number 1 with respect to Δ and the points of Δ have winding number 0 with respect to Γ. Then

$$\begin{aligned}
f(a)\,g(a) &= \left(-\frac{1}{2\pi i}\right)^2 \int_\Gamma \int_\Delta (a - \lambda 1)^{-1}(a - \mu 1)^{-1} f(\lambda) g(\mu)\, d\mu\, d\lambda \\
&= \left(-\frac{1}{2\pi i}\right)^2 \int_\Gamma (a - \lambda 1)^{-1} f(\lambda)\left(\int_\Delta (\lambda - \mu)^{-1} g(\mu)\, d\mu\right) d\lambda \\
&\quad - \left(-\frac{1}{2\pi i}\right)^2 \int_\Delta (a - \mu 1)^{-1} g(\mu)\left(\int_\Gamma (\lambda - \mu)^{-1} f(\lambda)\, d\lambda\right) d\mu \\
&= -\frac{1}{2\pi i}\int_\Gamma (a - \lambda 1)^{-1} f(\lambda)\, g(\lambda)\, d\lambda
\end{aligned}$$

where $\int_\Gamma (\lambda - \mu)^{-1} f(\lambda)\, d\lambda = 0$ for $\mu \in \Delta$ because the integrand is analytic at all points at which Γ has a non-zero winding number and

$$\int_\Delta (\lambda - \mu)^{-1} g(\mu)\, d\mu = -2\pi i\, g(\lambda)$$

for $\lambda \in \Gamma$ by the Cauchy integral formula.

The equations (i) can be obtained by taking Γ to be a large circle, expanding

$$(a - \lambda 1)^{-1} = -\lambda^{-1}(1 - \lambda^{-1}a)^{-1} = -\lambda^{-1}1 - \lambda^{-2}a - \lambda^{-3}a^2 - \ldots$$

and integrating term by term.

If $\phi \in \Phi_{\mathfrak{A}}$ then

$$\begin{aligned} f(a)\hat{}\,(\phi) &= -\frac{1}{2\pi i}\int_\Gamma \phi(a - \lambda 1)^{-1} f(\lambda)\, d\lambda \\ &= -\frac{1}{2\pi i}\int_\Gamma (\phi(a) - \lambda)^{-1} f(\lambda)\, d\lambda \\ &= f(\phi(a)) \\ &= (f \circ \hat{a})(\phi) \end{aligned}$$

by the Cauchy formula since $\phi(a) \in \sigma(a)$ so Γ has winding number 1 with respect to $\phi(a)$. The other part of (ii) follows from the relation $\hat{a}(\Phi_{\mathfrak{A}}) = \sigma(a)$.

For (iii) first of all note that because $f \mapsto f(a)$ is an algebra homomorphism which sends 1 to 1, $(f - \lambda 1)^{-1}(a) = (f(a) - \lambda 1)^{-1}$ for $\lambda \notin \sigma(f(a)) = f(\sigma(a))$. Then choose an admissible contour Δ for f in $f^{-1}(M)$ where M is an open neighbourhood of $f(\sigma(a))$ in which g is analytic and an admissible contour Γ for g and $f(a)$ such that each point of $f(\Delta)$ has winding number 1 with respect to Γ. Then Δ is an admissible contour for $g \circ f$ and

$$\begin{aligned} g(f(a)) &= \left(-\frac{1}{2\pi i}\right)^2 \int_\Gamma \int_\Delta (a - \mu 1)^{-1}(f(\mu) - \lambda 1)^{-1} g(\lambda)\, d\mu\, d\lambda \\ &= -\frac{1}{2\pi i}\int_\Delta (a - \mu 1)^{-1} g(f(\mu))\, d\mu \\ &= (g \circ f)(a) \end{aligned}$$

since for $\mu \in \Gamma$ the winding number of Δ with respect to $f(\mu)$ is 1. ■

If $\mathfrak{A}$ has no unit this construction can be made in $\mathfrak{A}^1$. If ϕ is the multiplicative linear functional $(a, \lambda) \mapsto \lambda$ then $f(a) \in \mathfrak{A}$ if and only if $0 = \phi(f(a)) = f(\phi(a)) = f(0)$. Thus if $f \in \mathcal{O}(\sigma(a))$, $f(0) = 0$ then $f(a) \in \mathfrak{A}$.

5. REGULARITY AND THE HULL-KERNEL TOPOLOGY

Throughout this section we shall assume that $\mathfrak{A}$ is a commutative Banach algebra with unit 1. The case in which $\mathfrak{A}$ has no 1 can be given a similar treatment by going to $\mathfrak{A}^1$ but we assume a unit to simplify matters.

If Ω is a compact Hausdorff space, E is a closed subset of Ω and J is a closed ideal in $\mathfrak{A} = C(\Omega)$ then

$$J_E = \{f; f \in \mathfrak{A}, f(\omega) = 0 \text{ for all } \omega \in E\}$$

is a closed ideal in $\mathfrak{A}$,

$$E_J = \{\omega; \omega \in \Omega, f(\omega) = 0 \text{ for all } f \in J\}$$

is a closed set in Ω, $J_{E_J} = J$, $E_{J_E} = E$ and $E \leftrightarrow J_E$ is a one to one correspondence between the closed subsets of Ω and the closed ideals of $\mathfrak{A}$.

Definition 5.1. For a general commutative Banach algebra if $E \subseteq \Phi_{\mathfrak{A}}$ and J is an ideal in $\mathfrak{A}$ then the *kernel* $k(E)$ of E is

$$\begin{aligned} k(E) &= \{a; a \in \mathfrak{A}, \hat{a}(\phi) = 0 \text{ for all } \phi \in E\} \\ &= \bigcap_{\phi \in E} \operatorname{Ker} \phi \end{aligned}$$

and the *hull* $h(J)$ of J is

$$h(J) = \{\phi; \phi \in \Phi_{\mathfrak{A}}, \hat{a}(\phi) = 0 \text{ for all } a \in J\}.$$

$k(E)$ is a closed ideal in $\mathfrak{A}$ and $h(J)$ is a closed subset of $\Phi_{\mathfrak{A}}$.

This does not always set up a one to one correspondence between the closed sets in $\Phi_{\mathfrak{A}}$ and the closed ideals in $\mathfrak{A}$; if Δ is the closed unit disc in $\mathbb{C}$ and $\mathfrak{A}$ is the closed subalgebra of $C(\Delta)$ consisting of functions analytic at all interior points of Δ then $\Phi_{\mathfrak{A}}$ is (homeomorphic with) Δ yet if $E \subseteq \Delta$ has a point of accumulation in the interior of Δ then $k(E) = \{0\}$. However

LEMMA 5.2. *If $E \subset \Phi_{\mathfrak{A}}$ and J is an ideal in $\mathfrak{A}$ then*

$$hkh(J) = h(J), \qquad khk(E) = k(E)$$

Proof. As the proofs are similar we give only one. For any subset E of $\Phi_{\mathfrak{A}}$ $hk(E) \supseteq E$ so $hkh(J) \supseteq h(J)$. If J_1, J_2 are two ideals in $\mathfrak{A}$ with $J_1 \subseteq J_2$ then $h(J_1) \supseteq h(J_2)$. Applying this to the inequality $kh(J) \supseteq J$ we get $hkh(J) \subseteq h(J)$. ∎

This lemma shows that there is a one to one correspondence between the range of h and the range of k. For some algebras such as group algebras the question of how far h, k depart from being one to one is important. The *problem of synthesis* for a closed set E in $\Phi_{\mathfrak{A}}$ is the question of whether $h^{-1}(E)$ contains more than 1 point. The range of h defines a topology on $\Phi_{\mathfrak{A}}$.

THEOREM 5.3. *The set $\mathscr{F} = \{h(J); J$ an ideal in $\mathfrak{A}\}$ of subsets of $\Phi_{\mathfrak{A}}$, that is the set of all sets E in $\Phi_{\mathfrak{A}}$ with $hk(E) = E$, form the closed sets in a topology on $\Phi_{\mathfrak{A}}$ which is coarser than the Gelfand topology.*

Proof. $\varnothing = h(\mathfrak{A})$ so $\varnothing \in \mathscr{F}$. $\Phi_{\mathfrak{A}} = h(0)$ so $\Phi_{\mathfrak{A}} \in \mathscr{F}$.

If J_1, J_2 are ideals in $\mathfrak{A}$ then so is $J_1 \cap J_2$ and, by the monotonicity of h, $h(J_1) \subseteq h(J_1 \cap J_2)$. Similarly $h(J_2) \subseteq h(J_1 \cap J_2)$ so $h(J_1) \cup h(J_2) \subseteq h(J_1 \cap J_2)$. If $\phi \notin h(J_1) \cap h(J_2)$ and $a_i \in J_i$ with $\phi(a_i) \neq 0$ $(i = 1, 2)$ then $a_1 a_2 \in J_1 \cap J_2$ and $\phi(a_1 a_2) \neq 0$ so $\phi \notin h(J_1 \cap J_2)$. Thus $h(J_1) \cup h(J_2) = h(J_1 \cap J_2)$. Hence $\mathscr{F}$ is closed under finite unions.

If J_α, $\alpha \in \mathrm{A}$, are ideals in $\mathfrak{A}$ then

$$\bigcap_\alpha h(J_\alpha) = h(\sum_\alpha J_\alpha)$$

where $\sum J_\alpha$ is the linear span of $\bigcup_\alpha J_\alpha$, so $\mathscr{F}$ is closed under arbitrary intersections.

Since $h(J)$ is always closed in the Gelfand topology the topology induced by $\mathscr{F}$ is coarser. ∎

Definition 5.4. The topology induced by $\mathscr{F}$ is called the *hull-kernel* topology. $\mathfrak{A}$ is said to be *regular* if the hull-kernel and the Gelfand topology coincide.

The hull-kernel topology is a T_1-topology, that is, one-point sets are closed (because they are $h(J)$ for a maximal ideal, by 3.6) and the identity map is continuous from $\Phi_{\mathfrak{A}}$ with the hull-kernel topology onto $\Phi_{\mathfrak{A}}$ with the Gelfand topology. As $\Phi_{\mathfrak{A}}$ is compact in the Gelfand topology, $\mathfrak{A}$ is regular if and only if the hull-kernel topology is Hausdorff.

One important use of regularity is that it enables us to check whether a complex valued function on $\Phi_{\mathfrak{A}}$ is in $\mathfrak{A}$ by considering the problem locally. The result used is

THEOREM 5.5. *Let $\mathfrak{A}$ be regular, $f \in C(\Phi_{\mathfrak{A}})$. If for each $\phi \in \Phi_{\mathfrak{A}}$ there exists $a \in \mathfrak{A}$ with $\hat{a} = f$ in a neighbourhood of ϕ then there exists $b \in \mathfrak{A}$ with $\hat{b} = f$ throughout $\Phi_{\mathfrak{A}}$.*

6. NON-COMMUTATIVE BANACH ALGEBRAS

(Our results here apply also to commutative algebras so really "non-commutative" could be replaced by "not-necessarily-commutative" throughout.)

The results and definitions concerning spectra, the function $\lambda \mapsto (a - \lambda 1)^{-1}$ and the functional calculus go through unchanged, and with the same proofs, provided the inverse b of an element a is required to satisfy $ab = 1 = ba$. Some of these results can also be extended to one-sided inverses but we shall not consider these. One way of seeing that the results we are concerned with carry over is to observe that all calculations are performed in the closed algebra generated by 1, a and the $(a - \lambda 1)^{-1}$ for $\lambda \in \mathbb{C}\backslash\sigma(a)$, a commutative Banach algebra. The results which carry over unchanged are 3.2, 3.3, 4.1 and 4.2 (except for the reference to the Gelfand transform). The material in 3.4, 3.5, 3.7 and 3.8 also carries over but is of little importance as the kernel of the Gelfand transform is a closed two-sided ideal such that the quotient algebra is commutative so that these results relate to this quotient algebra; and it is unusual for such a quotient to exist. In order to get a satisfactory theory for non-commutative algebras it is necessary to replace the notion of multiplicative linear functional by that of irreducible representation.

7. MODULES AND IRREDUCIBLE REPRESENTATIONS

Definition 7.1. Let $\mathfrak{A}$ be a Banach algebra and $\mathfrak{X}$ a Banach space. $\mathfrak{X}$ is a *left Banach* $\mathfrak{A}$*-module* if there is an associative bilinear product $(a, x) \mapsto ax$ from $\mathfrak{A} \times \mathfrak{X}$ into $\mathfrak{X}$ with

$$\|ax\| \leqslant K \| a \| \| x \| \qquad a \in \mathfrak{A}, \quad x \in \mathfrak{X}$$

for some $K > 0$. (The associative law we have in mind is

$$(ab)x = a(bx) \qquad a, b \in \mathfrak{A}, \quad x \in \mathfrak{X}.)$$

Definition 7.2. A *representation* of $\mathfrak{A}$ on $\mathfrak{X}$ is a continuous algebra homomorphism of $\mathfrak{A}$ into $\mathscr{L}(\mathfrak{X})$.

7.1 and 7.2 are different ways of describing the same system. If $\mathfrak{X}$ is a left Banach $\mathfrak{A}$-module then defining $\phi(a)x = ax$ we have a representation of $\mathfrak{A}$ on $\mathfrak{X}$ and the same formula used the other way shows how to construct a module from a representation. We shall use module notation as it is more convenient although the representation approach is traditional. We also propose to consider the case in which $\mathfrak{A}$ has no identity at the same time as the case in which it has. For the remainder of this section $\mathfrak{A}$ is a Banach algebra.

If J is a closed left ideal in $\mathfrak{A}$ then $\mathfrak{A}/J$ is a Banach space and $a(b + J) = ab + J$ defines a left $\mathfrak{A}$-module structure on $\mathfrak{A}/J$. Not all left $\mathfrak{A}$-modules arise in this way but all irreducible ones do.

Definition 7.3. A left $\mathfrak{A}$-module $\mathfrak{X}$ is *irreducible* if $\mathfrak{A}\mathfrak{X} \neq \{0\}$ and $\mathfrak{X}$ contains no proper submodules (that is no linear subspace $\mathfrak{Y}$ with $ay \in \mathfrak{Y}$ for all $a \in \mathfrak{A}$, $y \in \mathfrak{Y}$, $\mathfrak{Y} \neq \mathfrak{X}$ and $\mathfrak{Y} \neq \{0\}$). Note that we do not even allow $\mathfrak{X}$ to contain nonclosed submodules.

Definition 7.4. A left ideal J in $\mathfrak{A}$ is *modular* if and only if there is an element u of $\mathfrak{A}$ with $a - au \in J$ for all $a \in \mathfrak{A}$.

It is easy to see that if $\mathfrak{A}$ has a right unit then every left ideal is modular and conversely (consider the ideal $\{0\}$). Also any left ideal containing a modular left ideal is modular (with the same u).

THEOREM 7.5. *A left ideal J in $\mathfrak{A}$ is modular if and only if it is of the form $J' \cap \mathfrak{A}$ for some ideal J' of $\mathfrak{A}^1$ not contained in $\mathfrak{A}$.*

Proof. If $u \in \mathfrak{A}$ has $a - au \in J$ for all $a \in \mathfrak{A}$ put

$$J' = \{s : s \in \mathfrak{A}^1, su \in J\}.$$

Then J' is a left ideal in $\mathfrak{A}^1$ containing $1 - u \notin \mathfrak{A}$. Let $a \in \mathfrak{A}$. If $a \in J'$ then $au \in J$ so $a = a - au + au \in J$ and conversely. Thus $J' \cap \mathfrak{A} = J$.

If J' is a left ideal in $\mathfrak{A}^1$ not contained in $\mathfrak{A}$ then $J' + \mathfrak{A}$ is a subspace of $\mathfrak{A}^1$ properly containing $\mathfrak{A}$, a subspace of codimension 1, so $J' + \mathfrak{A} = \mathfrak{A}^1$. Let $u \in \mathfrak{A}$ with $1 - u \in J'$. If $a \in \mathfrak{A}$ than $a - au = a(1 - u) \in J' \cap \mathfrak{A}$ for each $a \in \mathfrak{A}$ so $J' \cap \mathfrak{A}$ is modular. ∎

COROLLARY 7.6. *If J is a maximal proper modular left ideal in $\mathfrak{A}$ then J is closed.*

Proof. Let I be a left ideal in $\mathfrak{A}^1$ properly containing J', constructed from J as above. Let $s \in I \backslash J'$. Then $su \notin J$, $su \in \mathfrak{A}$ and since $s - su \in J' \subseteq I$ we have $su \in I \cap \mathfrak{A}$. $I \cap \mathfrak{A}$ is thus a left ideal in $\mathfrak{A}$ properly containing J (because $su \in (I \cap \mathfrak{A}) \backslash J$) and hence $I \cap \mathfrak{A} = \mathfrak{A}$. As $I \supseteq J'$ we have $I \supseteq J' + \mathfrak{A} = \mathfrak{A}^1$ and so J' is maximal. The first part of the proof of 3.6 applies to show J' closed and hence $J = J' \cap \mathfrak{A}$ is closed. ∎

THEOREM 7.7. *If J is a maximal proper modular left ideal in $\mathfrak{A}$ then $\mathfrak{A}/J$ is an irreducible left Banach $\mathfrak{A}$-module. Conversely if $\mathfrak{X}$ is an irreducible left Banach $\mathfrak{A}$-module then there is a maximal proper modular left ideal J in $\mathfrak{A}$ and a linear homeomorphism U of $\mathfrak{A}/J$ onto $\mathfrak{X}$ with $U(ab + J) = aU(b + J)$ $(a, b \in \mathfrak{A})$.*

Proof. J is closed so $\mathfrak{A}/J$ is a left Banach $\mathfrak{A}$-module. As $J \neq \mathfrak{A}$ we have $u \in \mathfrak{A} \backslash J$ so $u^2 \in \mathfrak{A} \backslash J$ and so $\mathfrak{A}(\mathfrak{A}/J) \neq \{0\}$. Let q be the quotient map $\mathfrak{A} \to \mathfrak{A}/J$ and suppose $\mathfrak{Y}$ is a submodule of $\mathfrak{A}/J$. Then $q^{-1}(\mathfrak{Y})$ is a left ideal in $\mathfrak{A}$ containing J so $q^{-1}(\mathfrak{Y}) = J$ or $q^{-1}(\mathfrak{Y}) = \mathfrak{A}$. Hence $\mathfrak{Y} = q \cdot q^{-1}(\mathfrak{Y}) = \{0\}$ or $\mathfrak{A}/J$, and $\mathfrak{A}/J$ is irreducible. Conversely let ξ be a vector in $\mathfrak{X}$ with $\mathfrak{A}\xi \neq \{0\}$ (such exist, otherwise $\mathfrak{A}\mathfrak{X} = \{0\}$). Then $\mathfrak{A}\xi$ is a submodule of $\mathfrak{X}$ and so is $\mathfrak{X}$ by irreducibility. The map U_0; $a \to a\xi$ is a continuous map of $\mathfrak{A}$ onto $\mathfrak{X}$ with Ker $U_0 = J = \{a : a \in \mathfrak{A}, a\xi = 0\}$ where J is clearly a closed left ideal and $U_0 = Uq$ where q is the quotient map $\mathfrak{A} \to \mathfrak{A}/J$ and U is a linear homeomorphism of $\mathfrak{A}/J$ onto $\mathfrak{X}$. As $U_0(ab) = aU_0(b)$ we see $U(ab + J) = aU(b + J)$. If I is a left ideal in $\mathfrak{A}$ containing J then $U_0 I$ is a submodule of $\mathfrak{X}$ so $U_0 I = \{0\}$ or $\mathfrak{X}$. As $I = U_0^{-1} UI$ this shows $I = J$ or $\mathfrak{A}$. J is modular because as $\mathfrak{A}\xi = \mathfrak{A}$ there is $u \in \mathfrak{A}$ with $u\xi = \xi$ and hence $au\xi = a\xi$ for all $a \in \mathfrak{A}$. ■

THEOREM 7.8. *Let $\mathfrak{X}$ be an irreducible left $\mathfrak{A}$-module and let $\xi_1, \ldots, \xi_n$ be linearly independent in $\mathfrak{X}$, $\eta_1, \ldots, \eta_n \in \mathfrak{X}$. Then there is $a \in \mathfrak{A}$ with $a\xi_i = \eta_i$ $(i = 1, 2, \ldots)$.*

Proof. We prove the result by induction on n. Consider first the case $n = 1$. $\{\xi; \xi \in \mathfrak{X}, a\xi = 0 \text{ for all } a \in \mathfrak{A}\}$ is a submodule of $\mathfrak{X}$ and so is $\{0\}$ or $\mathfrak{X}$. If it were $\mathfrak{X}$ we would have $\mathfrak{A}\mathfrak{X} = \{0\}$ so it must be $\{0\}$. Thus, as $\xi_1 \neq 0$, $\mathfrak{A}\xi_1 = \mathfrak{X}$ so there is $a \in \mathfrak{A}$ with $a\xi_1 = \eta_1$.

Now suppose $n \geqslant 2$ and the theorem is true with n replaced by $n - 1$. If for each i $(1 \leqslant i \leqslant n)$ there is $a_i \in \mathfrak{A}$ with $a_i\xi_i \neq 0$, $a_i\xi_j = 0$ $i \neq j$ then choose $b_i \in \mathfrak{A}$ with $b_i a_i \xi_i = \eta_i$ and putting $a = \sum b_i a_i$ we see $a\xi_i = \eta_i$ $(i = 1, \ldots, n)$. Thus suppose that for one value of i the choice of a_i is not possible; for convenience we suppose it fails for $i = 1$. We are supposing

$$\text{If } a \in \mathfrak{A} \text{ and } a\xi_2 = a\xi_3 \ldots = a\xi_n = 0 \text{ then } a\xi_1 = 0. \tag{*}$$

Put

$$I = \{a; a \in \mathfrak{A}, a\xi_2 = a\xi_3 \ldots = a\xi_n = 0\}$$

$$J = \{a; a \in \mathfrak{A}, a\xi_3 = \ldots = a\xi_n = 0\}$$

(taking $J = \mathfrak{A}$ when $n = 2$) so I, J are closed left ideals in $\mathfrak{A}$. Define T; $\mathfrak{X} \to \mathfrak{X}$ by $T\eta = a\xi_1$ $(\eta \in \mathfrak{X})$ where $a \in J$ has $a\xi_2 = \eta$. The inductive hypothesis ensures that for each η such an a exists and condition (*) shows that $T\eta$ depends on η and not on the choice of a. T is linear and if $b \in \mathfrak{A}$ then

$$T(b\eta) = T(ba\xi_2) = ba\xi_1 = bT(\eta),$$

where $a \in J$ has $a\xi_2 = \eta$. The map $a \mapsto a\xi_2$ maps J continuously onto $\mathfrak{X}$ (by the inductive hypothesis), has kernel I and so can be written as the product of the quotient map $J \to J/I$ and a linear homeomorphism of J/I onto $\mathfrak{X}$. If $\zeta_j \to 0$ in $\mathfrak{X}$ then the corresponding sequence in J/I has a zero limit and so is the image, under the quotient map, of a sequence $\{a_j\}$ in J with $a_j \to 0$. Thus

$$T(\zeta_j) = T(a_j\xi_2) = a_j\xi_1 \to 0$$

showing that T is continuous.

Put

$$\mathfrak{C} = \{S;\ S \in \mathscr{L}(\mathfrak{X}),\ S(a\eta) = aS(\eta), a \in \mathfrak{A},\ \eta \in \mathfrak{X}\},$$

a closed subalgebra of $\mathscr{L}(\mathfrak{X})$ containing T and I. If $S \in \mathfrak{C}$ then Ker S and Im S are submodules of $\mathfrak{X}$ and so are each either $\{0\}$ or $\mathfrak{X}$. If $S \neq 0$ then Ker $S = \{0\}$, Im $S = \mathfrak{X}$ so that, by the open mapping theorem, S has an inverse in $\mathscr{L}(\mathfrak{X})$. Let $\lambda \in \sigma(T)$. Then $T - \lambda I$ is in $\mathfrak{C}$ and has no inverse in $\mathscr{L}(\mathfrak{X})$ and so $T - \lambda I = 0$. Thus $a\xi_1 = T(a\xi_2) = \lambda a\xi_2$ for all $a \in J$, that is $a(\xi_1 - \lambda\xi_2) = 0$ whenever $a \in \mathfrak{A}$ has $a\xi_3 = \ldots = a\xi_n = 0$. Thus by the inductive hypothesis $\xi_1 - \lambda\xi_2, \xi_3, \ldots, \xi_n$ are dependent, contradicting our assumption that $\xi_1, \xi_2, \ldots, \xi_n$ are linearly independent. ■

Definition 7.9. The *radical* of a Banach algebra $\mathfrak{A}$ is

$$\{a; a \in \mathfrak{A}, a\xi = 0 \text{ for all } \xi \in \mathfrak{X}, \text{ for every irreducible left } \mathfrak{A}\text{-module } \mathfrak{X}).$$

By 7.7 the radical is also the intersection of the maximal proper modular left ideals of $\mathfrak{A}$. $\mathfrak{A}$ is *semi-simple* if its radical is $\{0\}$.

This definition agrees with 3.10 when $\mathfrak{A}$ is commutative, as if $\phi \in \Phi_{\mathfrak{A}}$ then $a\xi = \phi(a)\xi$, $\xi \in \mathbb{C}$ defines $\mathbb{C}$ as an irreducible $\mathfrak{A}$-module and every irreducible $\mathfrak{A}$-module $\mathfrak{X}$ is of this form because it is one-dimensional (if not take ξ_1, ξ_2 independent and $a, b \in \mathfrak{A}$ with $a\xi_1 = \xi_1$, $a\xi_2 = 0$, $b\xi_1 = \xi_2$, $b\xi_2 = \xi_1$ so that $\xi_1 = ab\xi_2 = ba\xi_2 = 0$).

Results similar to those obtained in this section hold for right ideals, right modules and anti-representations (linear maps $\mathfrak{A} \to \mathscr{L}(\mathfrak{X})$ with $T(ab) = T(b)T(a)$; $a, b \in \mathfrak{A}$). The radical, as defined in 7.9 can be shown to be the same as that defined by similar formulae involving right modules and anti-representations.

8. UNIQUENESS OF NORM FOR BANACH ALGEBRAS

THEOREM 8.1. *Let $\mathfrak{X}$ be a Banach space and an irreducible left module over the Banach algebra $\mathfrak{A}$. If the product $(a, \xi) \mapsto a\xi$ is continuous in ξ then it is continuous in a, and $\mathfrak{X}$ is a left Banach $\mathfrak{A}$-module.*

Proof. Theorem 7.8 applies in our case but the proof must be expanded. In the first place I, J are intersections of maximal modular ideals and so are closed. The other change necessary is in the proof that T is a scalar multiple of the identity. The map $a \to a\xi_2$ gives a one to one linear map α of J/I onto $\mathfrak{X}$. With module operations $a(j + I) = aj + I$, J/I is a Banach $\mathfrak{A}$-module and $\alpha(am) = a\,\alpha(m)$, $a \in \mathfrak{A}$, $m \in J/I$. Thus J/I is irreducible because $\mathfrak{X}$ is and we can argue as before with $\alpha^{-1}T\alpha$ and J/I replacing T and $\mathfrak{X}$ to show that $\alpha^{-1}T\alpha$ is a scalar multiple of the identity in J/I which implies that T is a scalar multiple of the identity on $\mathfrak{X}$. The remainder of the proof is as before.

If $\mathfrak{X}$ is finite dimensional put

$$J_\xi = \{a; a \in \mathfrak{A}, a\xi = 0\}, \qquad \xi \in \mathfrak{X}.$$

Then J_ξ is a maximal modular left ideal in $\mathfrak{A}$ and so is closed. The map $a \mapsto a\xi$ has kernel J_ξ and so gives a one to one map of the Banach space $\mathfrak{A}/J_\xi$ into the finite dimensional space $\mathfrak{X}$. As linear maps between finite dimensional Banach spaces are continuous the map $a \mapsto a\xi$, being a quotient map followed by a map from a finite dimensional Banach space into another, is continuous.

If $\mathfrak{X}$ is infinite dimensional and $\xi_0 \in \mathfrak{X}$ with $\xi_0 \neq 0$ and $a \mapsto a\xi_0$ continuous then, for each $b \in \mathfrak{A}$, the map $a \mapsto (ab)\xi_0$ is continuous. As $\mathfrak{A}\xi_0 = \mathfrak{X}$ we see that if $a \mapsto a\xi$ is continuous for one non-zero ξ in $\mathfrak{X}$ then it is continuous for all non-zero ξ in $\mathfrak{X}$. Thus suppose that if $\xi \neq 0$ then $a \mapsto a\xi$ is discontinuous. Let $\xi_1, \xi_2, \ldots$ be a linearly independent sequence of vectors in $\mathfrak{X}$ with $\|\xi_i\| = 1$ for all i. By induction choose a sequence $\{a_i\}$ from $\mathfrak{A}$ with

(i) $a_i\xi_j = 0 \qquad j < i$

(ii) $\|a_i\xi_i\| \geqslant i + \|\sum_{j<i} a_j\xi_i\|$

(iii) $\|a_i\| < 2^{-i}$.

The choice of a_1 is trivial as (i) is vacuous. When $a_1, \ldots, a_{n-1}$ have been chosen take $b_n \in \mathfrak{A}$ with $b_n\xi_j = 0$ $(j = 1, \ldots, n-1)$ and $b_n\xi_n = \xi_n$. Choose a'_n with

$$\|a'_n\xi_n\| \geqslant i + \|\sum_{j<n} a_j\xi_i\|$$

and

$$\|a'_n\| < 2^{-n}\|b_n\|^{-1},$$

such a choice being possible by the discontinuity of $a \mapsto a\xi_n$. Finally put $a_n = a'_n b_n$.

Define $c_n = \sum_{j>n} a_j$, the series being absolutely convergent by (iii). As $a_j \in J_{\xi_i}$ for $j > i$ and J_{ξ_i} is closed, $c_n \in J_{\xi_n}$. Then

$$\begin{aligned} \sup_{\|\xi\| \leqslant 1} \|c_0\xi\| &\geqslant \|c_0\xi_n\| \\ &= \|\sum_{j<n} a_j\xi_n + a_n\xi_n + c_n\xi_n\| \\ &\geqslant \|a_n\xi_n\| - \|\sum_{j<n} a_j\xi_n\| \\ &\geqslant n. \end{aligned}$$

However $\xi \mapsto c_0\xi$ is continuous, so the supremum is finite. This contradiction shows $a \mapsto a\xi$ is continuous for all $\xi \in \mathfrak{X}$.

The uniform boundedness theorem then shows $(a, \xi) \mapsto a\xi$ is jointly continuous (the maps $a \mapsto a\xi$, $\|a\| \leqslant 1$ form a pointwise bounded set and hence are uniformly bounded). ■

THEOREM 8.2. *Let $\mathfrak{A}$ be a semi-simple Banach algebra with norm $\| \ \|$. Suppose $| \ |$ is another norm in $\mathfrak{A}$ making $\mathfrak{A}$ a Banach algebra. Then $| \ |$ and $\| \ \|$ are equivalent, that is they determine the same norm topology.*

Proof. We use the closed graph theorem. Suppose $\{a_n\}$ is a sequence in $\mathfrak{A}$ which converges to 0 in $\| \ \|$ and to b in $| \ |$. If $\mathfrak{X}$ is an irreducible Banach $\mathfrak{A}$-module (with respect to $\| \ \|$) and $\xi \in \mathfrak{X}$, then $a \mapsto a\xi$ is continuous in $| \ |$ by 8.1 so $b\xi = \lim a_n\xi = 0$. Thus, because $\mathfrak{A}$ is semi-simple, $b = 0$ and the closed graph theorem shows that the identity map, considered as a map $(\mathfrak{A}, \| \ \|) \mapsto (\mathfrak{A}, | \ |)$ is continuous. By the open mapping theorem it is a homeomorphism. ■

9. APPROXIMATE UNITS AND FACTORISATION

Throughout this section $\mathfrak{A}$ will denote a Banach algebra. Our results will be trivial if $\mathfrak{A}$ has a unit.

Definition 9.1. $\mathfrak{A}$ has a *bounded left approximate unit* (or *identity*) if there is a net $\{e_\alpha\}$ in $\mathfrak{A}$ and $K > 0$ with $\| e_\alpha \| \leqslant K$ for all α and $e_\alpha a \to a$ for all $a \in \mathfrak{A}$.

Equivalently, if for each finite subset $a_1, \ldots, a_n$ of $\mathfrak{A}$ and each $\varepsilon > 0$ there is $e \in \mathfrak{A}$ with $\| ea_i - a_i \| < \varepsilon$ and $\|e\| < K$.

THEOREM 9.2. *Let $\mathfrak{A}$ have a bounded left approximate unit and let $\mathfrak{X}$ be a left Banach $\mathfrak{A}$-module. Then*

$$\{ax;\ a \in \mathfrak{A}, x \in \mathfrak{X}\}$$

is a closed submodule.

Proof. The closed linear span $\mathfrak{X}_0$ of $\{ax; a\in\mathfrak{A},\ x\in\mathfrak{X}\}$ is $\{x; x\in\mathfrak{X},\ e_\alpha x \to x\}$. Let $x \in \mathfrak{X}_0$. We shall define a sequence e_n of elements of $\mathfrak{A}$ with $\| e_n \| \leqslant K$ so that

$$\begin{aligned} h_n &= \left[\left(1 + \frac{1}{K}\right)1 - \frac{1}{K}e_n\right]^{-1} \cdots \left[\left(1 + \frac{1}{K}\right)1 - \frac{1}{K}e_1\right]^{-1} \\ &= \left(1 + \frac{1}{K}\right)^{-n} 1 + b_n \qquad (b_n \in \mathfrak{A}) \end{aligned}$$

converges in $\mathfrak{A}^1$ to a limit b and $y_n = h_n^{-1}x$ converges in $\mathfrak{X}$ (in fact in $\mathfrak{X}_0$) to an element y. Note that, because $\|e_n\| \leqslant K$,

$$1 - \frac{1}{K}\left(1 + \frac{1}{K}\right)^{-1} e_n,$$

and hence

$$\left(1+\frac{1}{K}\right)1-\frac{1}{K}e_n,$$

is regular. Using the series expansions in 3.2 we see

$$\left[\left(1+\frac{1}{K}\right)1-\frac{1}{K}e_n\right]^{-1}=\left(1+\frac{1}{K}\right)^{-1}1+\text{an element of }\mathfrak{A}.$$

Thus the above formulae will define the elements h_n, b_n in the appropriate spaces. $\mathfrak{X}$ is a left Banach $\mathfrak{A}^1$-module if we put $1x = x$, $x \in \mathfrak{X}$. Once we have this, we have $by = \lim h_n y_n = x$.

The e_n are defined inductively, e_{n+1} being chosen so that

$$\| e_{n+1} \| \leqslant K$$

$$\| e_{n+1} b_n - b_n \| < K\left(1+\frac{1}{K}\right)^{-n-1}$$

$$\| x - e_{n+1} x \| < \| h_n^{-1} \|^{-1} K\left(1+\frac{1}{K}\right)^{-n-1}.$$

e_1 is defined by taking $h_0 = 1$, $b_0 = 0$. As $h_n - b_n \to 0$, to show h_n convergent we need only show that the b_n form a Cauchy sequence. We have

$$\begin{aligned} b_{n+1} &= h_{n+1} - \left(1+\frac{1}{K}\right)^{-n-1} 1 \\ &= \left[\left(1+\frac{1}{K}\right)1-\frac{1}{K}e_{n+1}\right]^{-1}\left[\left(1+\frac{1}{K}\right)^{-n}1+b_n\right]-\left(1+\frac{1}{K}\right)^{-n-1}1 \end{aligned}$$

so that

$$\begin{aligned} \| b_{n+1} - b_n \| = \Bigg\| \left[\left(1+\frac{1}{K}\right)1-\frac{1}{K}e_{n+1}\right]^{-1}\Bigg[\frac{1}{K}\left(1+\frac{1}{K}\right)^{-n-1} e_{n+1} \\ + \frac{1}{K}e_{n+1}b_n - \frac{1}{K}b_n\Bigg]\Bigg\| \end{aligned}$$

where, as in 3.2, expanding $[1-(1/K+1)\,e_{n+1}]^{-1}$ as a power series and using $\| e_{n+1} \| \leqslant K$, we have

$$\left\| \left[\left(1+\frac{1}{K}\right)1-\frac{1}{K}e_{n+1}\right]^{-1} \right\| \leqslant K.$$

As

$$\left\| \frac{1}{K}\left(1+\frac{1}{K}\right)^{-n-1} e_{n+1} \right\| \leqslant \left(1+\frac{1}{K}\right)^{-n-1}$$

and

$$\frac{1}{K}\| e_{n+1}b_n - b_n \| \leqslant \left(1+\frac{1}{K}\right)^{-n-1}$$

we have

$$\| b_{n+1} - b_n \| \leqslant 2K\left(1+\frac{1}{K}\right)^{-n-1}$$

Thus

$$\begin{aligned}
\| b_n - b_m \| &\leqslant \| b_n - b_{n+1} \| + \| b_{n+1} - b_{n+2} \| \ldots + \| b_{m-1} - b_m \| \\
&\leqslant 2K\left(1+\frac{1}{K}\right)^{-n-1}\left(1+\left(1+\frac{1}{K}\right)^{-1}+\ldots+\left(1+\frac{1}{K}\right)^{n-m+1}\right) \\
&\leqslant 2K(K+1)\left(1+\frac{1}{K}\right)^{-n-1} \\
&\to 0 \text{ as } n \to \infty.
\end{aligned}$$

This shows that the b_n form a Cauchy sequence in $\mathfrak{A}$.
Also

$$y_{n+1} = h_n^{-1}\left[\left(1+\frac{1}{K}\right)1 - \frac{1}{K}e_{n+1}\right]x$$

so

$$\begin{aligned}
\| y_{n+1} - y_n \| &= \left\| h_n^{-1}\left[\frac{1}{K}x - \frac{1}{K}e_{n+1}x\right]\right\| \\
&\leqslant \frac{1}{K}\| h_n^{-1} \| \, \| x - e_{n+1}x \| \\
&\leqslant \left(1+\frac{1}{K}\right)^{-n-1}
\end{aligned}$$

and we see, as for the b_n, that the y_n form a Cauchy sequence. ∎

Corollary 9.3. *If C is a compact subset of $\mathfrak{X}_0$ then there is $a \in \mathfrak{A}$ and a compact subset C' of $\mathfrak{X}_0$ with $C = aC'$.*

Proof. Let $\mathfrak{Y}$ be the Banach space of continuous functions $C \mapsto \mathfrak{X}_0$ with $\|f\| = \sup_{c \in C} \|f(c)\|$. $\mathfrak{Y}$ becomes a left Banach $\mathfrak{A}$-module with the operation $(af)(c) = a(f(c))$, $a \in \mathfrak{A}$, $c \in C$, $f \in \mathfrak{Y}$. If $\{e_\alpha\}$ is a bounded approximate identity in $\mathfrak{A}$ then $e_\alpha x \to x$ for each $x \in \mathfrak{X}_0$ and, as it is bounded, $e_\alpha x \to x$ uniformly on compact subsets of $\mathfrak{X}_0$. If $f \in \mathfrak{Y}$ then Im f, being the continuous image of C, is compact so $e_\alpha f(c) \to f(c)$ uniformly on C. Hence $\mathfrak{Y}_0 = \mathfrak{Y}$. Let $j \in \mathfrak{Y}$ be defined by $j(c) = c$. Then, by 9.2, there is $a \in \mathfrak{A}$ and $k \in \mathfrak{Y}$ with $j = ak$ and, in particular $C = a \operatorname{Im} k$. Put $C' = \operatorname{Im} k$. ∎

If x_n is a sequence in $\mathfrak{X}_0$ converging to 0 then we can apply the above with $C = \{0, x_1, x_2, \ldots\}$ and put $y_n = k(x_n)$. Then $ay_n = x_n$, $ak(0) = 0$, and $y_n \to k(0)$. Put $z_n = y_n - k(0)$. Then $az_n = x_n$ and $z_n \to 0$.

3. Introduction to Cohomology in Banach Algebras

B. E. JOHNSON

Department of Mathematics,
University of Newcastle, England

1. Extensions of Banach Algebras 84
2. Derivations and Automorphisms of Banach Algebras 87
3. The Hochschild Cohomology Groups 89
4. Cohomology and Fixed Points 90
5. Reflexive Modules and Modules with a Reflexive Auxilary Norm 91
6. Amenable Groups and Algebras 94
7. Modules which are not Dual Modules 95
8. Helemskiĭ's Cohomology Theory 96
References 99

1. EXTENSIONS OF BANACH ALGEBRAS

If completely developed, the theory of Banach algebras might consist of three parts; the theory of semi-simple Banach algebras, the theory of radical Banach algebras and then a study of the way in which these can be combined to form the most general Banach algebra. We shall consider the last of these in the special case in which $xy = 0$ for all x, y in the radical. A theory for nilpotent radicals can be based on this, but the general case is not dealt with. As a reference for this section, we cite [**6**].

Let $\mathfrak{B}$ be a Banach algebra, $\mathfrak{X}$ a closed two sided ideal with $xy = 0$ whenever $x, y \in \mathfrak{X}$ (we do not need to assume that $\mathfrak{X}$ is the radical of $\mathfrak{A}$). Put $\mathfrak{A} = \mathfrak{B}/\mathfrak{X}$. Then $\mathfrak{X}$ is a (two-sided) Banach $\mathfrak{A}$-module if we define

$$ax = bx \qquad a \in \mathfrak{A}, \quad x \in \mathfrak{X}$$

$$xa = xb$$

where $b \in \mathfrak{B}$ is such that $q(b) = a$, q the quotient map $\mathfrak{B} \to \mathfrak{A}$. By saying that $\mathfrak{X}$ is a two-sided Banach $\mathfrak{A}$-module we mean that it is a left and a right Banach $\mathfrak{A}$-module and that the further associative law

$$a(xa') = (ax)a' \qquad a, a' \in \mathfrak{A}, \quad x \in \mathfrak{X}$$

holds.

If $\mathfrak{X}$ has a Banach space complement in $\mathfrak{B}$ then $\mathfrak{B}$ is linearly homeomorphic with $\mathfrak{A} \oplus \mathfrak{X}$ and the multiplication on $\mathfrak{B}$ gives, via this homeomorphism, a multiplication on $\mathfrak{A}$ making it a Banach algebra (or at least a Banach space with continuous multiplication which on renorming can be made into a Banach algebra). $\mathfrak{X}$ is an ideal in $\mathfrak{A} \oplus \mathfrak{X}$ and the projection $\mathfrak{A} \oplus \mathfrak{X} \to \mathfrak{A}$ is an algebra homomorphism. This shows that the product $(a, 0)\,(b, 0)$ has first coordinate ab and so is $(ab, T(a, b))$. As this product is bilinear and continuous, T is a continuous bilinear map $\mathfrak{A} \times \mathfrak{A} \to \mathfrak{X}$. We have, from the homeomorphism between $\mathfrak{A} \oplus \mathfrak{X}$ and $\mathfrak{B}$,

$$(a, 0)\,(0, x) = (0, ax) \quad \text{and} \quad (0, x)\,(a, 0) = (0, xa),$$

$a \in \mathfrak{A}$, $x \in \mathfrak{X}$. Thus the product in $\mathfrak{A} \oplus \mathfrak{X}$ is

$$(a, x)\,(b, y) = (ab, ay + xb + T(a, b)). \tag{i}$$

The associative law shows

$$aT(b, c) - T(ab, c) + T(a, bc) - T(a, b)c = 0 \qquad a, b, c \in \mathfrak{A}. \tag{ii}$$

Conversely starting with a Banach algebra $\mathfrak{A}$, a two-sided Banach $\mathfrak{A}$-module $\mathfrak{X}$ and a continuous bilinear map T which satisfies (ii) then (i) defines a continuous multiplication on $\mathfrak{A} \oplus \mathfrak{X}$ making $\mathfrak{A} \oplus \mathfrak{X}$ an extension of $\mathfrak{A}$ by $\mathfrak{X}$ according to the following definition.

Definition. Let $\mathfrak{A}$ be a Banach algebra and $\mathfrak{X}$ a two sided Banach $\mathfrak{A}$-module. Then an *extension of* $\mathfrak{A}$ *by* $\mathfrak{X}$ is a Banach algebra $\mathfrak{B}$ containing an ideal $\mathfrak{X}_0$ homeomorphic with $\mathfrak{X}$, with $xy = 0$ for all $x, y \in \mathfrak{X}_0$, such that $\mathfrak{B}/\mathfrak{X}_0$ is homeomorphic and algebraically isomorphic with $\mathfrak{A}$ and the $\mathfrak{A}$-module structure induced on $\mathfrak{X}$ by the multiplication in $\mathfrak{B}$ is the original one.

We have seen how an extension can be constructed from T and how, given an extension and a complement of $\mathfrak{X}$, T can be constructed. We now consider how the function T is affected by the choice of complement. Once we have chosen one complement $\mathfrak{B}$ is isomorphic with $\mathfrak{A} \oplus \mathfrak{X}$ so we look to see what happens to T if we replace the canonical complement of $\mathfrak{X}$ in $\mathfrak{A} \oplus \mathfrak{X}$ by some other. If $S \in \mathscr{L}(\mathfrak{A}, \mathfrak{X})$, the space of bounded linear operators $\mathfrak{A} \to \mathfrak{X}$ then $\{(a, Sa);\ a \in \mathfrak{A}\}$ is a closed complement of $\mathfrak{X}$ in $\mathfrak{A} \oplus \mathfrak{X}$ and every such complement is obtained in this way. The bilinear map T' associated with such a choice of complement is defined by

$$(a, Sa)\,(b, Sb) - (ab, S(ab)) = (0, T'(a, b))$$

and as $(a, Sa)\,(b, Sb) = (ab, as(b) + S(a)b + T(a, b))$ we see

$$T'(a, b) - T(a, b) = aS(b) - S(ab) + S(a)b, \qquad a, b \in \mathfrak{A}. \tag{iii}$$

Thus if T, T' come from the same extension of $\mathfrak{A}$ by $\mathfrak{X}$ by choosing different complements then they are connected by (iii). It is easy to reverse the argument showing that if T, T' are connected by (iii) then they come from the same extension of $\mathfrak{A}$ by $\mathfrak{X}$ but are associated with different complements. The set of bilinear maps from $\mathfrak{A}$ into $\mathfrak{X}$ which satisfy (ii) form a linear space $Z(\mathfrak{A}, \mathfrak{X})$. Taking $T \sim T'$ when there is $S \in \mathscr{L}(\mathfrak{A}, \mathfrak{X})$ such that (iii) is satisfied we have an equivalence relation on $Z(\mathfrak{A}, \mathfrak{X})$, the equivalence classes are in fact the cosets of the subspace $N(\mathfrak{A}, \mathfrak{X})$ of bilinear maps T of the form

$$T(a, b) = aS(b) - S(ab) + S(a)b \qquad a, b \in \mathfrak{A} \tag{iv}$$

for some $S \in \mathscr{L}(\mathfrak{A}, \mathfrak{X})$. Note that such a T always satisfies (ii).

Definition. Two extensions $\mathfrak{B}$, $\mathfrak{B}'$ of $\mathfrak{A}$ by $\mathfrak{X}$ are *equivalent* if there is an algebra isomorphism k of $\mathfrak{B}$ onto $\mathfrak{B}'$ so that the diagram

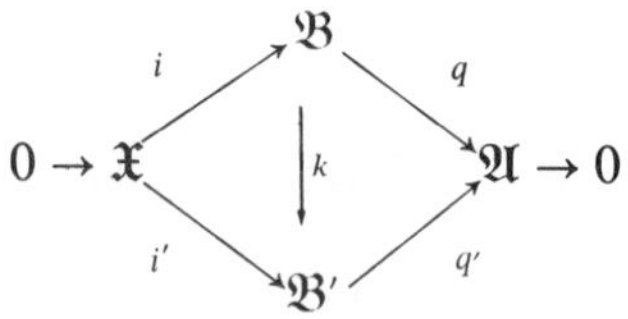

commutes, where i, i' are the injections and q, q' the quotient maps for the extensions. Note that k is necessarily continuous.

Our above discussion shows that there is a one to one correspondence between equivalence classes of extensions of $\mathfrak{A}$ by $\mathfrak{X}$ in which the copy of $\mathfrak{X}$ is complemented and elements of the group $\mathscr{H}^2(\mathfrak{A}, \mathfrak{X}) = Z(\mathfrak{A}, \mathfrak{X})/N(\mathfrak{A}, \mathfrak{X})$.

If we construct the extension of $\mathfrak{A}$ by $\mathfrak{X}$ corresponding to the zero element of $Z(\mathfrak{A}, \mathfrak{X})$ then the canonical copy of $\mathfrak{A}$ in $\mathfrak{A} \oplus \mathfrak{X}$ is a subalgebra. If we construct the extension corresponding to an element of $N(\mathfrak{A}, \mathfrak{X})$ then, in $\mathfrak{A} \oplus \mathfrak{X}$ there is a closed subalgebra complementary to $\mathfrak{X}$. Thus the zero of $\mathscr{H}^2$ corresponds to those extensions in which $\mathfrak{X}$ has a complementary subalgebra. If $\mathfrak{A} \oplus \mathfrak{X}$ is the extension of $\mathfrak{A}$ by $\mathfrak{X}$ corresponding to $T = 0$, $S \in \mathscr{L}(\mathfrak{A}, \mathfrak{X})$ and $\{(a, Sa), a \in \mathfrak{A}\}$ is a closed subalgebra of $\mathfrak{A} \oplus \mathfrak{X}$ then

$$(ab, S(ab)) = (a, Sa)(b, Sb) = (ab, aSb + (Sa)b)$$

so $S(ab) = aSb + (Sa)b$. Thus the subalgebras complementary to $\mathfrak{X}$ are in one to one correspondence with the elements of $\mathscr{L}(\mathfrak{A}, \mathfrak{X})$ satisfying this formula.

Note. In some cases the assumption that $\mathfrak{X}$ have a closed complement in $\mathfrak{B}$ is automatically satisfied. This will be so if $\mathfrak{X}$ is finite-dimensional or if $\mathfrak{A}$ is linearly homeomorphic with an l^1-space.

Example. If $\mathfrak{A}$ is the algebra of all power series $\sum_{i+j\geq 2} a_{ij}z^i w^j$ in the variables z, w with $\|a\| = \sum |a_{ij}| < \infty$ as norm and usual power series multiplication then $\mathfrak{X} = \mathbb{C}$ is an $\mathfrak{A}$-module if we put $ax = 0 = xa$, $a \in \mathfrak{A}$, $x \in \mathfrak{X}$. Define $T(a, b) = a_{11}b_{11}$. Formula (ii) in our case is $T(ab, c) = T(a, bc)$ which holds because each side is always zero. If $T' \in N(\mathfrak{A}, \mathfrak{X})$ then $T'(zw, zw) = T'(z^2, w^2)$ and since T does not satisfy this, $T \in Z(\mathfrak{A}, \mathfrak{X}) \backslash N(\mathfrak{A}, \mathfrak{X})$. By adjoining a unit to $\mathfrak{A}$ and defining $T(a, 1) = 0 = T(1, a)$, $1x = x = x1$, $a \in \mathfrak{A}$, $x \in \mathfrak{X}$, we get an example in which the algebra has a unit.

2. DERIVATIONS AND AUTOMORPHISMS OF BANACH ALGEBRAS

Definition. The *tensor norm* on the algebraic tensor product of two Banach spaces $\mathfrak{X}$, $\mathfrak{Y}$ is

$$\|t\| = \inf\left\{\sum_{i=1}^{n} \|x_i\| \, \|y_i\| ; x_1, \ldots, x_n \in \mathfrak{X}, y_1, \ldots, y_n \in \mathfrak{Y}, \sum x_i \otimes y_i = t\right\}.$$

$\mathfrak{X} \hat{\otimes} \mathfrak{Y}$ is the completion of $\mathfrak{X} \otimes \mathfrak{Y}$ in this norm.

If T is continuous $\mathfrak{X} \hat{\otimes} \mathfrak{Y} \to \mathfrak{Z}$, where $\mathfrak{Z}$ is a third Banach space then $T^0(x, y) = T(x \otimes y)$ is a continuous bilinear map. Conversely if T^0 is a continuous bilinear map $\mathfrak{X} \times \mathfrak{Y} \to \mathfrak{Z}$ then $T(\sum x_i \otimes y_i) = T^0(x_i, y_i)$ defines a continuous map $\mathfrak{X} \otimes \mathfrak{Y} \to \mathfrak{Z}$ extending to $\mathfrak{X} \hat{\otimes} \mathfrak{Y}$ by continuity. This connection between bilinear maps and maps from the tensor product space is the chief value of the tensor product and can be used to characterize it. If $S \in \mathscr{L}(\mathfrak{X})$, $T \in \mathscr{L}(\mathfrak{Y})$ then $(S \otimes T)(x \otimes y) = S(x) \otimes T(y)$ defines $S \otimes T \in \mathscr{L}(\mathfrak{X} \hat{\otimes} \mathfrak{Y})$.

Definition. Let $\mathfrak{A}$ be a Banach algebra and $\mathfrak{X}$ a Banach $\mathfrak{A}$-module. A *derivation* from $\mathfrak{A}$ into $\mathfrak{X}$ is a linear map with $D(ab) = aD(b) + D(a)b$, $a, b \in \mathfrak{A}$. If $x \in \mathfrak{X}$ then

$$D(a) = ax - xa, \qquad a \in \mathfrak{A}$$

defines a derivation. Derivations of this kind are called *inner*.

The condition for the complement of $\mathfrak{X}$ in $\mathfrak{A} \oplus \mathfrak{X}$ to be a subalgebra obtained at the end of last section is that S should be a derivation.

Definition. An *automorphism* of a Banach algebra $\mathfrak{A}$ is a continuous one to one linear multiplicative map of $\mathfrak{A}$ onto $\mathfrak{A}$. Automorphisms are connected with derivations as follows.

THEOREM. [2, p. 313]. *If $\mathfrak{A}$ is a Banach algebra and D is a derivation of $\mathfrak{A}$ into*

$\mathfrak{A}$ *then* $\exp D$ *is an automorphism. If* α *is an automorphism of* $\mathfrak{A}$ *with*

$$\sigma(\alpha) \subset \left\{z; z \in \mathbb{C}, -\frac{2\pi}{3} < \arg z < \frac{2\pi}{3}\right\}$$

then α *is a derivation.*

$\sigma(\alpha)$ is the spectrum of α in $\mathscr{L}(\mathfrak{A})$. $\operatorname{Exp} D$ and $\log \alpha$ are calculated in this algebra, log being calculated from a principal branch of the logarithm.

Proof. If D is a derivation we have the Leibnitz rule

$$D^n(ab) = aD^n(b) + \binom{n}{1} D(a)D^{n-1}(b) + \binom{n}{2} D^2(a)D^{n-2}(b) + \dots$$

which follows from the derivation formula exactly as in elementary calculus. Using it, it is easy to show that

$$\alpha(ab) = \left(I + D + \frac{D^2}{2!} + \frac{D^3}{3!} + \dots\right)(ab) = \alpha(a)\alpha(b).$$

Thus α is continuous, linear and multiplicative. By the functional calculus $\exp D \exp -D = I$, so α has the inverse $\exp -D$.

We prove the converse result only in the case

$$\sigma(\alpha) \subset \left\{z{:}z \in \mathbb{C}, -\frac{\pi}{2} < \arg z < \frac{\pi}{2}\right\} = \Delta.$$

Suppose a, b are commuting elements of a Banach algebra with $\sigma(a) \subset \Delta$, $\sigma(b) \subset \Delta$. It follows from the two-variable functional calculus and the identity $\log zw = \log z + \log w$, valid if $(z, w) \in \Delta \times \Delta$, that $\log a + \log b = \log ab$, where $\sigma(ab) \subset \Delta^2$ by the spectral mapping theorem, so $\log ab$ exists.

The product on $\mathfrak{A}$ is bilinear and so gives a map $\pi{:}\, \mathfrak{A} \hat{\otimes} \mathfrak{A} \to \mathfrak{A}$ with $\pi(a \otimes b) = ab$. The formula $\alpha(ab) = \alpha(a)\,\alpha(b)$ can be rewritten $\alpha \circ \pi = \pi \circ (\alpha \otimes \alpha)$. In the same way, for $n = 0, 1, 2, \dots, \alpha^n \circ \pi = \pi \circ (\alpha \otimes \alpha)^n$ and hence for any polynomial P, $P(\alpha) \circ \pi = \pi \circ P(\alpha \otimes \alpha)$. Approximating $\log z$ by polynomials in a neighbourhood of $\sigma(\alpha)$ and using $\sigma(\alpha \otimes \alpha) = \sigma(\alpha)^2$ (because $\alpha \otimes \alpha$ is the product of the commuting operators $\alpha \otimes I$ and $I \otimes \alpha$, each with spectrum $\sigma(\alpha)$) we get

$$\log \alpha \circ \pi = \pi \circ \log(\alpha \otimes \alpha) = \pi \circ (\log \alpha \otimes I + \log I \otimes \alpha) = \pi \circ [(\log \alpha) \otimes I + I \otimes \log \alpha].$$

However $\log \alpha \circ \pi = \pi \circ [(\log \alpha) \otimes I + I \otimes \log \alpha]$ is just the statement that $\log \alpha$ is a derivation. ■

Example. If $\mathfrak{A} = \mathbb{C}^3$ with coordinatewise multiplication $(ab)_i = a_i b_i$ and

$\|a\| = \max |a_i|$, that is $\mathfrak{A}$ is $C(\Omega)$ when $\Omega = \{1, 2, 3\}$ and $\alpha(a)_j = a_k$ where $j \equiv k + 1 \bmod 3$ then $\sigma(\alpha)$ consists of the three cube roots of 1 and $\mathfrak{A}$ has no non-zero derivations (for an idempotent e, $D(e) = D(e^2) = 2eD(e)$ so $eD(e) = 2eD(e)$, $eDe = 0$ and $D(e) = 0$). This shows that the value $\frac{2}{3}\pi$ in the Theorem is best possible.

The main value of the Theorem is that it shows that the derivations are the Lie algebra of the automorphism group of $\mathfrak{A}$ and consequently determine its connected component of I.

3. THE HOCHSCHILD COHOMOLOGY GROUPS

Throughout this section $\mathfrak{A}$ will be a Banach algebra and $\mathfrak{X}$ a two-sided Banach $\mathfrak{A}$-module. $\mathscr{L}^n(\mathfrak{A}, \mathfrak{X})$ is the set of continuous n-linear maps from $\mathfrak{A}$ into $\mathfrak{X}$. $\mathscr{L}^0(\mathfrak{A}, \mathfrak{X})$ is taken to be $\mathfrak{X}$. We define $\delta^n \colon \mathscr{L}^{n-1}(\mathfrak{A}, \mathfrak{X}) \to \mathscr{L}^n(\mathfrak{A}, \mathfrak{X})$ by

$$\begin{aligned}(\delta^n T)(a_1, \ldots, a_n) &= a_1 T(a_2, \ldots, a_n) - T(a_1 a_2, a_3 \ldots, a_n)\\ &\quad + T(a_1, a_2 a_3, \ldots, a_n) \ldots + (-1)^{n-1} T(a_1, \ldots, a_{n-1} a_n)\\ &\quad + (-1)^n T(a_1, \ldots, a_{n-1}) a_n,\end{aligned}$$

where $T \in \mathscr{L}^{n-1}(\mathfrak{A}, \mathfrak{X})$, $a_1, \ldots, a_n \in \mathfrak{A}$.

We have $\delta^{n+1}\delta^n = 0$, that is $\operatorname{Ker} \delta^{n+1} \supseteq \operatorname{Im} \delta^n$. The quotient space $\operatorname{Ker} \delta^{n+1}/\operatorname{Im} \delta^n$ is denoted by $\mathscr{H}^n(\mathfrak{A}, \mathfrak{X})$.

We now examine $\mathscr{H}^1$ and $\mathscr{H}^2$ in detail. As $(\delta^2 S)(a, b) = aS(b) - S(ab) + S(a)b$ we see that $\operatorname{Ker} \delta^2$ is the set of continuous derivations from $\mathfrak{A}$ into $\mathfrak{X}$. $\operatorname{Im} \delta^1$ is the set of maps $S; \mathfrak{A} \to \mathfrak{X}$ of the form $S(a) = ax - xa$, $a \in \mathfrak{A}$, that is the set of inner derivations. Thus $\mathscr{H}^1(\mathfrak{A}, \mathfrak{X})$ is the set of continuous derivations from $\mathfrak{A}$ into $\mathfrak{X}$ modulo the inner derivations. In particular $\mathscr{H}^1(\mathfrak{A}, \mathfrak{X}) = 0$ means that every derivation is inner.

As $\delta^3 T(a, b, c) = aT(b, c) - T(ab, c) + T(a, bc) - T(a, b)c$, $\operatorname{Ker} \delta^3$ is the set of elements of $\mathscr{L}^2(\mathfrak{A}, \mathfrak{X})$ which satisfy equation (ii) of Section 1 and $\operatorname{Im} \delta^2$ is the set of elements of $\mathscr{L}^2(\mathfrak{A}, \mathfrak{X})$ given by (iv). Thus the definition of $\mathscr{H}^2(\mathfrak{A}, \mathfrak{X})$ given here agrees with that given in Section 1.

It is sometimes convenient to be able to write $\mathscr{H}^n(\mathfrak{A}, \mathfrak{X})$ as $\mathscr{H}^q$ of something else with $q < n$. $\mathscr{L}^p$ becomes an $\mathfrak{A}$-module if

$$\begin{aligned}(aT)(a_1, \ldots, a_p) &= aT(a_1, \ldots, a_p),\\ (Ta)(a_1, \ldots, a_p) &= T(aa_1, a_2, \ldots, a_p)\\ &\quad - T(a, a_1 a_2, \ldots, a_p) + T(a, a_1, a_2 a_3, \ldots, a_p)\\ &\quad + (-1)^{p-1} T(a, a_1, \ldots, a_{p-1} a_p) + (-1)^p T(a, a_1, \ldots, a_{p-1}) a_p.\end{aligned}$$

The elementary isomorphisms $\mathscr{L}^{n+p}(\mathfrak{A}, \mathfrak{X}) \approx \mathscr{L}^n(\mathfrak{A}, \mathscr{L}^p(\mathfrak{A}, \mathfrak{X}))$ given by

$$[T(a_1, \ldots, a_n)](a_{n+1}, \ldots, a_{n+p}) = T(a_1, \ldots, a_{n+p})$$

then make the diagram

$$\begin{array}{ccccccc} 0 \to \mathscr{L}^p(\mathfrak{A}, \mathfrak{X}) & \to & \mathscr{L}(\mathfrak{A}, \mathscr{L}^p(\mathfrak{A}, \mathfrak{X})) & \to & \mathscr{L}^2(\mathfrak{A}, \mathscr{L}^p(\mathfrak{A}, \mathfrak{X})) & \to & \ldots \\ \downarrow & & \downarrow & & \downarrow & & \\ \mathscr{L}^p(\mathfrak{A}, \mathfrak{X}) & \to & \mathscr{L}^{p+1}(\mathfrak{A}, \mathfrak{X}) & \to & \mathscr{L}^{p+2}(\mathfrak{A}, \mathfrak{X}) & \to & \end{array}$$

commute and show $\mathscr{H}^{n+p}(\mathfrak{A}, \mathfrak{X}) \approx \mathscr{H}^n(\mathfrak{A}, \mathscr{L}^p(\mathfrak{A}, \mathfrak{X}))$.

4. COHOMOLOGY AND FIXED POINTS

Let $\mathfrak{A}$ be a Banach algebra with a unit and containing a bounded subset G which forms a group under multiplication and such that the closed linear span of G is dense in $\mathfrak{A}$. If $\mathfrak{A}$ is a B^*-algebra then we can take G to be the set of unitary elements of $\mathfrak{A}$, though in some cases a smaller group is more convenient to work with. If $\mathfrak{A}$ is the group algebra of a discrete group then G can be taken to be the set of functions having the value one at one point and zero everywhere else. Suppose $\mathfrak{X}$ is a Banach $\mathfrak{A}$-module with $1x = x = x1$ $(x \in \mathfrak{X})$ and D is a continuous derivation of $\mathfrak{A}$ into $\mathfrak{X}$. Then put

$$g \circ x = gxg^{-1} \qquad x \in \mathfrak{X}, \quad g \in G$$

$$\Phi(g) = D(g)g^{-1}.$$

The function Φ; $G \to \mathfrak{X}$ satisfies

$$\begin{aligned} \Phi(gh) &= D(gh)h^{-1}g^{-1} \\ &= D(g)g^{-1} + gD(h)h^{-1}g^{-1} \\ &= \Phi(g) + g \circ \Phi(h). \end{aligned} \tag{i}$$

If D is inner, $D = \delta^1 x$ say, then

$$\Phi(g) = g \circ x - x \qquad g \in G. \tag{ii}$$

Conversely if $x \in \mathfrak{X}$ is such that (ii) holds then $D = \delta^1 x$ on G, hence on linear combinations of elements of G, since each side is a linear operator on $\mathfrak{A}$, and so on the whole of $\mathfrak{A}$ by continuity. Thus D is inner.

Let $A_g y = g \circ y - \Phi(g)$ $(g \in G,\ y \in \mathfrak{X})$; then for each g in G, A_g is an affine map (that is $A_g(\sum \lambda_i x_i) = \sum \lambda_i A_g(x_i)$ when $\sum \lambda_i = 1$) with $A_g A_h = A_{gh}$, $A_1 =$ the identity on $\mathfrak{X}$. Thus the A_g form a group of continuous affine transformations on $\mathfrak{X}$. Moreover $A_g(-\Phi(h)) = -g \circ \Phi(h) - \Phi(g) = -\Phi(gh)$ so $\{-\Phi(g) : g \in G\}$ is closed under the maps A_h, $h \in G$.

If x is a common fixed point for all the A_g then $x = A_g x = g \circ x - \Phi(g)$

so $\Phi(g) = g \circ x - x\,(g \in G)$. Conversely if (ii) holds then x is a common fixed point for the A_g.

In this section we have seen how problems about inner derivations can be transformed into the problem of solving the functional equation (i) and so into problems concerning fixed points. In the next two sections we show how these problems can be solved, in one case by imposing hypotheses on the space and in the second by imposing hypotheses on the group.

5. REFLEXIVE MODULES AND MODULES WITH A REFLEXIVE AUXILIARY NORM

Fixed point theorems for groups of affine maps involve compactness assumptions. If $\mathfrak{X}$ is reflexive then, as $\{\Phi(z); z \in G\}$ is a bounded set (this is where the boundedness of G enters essentially), its weakly closed convex cover is a compact convex A_g-invariant set and fixed point theorems apply. It is possible to separate the contribution that reflexity makes to this theorem from the contribution made by the compactness, and we now do this.

If $\mathfrak{Y}$ is a Banach $\mathfrak{A}$-module then $\mathfrak{Y}^*$ is also a Banach $\mathfrak{A}$-module if we define

$$(af, y) = (f, ya)$$

$$(fa, y) = (f, ay) \qquad a \in \mathfrak{A} \quad f \in \mathfrak{Y}^*, \quad y \in \mathfrak{Y}.$$

$\mathfrak{Y}^*$ is then an $\mathfrak{A}$-module which is a dual space and the module actions $f \mapsto af$, $f \mapsto fa$ are continuous in the weak* topology. This characterizes modules which are the duals of other modules; we refer to such a module as a *dual* $\mathfrak{A}$-module. Application of the Ryll–Nardzewski fixed point theorem [**8**] then shows

THEOREM. *If $\mathfrak{X}$ is a dual $\mathfrak{A}$-module, where $\mathfrak{A}$ is a Banach algebra which satisfies the conditions at the beginning of Section* 4, *and if $|\;|$ is an auxilliary norm on $\mathfrak{A}$ satisfying the following conditions*

(i) *for some K* $\quad |x| \leqslant K\|x\| \qquad x \in \mathfrak{X}$,

(ii) *for some L* $\quad |g \circ x| \leqslant L|x| \qquad g \in G, x \in \mathfrak{X}$,

(iii) *the $|\;|$ unit ball in $\mathfrak{X}$ is weak*-closed,*

(iv) *the completion of $\mathfrak{X}$ in $|\;|$ is reflexive,*

then $\mathscr{H}^1(\mathfrak{A}, \mathfrak{X}) = 0$.

We shall prove a simpler theorem of the same kind.

THEOREM. *Suppose the auxilliary norm satisfies the following conditions in place of* (i)–(iv) *above*

(i) *for some* K $\quad |x| \leqslant K\|x\| \qquad x \in \mathfrak{X}$

(ii) *for some* L $\quad |g \bigcirc x| \leqslant L|x| \qquad g \in G, x \in \mathfrak{X}$

(iii) $|\ |$ *is uniformly convex.*

Then $\mathscr{H}^1(\mathfrak{A}, \mathfrak{X}) = 0$.

By saying that $|\ |$ is *uniformly convex* we mean that if $\{x_n\}, \{y_n\}$ are sequences in $\mathfrak{X}$ with $|x_n| \leqslant N$, $|y_n| \leqslant N$, $|x_n + y_n| \to 2N$ then $|x_n - y_n| \to 0$. The completion of a uniformly convex space is uniformly convex and hence reflexive.

Proof. We consider in detail only the case in which G has two generators a, b. The space $l^\infty(1, \infty)$ has a generalised limit, that is a linear functional lim such that

(i) If $a_n \geqslant 0$ for all n then $\lim a_n \geqslant 0$.

(ii) If $a_n \to \alpha$ as $n \to \infty$ then $\lim a_n = \alpha$.

(iii) $\lim a_n = \lim a_{n+1}$.

The existence of such a functional is shown in [**1**, p. 33] and we shall consider similar functionals in the next section.

Suppose $\mathfrak{X}$ is the dual of the $\mathfrak{A}$-module $\mathfrak{Y}$, D is a continuous derivation $\mathfrak{A} \to \mathfrak{X}$ and Φ is the function constructed from D in Section 4. Define $x \in \mathfrak{X}$ by

$$(x, y) = -\lim 2^{-n} \sum_{g \in G_n} (\Phi(g), y)$$

where G_n is the set of n-fold products of a's and b's so that G_n contains 2^n elements and $y \bigcirc g = y^{-1}gy$. It is easy to check that, for each y, lim is applied to a bounded sequence and that the resulting number is linear and continuous in y, thus defining $x \in \mathfrak{X}$.

We have

$$\begin{aligned}
(a \bigcirc x - x + b \bigcirc x - x, y) &= (x, y \bigcirc a) + (x, y \bigcirc b) - 2(x, y) \\
&= -\lim 2^{-n} \sum_{g \in G_n} (a \bigcirc \Phi(g) + b \bigcirc \Phi(g), y) - 2(x, y) \\
&= -\lim \Big[2^{-n} \sum_{g \in G_n} (\Phi(ag) + \Phi(bg), y) - (\Phi(a) + \Phi(b), y)\Big] - 2(x, y) \\
&= -\lim 2^{-n} \sum_{g \in G_{n+1}} (\Phi(g), y) + (\Phi(a) + \Phi(b), y) - 2(x, y) \\
&= (\Phi(a) + \Phi(b), y)
\end{aligned}$$

because $\lim 2^{-n} \sum_{g \in G_{n+1}} (\Phi(g), y) = 2(x, y)$ by condition (iii) for "lim." Thus

$\Phi(a) - ((a \circ x) - x) = -[\Phi(b) - (b \circ x - x)]$. Define, for $g \in G$,

$$\Psi(g) = \Phi(g) - g \circ x + x.$$

Then Ψ satisfies condition (i) of Section 4, the range of Ψ is bounded with respect to $\| \ \|$ and $\Psi(a) = -\Psi(b)$.

Let $N = \sup\{|\Psi(g)|; g \in G\} < \infty$ and suppose $|\Psi(h_i)| \to N$ as $i \to \infty$. Then

$$\Psi(h_i a) = \Psi(h_i) + h_i \circ \Psi(a)$$

$$\Psi(h_i b) = \Psi(h_i) - h_i \circ \Psi(a)$$

so that $|\Psi(h_i a)| \leqslant N$, $|\Psi(h_i b)| \leqslant N$, $|\Psi(h_i a) + \Psi(h_i b)| = 2|\Psi(h_i)| \to 2N$ and, by uniform convexity, $|h_i \circ \Psi(a)| \to 0$. However $|\Psi(a)| = |h_i^{-1} \circ h_i \circ \Psi(a)| \leqslant |h_i \circ \Psi(a)| \to 0$ so $\Psi(a) = 0$ and hence $\Psi(b) = 0$. It follows from the relationship $\Psi(gh) = \Psi(g) + g \circ \Psi(h)$ that the set of elements at which Ψ is zero is a subgroup of G, so Ψ is zero on the group generated by a,b. Hence $D = \delta^1 x$.

The case for a group with n generators is treated similarly. If G is not finitely generated then for each finitely generated subgroup H we obtain an element x_H with $\Psi(h) = h \circ x_H - x_H$ for all $h \in H$. These elements x_H form a bounded net and we have $D = \delta^1 x$ for any w^* limit point of this net. ■

Example. If G is a discrete group then every derivation from $l^1(G)$ into $l^1(G)$ is inner. As explained in Section 4, the standard basis for $l^1(G)$ is a bounded group in $l^1(G)$ and is in fact isomorphic with G under the obvious one to one correspondence. We shall show that the l^2 norm $\| \ \|_2$ on $l^1(G)$ is an auxiliary norm which satisfies the conditions of our Theorem so that this result is an immediate consequence of it.

The inequality $\|a\|_2 \leqslant \|a\|_1$ is well known. $g \circ x$ is obtained from x by translating on the right and on the left, so $\|g \circ x\|_2 = \|x\|_2$. To see that $\| \ \|_2$ is uniformly convex we consider the equality

$$\|x + y\|_2^2 + \|x - y\|_2^2 = 2\|x\|_2^2 + 2\|y\|_2^2$$

valid in any inner product space. If $\|x_n\|_2 \leqslant N$, $\|y_n\|_2 \leqslant N$, $\|x_n + y_n\|_2 \to 2N$ then $0 \leqslant \|x_n - y_n\|_2^2 \leqslant 4N^2 - \|x_n + y_n\|_2^2 \to 0$ so $\|x_n - y_n\|_2 \to 0$.

Problem. If G is a locally compact group, $\mu \in M(G)$ then $Df = f*\mu - \mu*f$ defines a derivation on $L^1(G)$. Are there any others? This result is known in special cases but the general problem has not been settled.

The methods we have used do not give any information about $\mathscr{H}^n(\mathfrak{A}, \mathfrak{X})$, $n > 1$, even when $\mathfrak{X}$ is reflexive, because the Reduction of Dimension Theorem turns $\mathscr{H}^2(\mathfrak{A}, \mathfrak{X})$ into $\mathscr{H}^1(\mathfrak{A}, \mathscr{L}(\mathfrak{A}, \mathfrak{X}))$ and $\mathscr{L}(\mathfrak{A}, \mathfrak{X})$ is not reflexive even if $\mathfrak{X}$ is.

6. AMENABLE GROUPS AND ALGEBRAS

Let G be a (discrete) group and $l^\infty(G)$ the Banach space of bounded complex valued functions on G with the sup norm. A *mean* on $l^\infty(G)$ is a positive linear functional M with $M(1) = 1$ where 1 is the constant function with value 1. The mean is *translation invariant* if $M(f) = M(\tau_g f)$ for $f \in l^\infty(G)$, $g \in G$ where τ_g is the translation $\tau_g(f)(h) = f(g^{-1}h)$. A group is *amenable* if it has a translation invariant mean. It is easy to see that, if G is finite, then $M(f) = |G|^{-1} \sum_{g \in G} f(g)$ is a translation invariant mean, so finite groups are amenable. Products, subgroups and quotients of amenable groups are amenable. If a group has a composition series

$$\{e\} = G_0 \lhd G_1 \lhd G_2 \ldots \lhd G_n = G,$$

where $G_i \lhd G_{i+1}$ means G_i is a normal subgroup of G_{i+1}, and G_{i+1}/G_i is amenable, $0 \leqslant i < n$, then G is amenable. This is true even if n is a transfinite ordinal so that the series is a transfinite one.

The group $\mathbb{Z}$ of integers is amenable—to see this let $\mathscr{U}$ be a free ultrafilter on $\mathbb{Z}^+$ and define

$$M(a) = \lim_{n \in \mathscr{U}} \frac{1}{2n+1} \sum_{i=-n}^{n} a_i.$$

It is easy to see that M is a mean and since

$$\left| \frac{1}{2n+1} \sum_{i=-n}^{n} a_i - \frac{1}{2n+1} \sum_{i=-n}^{n} a_{i+m} \right| = \left| \frac{1}{2n+1} \left(\sum_{i=-n}^{-n+m-1} - \sum_{i=n+1}^{n+m} \right) a_i \right|$$

$$\leqslant \frac{2m}{2n+1} \| a \|_\infty \to 0 \text{ as } n \to \infty$$

we have

$$\lim_{n \in \mathscr{U}} \frac{1}{2n+1} \sum_{i=-n}^{n} a_i = \lim_{n \in \mathscr{U}} \frac{1}{2n+1} \sum_{i=-n}^{n} a_{i+m}$$

so M is translation invariant. An alternative description of M is to take ρ, a state on the B^*-algebra $l^\infty(\mathbb{Z}^+)/c_0$ and let $M(a)$ be the value of ρ on the coset in $l^\infty(\mathbb{Z}^+)/c_0$ of the sequence

$$b_n = \frac{1}{2n+1} \sum_{i=-n}^{n} a_i.$$

Because the sequence $\{b'_n\}$ associated with a translate of $\{a_n\}$ has $b - b' \in c_0$ as above, we have $\rho(b) = \rho(b')$. It follows from this that any abelian group is amenable.

Some groups are not amenable, for example the free group on two or more generators is not amenable.

If G is a locally compact topological group then amenability is defined in terms of $L^\infty(G)$ rather than $l^\infty(G)$. A full discussion of this appears in [3].

THEOREM. *Let $\mathfrak{A}$ be a Banach algebra containing a bounded multiplicative group G with* span *G dense in $\mathfrak{A}$ which is amenable. Let $\mathfrak{X}$ be a dual $\mathfrak{A}$-module. Then* $\mathscr{H}^1(\mathfrak{A}, \mathfrak{X}) = 0$.

Proof. Let M be an invariant mean on G and suppose $\mathfrak{X}$ is the dual of the $\mathfrak{A}$-module $\mathfrak{Y}$. Let D be a continuous derivation $\mathfrak{A} \to \mathfrak{X}$ and Φ the map constructed from it in Section 4. Define $x \in \mathfrak{Y}^* = \mathfrak{X}$ by

$$(x, y) = -\underset{g}{M}(\Phi(g), y) \qquad y \in \mathfrak{Y},$$

where $\underset{g}{M}$ indicates that we treat $(\Phi(g), y)$ as a function of g and apply the invariant mean to it. Because $\{\Phi(g); g \in G\}$ is bounded, $g \mapsto (\Phi(g), y)$ is in $l^\infty(G)$, so (x, y) is well defined, and it is easy to see that it is continuous and linear in y. Thus x is well defined. Consider

$$\begin{aligned}(h \circ x - x, y) &= (x, y \circ h) - (x, y)\\ &= -\underset{g}{M}(h \circ \Phi(g), y) - (x, y)\\ &= -\underset{g}{M}(\Phi(hg), y) + \underset{g}{M}(\Phi(h), y) - (x, y)\\ &= (\Phi(h), y)\end{aligned}$$

because the first term is just M applied to a translate of $g \mapsto (\Phi(g), y)$, and so is (x, y) by translation invariance, and the second is M applied to a constant sequence. We thus have $(h \circ x - x, y) = (\Phi(h), y)$ for each $y \in \mathfrak{Y}$, that is $\Phi(b) = b \circ x - x$ for each $h \in G$. As before this implies $D = \delta^1 x$. ■

Motivated by this result we say that a Banach algebra $\mathfrak{A}$ is *amenable* if $\mathscr{H}^1(\mathfrak{A}, \mathfrak{X}) = 0$ for all dual $\mathfrak{A}$-modules $\mathfrak{X}$. We have seen that the group algebra of a discrete amenable group and a commutative C^*-algebra are amenable. The group algebra of any locally compact amenable group is amenable. The algebra of compact operators on most of the usual classical Banach spaces is amenable.

In this case the higher cohomology groups vanish as well by the Reduction of Dimension Theorem—if $\mathfrak{X}$ is a dual $\mathfrak{A}$-module so is $\mathscr{L}^p(\mathfrak{A}, \mathfrak{X}) \approx (\mathfrak{A} \hat{\otimes} \ldots \hat{\otimes} \mathfrak{A} \hat{\otimes} \mathfrak{Y})^*$.

7. MODULES WHICH ARE NOT DUAL MODULES

If $\mathfrak{X}$ is a not a dual module and $T \in \mathscr{L}^n(\mathfrak{A}, \mathfrak{X})$ with $\delta T = 0$ then the methods of the previous sections might apply to show $T = \delta S$ for some $S \in \mathscr{L}^n(\mathfrak{A}, \mathfrak{X}^{**})$.

If we require $S \in \mathscr{L}^n(\mathfrak{A}, \mathfrak{X})$ then considerable difficulties can arise. If $\mathfrak{A}$ is a commutative semi-simple infinite dimensional Banach algebra then we can always find an $\mathfrak{A}$-module $\mathfrak{X}$ with $\mathscr{H}^1(\mathfrak{A}, \mathfrak{X}) \neq 0$. To do this assume $\mathfrak{A}$ has a 1 so that its maximal ideal space $\Phi_{\mathfrak{A}}$ is compact. $C(\Phi_{\mathfrak{A}} \times \Phi_{\mathfrak{A}})$ is an $\mathfrak{A}$-module if we put

$$(aF)(s, t) = \hat{a}(s)\, F(s, t)$$

$$(Fa)(s, t) = \hat{a}(t)\, F(s, t) \qquad a \in \mathfrak{A},\ F \in C(\Phi_{\mathfrak{A}} \times \Phi_{\mathfrak{A}}),\ s, t \in \Phi_{\mathfrak{A}}.$$

The space

$$\mathfrak{X} = \{F : F \in C(\Phi_{\mathfrak{A}} \times \Phi_{\mathfrak{A}}),\ F(s, s) = 0,\ s \in \Phi_{\mathfrak{A}}\}$$

is a closed submodule and $(Da)\,(s, t) = \hat{a}(s) - \hat{a}(t)$ is a derivation from $\mathfrak{A}$ into $\mathfrak{X}$. If $D = \delta F$ for some $F \in \mathfrak{X}$ then $\hat{a}(s) - \hat{a}(t) = (\hat{a}(s) - \hat{a}(t))\, F(s, t)$ so $(F(s, t) - 1)$ $(\hat{a}(s) - \hat{a}(t)) = 0$ for $a \in \mathfrak{A}$, $s, t \in \Phi_{\mathfrak{A}}$. This implies $F(s, t) = 1$ if $s \neq t$ whereas $F \in \mathfrak{X}$ implies $F(s, t) = 0$ if $s = t$. Since $\Phi_{\mathfrak{A}}$ is infinite and compact this contradicts the continuity of F.

Similar problems arise for $\mathscr{H}^2$—if $\mathfrak{A}$ is as above and amenable then there is an $\mathfrak{A}$-module with $\mathscr{H}^2(\mathfrak{A}, \mathfrak{X}) \neq 0$. At $\mathscr{H}^3$ things begin to happen. For certain algebras, in particular c_0 and the group algebra of a compact topological group, we have $\mathscr{H}^3(\mathfrak{A}, \mathfrak{X}) = 0$ for all $\mathfrak{X}$ [7]. This raises a number of interesting questions which are unanswered so far.

For commutative algebras $\mathfrak{A}$ we can ask whether $\mathscr{H}^n(\mathfrak{A}, \mathfrak{X}) = 0$ for all Banach $\mathfrak{A}$-modules $\mathfrak{X}$ with $ax = xa$, $x \in \mathfrak{X}$, $a \in \mathfrak{A}$. In this case for group algebras and B^*-algebras we get $\mathscr{H}^1(\mathfrak{A}, \mathfrak{X}) = 0 = \mathscr{H}^2(\mathfrak{A}, \mathfrak{X})$ but what happens in higher dimensions is not known (reduction of dimension does not help as $\mathscr{L}^p(\mathfrak{A}, \mathfrak{X})$ does not have the property $ax = xa$ even when $\mathfrak{X}$ does).

8. HELEMSKIĬ'S COHOMOLOGY THEORY

The (algebraic) cohomology groups of an algebra A with coefficients in a module X, although originally defined by the formulae in Section 3, are now usually introduced in terms of projective resolutions and the functor Ext. In this section we shall outline this theory and show how it can be adapted to the case of Banach algebras. The ideas involved are due to Helemskiĭ [4].

Let A be an associative algebra with a 1. We assume the modules X we consider have $1x = x$, $x \in X$. A left A-module P is *projective* if the diagram of left A-modules

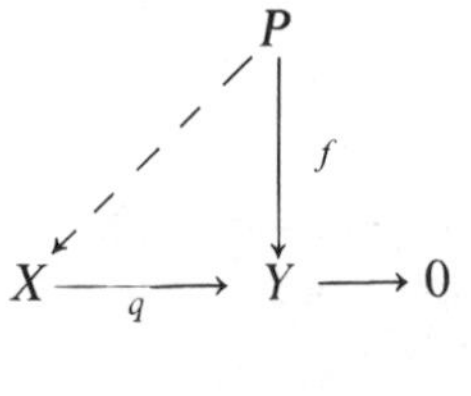

(i)

where the row is exact, can be completed as indicated. In other words if X is a left A-module, Z a submodule and f is a linear map $P \to X/Z$ with $f(ap) = af(p)$ where $p \in P$, $a \in A$, then there is a linear map $g; P \to X$ with $g(ap) = ag(p)$ for $a \in A$, $p \in P$ and $f = qg$, q being the quotient map. A map g from one module P into another X with $g(ap) = ag(p)$ is a *module homomorphism.*

A *projective resolution* of a left A-module X is an exact sequence of left A-modules

$$\ldots P_2 \xrightarrow{\phi_1} P_1 \xrightarrow{\phi_0} P_0 \xrightarrow{\varepsilon} X \longrightarrow 0 \tag{ii}$$

where the P_i are projective. The arrows represent module homomorphisms and saying that the sequence is exact means that, for each module in the chain, the range of the incoming map is the kernel of the outgoing map. To show that every left A-module W has a projective resolution we first show that $A \otimes W$, with module action $a(b \otimes w) = ab \otimes w$ for $a, b \in A$, $w \in W$, is a projective left A-module. If we are in the situation of the projective diagram (i) with $P = A \otimes W$ then we can find a linear map $g_0: W \to X$ with $qg_0(w) = f(1 \otimes w)$ and define $g: A \otimes W \to X$ by $g(a \otimes w) = ag_0(w)$ Once we have done this we construct the projective resolution by taking $P_0 = A \otimes W$, $\varepsilon(a \otimes w) = aw$, $P_1 = A \otimes \operatorname{Ker} \varepsilon$ *etc.*

If X, Y are two left A-modules then $\operatorname{Hom}_A(X, Y)$ is the linear space of all module homomorphisms from X into Y. Starting with the projective resolution (ii) for X we get a system

$$\ldots \xleftarrow{\phi_1^*} \operatorname{Hom}_A(P_1, Y) \xleftarrow{\phi_0^*} \operatorname{Hom}_A(P_0, Y) \xleftarrow{\varepsilon^*} \operatorname{Hom}_A(X, Y) \longleftarrow 0$$

where, for $T \in \operatorname{Hom}_A(P_n, Y)$,

$$(\phi_n^* T)(p) = T(\phi_n p)(p \in P_{n+1}).$$

Because

$$\phi_n^* \phi_{n-1}^* = (\phi_{n-1}\phi_n)^* = 0$$

we have $\operatorname{Ker} \phi_n^* \supseteq \operatorname{Im} \phi_{n-1}^*$ although the sequence need not be exact. We define

$$\operatorname{Ext}^n(X, Y) = \operatorname{Ker} \phi_n^* / \operatorname{Im} \phi_{n-1}^* \ (\operatorname{Ext}^0(X, Y) = \operatorname{Ker} \phi_0^*).$$

The first result to be proved is that the spaces $\operatorname{Ext}^n(X, Y)$ do not depend on the particular choice of projective resolution. The second main result is the Long Exact Sequence of Homology Theorem which says that if

$$0 \to X' \to X \to X'' \to 0$$

is an exact sequence of left A-modules, that is if X is a left A-module, X' a submodule and X'' the quotient module X/X', then maps can be defined to make the sequence

$$0 \to \mathrm{Ext}^0 (X'', Y) \to \ldots \to \mathrm{Ext}^{n-1} (X', Y) \to \mathrm{Ext}^n (X'', Y)$$
$$\to \mathrm{Ext}^n (X, Y) \to \mathrm{Ext}^n (X', Y) \to \mathrm{Ext}^{n+1} (X'', Y) \to \ldots$$

exact. This enables us to compute Ext (X, Y) in terms of Ext (X', Y) and Ext (X'', Y).

The obvious extension of this to Banach algebras does not work, as $\mathfrak{A} \hat{\otimes} \mathfrak{X}$ is not usually a projective $\mathfrak{A}$-module. In fact when $\mathfrak{A} = \mathbb{C}$ then a left Banach $\mathfrak{A}$-module is just a Banach space, $\mathfrak{A} \hat{\otimes} \mathfrak{X} = \mathfrak{X}$ and so $\mathfrak{A} \hat{\otimes} \mathfrak{X}$ is projective if and only if $\mathfrak{X}$ is projective in the category of Banach spaces. It is well known that some Banach spaces are not projective in this sense; in fact very few are. The extension is possible with the notion of relative cohomology. Let $\mathfrak{A}$ be a Banach algebra with a unit. A linear map ρ of one left $\mathfrak{A}$-module $\mathfrak{X}$ into another $\mathfrak{Y}$ is *admissible* if it is continuous, its kernel has a Banach space complement in $\mathfrak{X}$ and its range is closed and has a Banach space complement in $\mathfrak{Y}$. A *relatively projective* module is defined in the same way as a projective module except that we require q to be an admissible map and all the maps to be continuous. Similarly, for a relatively projective resolution the P_i are only required to be relatively projective $\mathfrak{A}$-modules and the maps ε_i and ϕ_i are required to be admissible. With these changes, and replacing $\mathrm{Hom}_A(P_i, Y)$ by the space of continuous module homomorphisms of P_i into $\mathfrak{Y}$, we define the spaces $\mathrm{Ext}^n_{\mathfrak{A}}(\mathfrak{X}, \mathfrak{Y})$. As before, it is easy to see that $\mathfrak{A} \hat{\otimes} \mathfrak{X}$ is a relatively projective $\mathfrak{A}$-module and that each left Banach $\mathfrak{A}$-module has a relatively projective resolution.

The connection between Ext and the cohomology groups which we have been considering is the isomorphism $\mathscr{H}^n(\mathfrak{A}, \mathfrak{X}) \approx \mathrm{Ext}^n_{\mathfrak{A} \hat{\otimes} \mathfrak{A}_e}(\mathfrak{A}, \mathfrak{X})$ which we proceed to explain. $\mathfrak{A}$ is a Banach algebra with a unit and $\mathfrak{X}$ is a two-sided Banach $\mathfrak{A}$-module. $\mathfrak{A}_e$ is the Banach algebra obtained by defining a multiplication $\bigcirc$ on the Banach space $\mathfrak{A}$ by $a \bigcirc b = ba \, (a, b \in \mathfrak{A})$. $\mathfrak{A} \hat{\otimes} \mathfrak{A}_e$ is the Banach space tensor product with multiplication

$$\begin{aligned}(a \otimes b)(a' \otimes b') &= aa' \otimes b \bigcirc b' \\ &= aa' \otimes b'b.\end{aligned}$$

$\mathfrak{X}$ is a left $\mathfrak{A} \hat{\otimes} \mathfrak{A}_e$-module if we put $(a \otimes b)x = axb, a, b \in \mathfrak{A}, x \in \mathfrak{X}$ and $\mathfrak{A}$ becomes a left $\mathfrak{A} \hat{\otimes} \mathfrak{A}_e$-module by the same formula with $\mathfrak{A}$ replacing $\mathfrak{X}$. Since $\mathfrak{A}$ and $\mathfrak{X}$ are left Banach $\mathfrak{A} \hat{\otimes} \mathfrak{A}_e$-modules, $\mathrm{Ext}^n_{\mathfrak{A} \hat{\otimes} \mathfrak{A}_e}(\mathfrak{A}, \mathfrak{X})$ is defined as above. To show the isomorphism we consider the relatively projective resolution of $\mathfrak{A}$ given by taking $P_n = \mathfrak{A} \hat{\otimes} \mathfrak{A} \ldots \hat{\otimes} \mathfrak{A}$, where there are $n+2$ $\mathfrak{A}$'s, $\varepsilon(a \otimes b) = ab$ and, for $a_0, a_1, \ldots, a_{n+2} \in \mathfrak{A}$,

$$\phi_n(a_0 \otimes \dots a_{n+2}) = a_0 a_1 \otimes \dots \otimes a_{n+2} - a_0 \otimes a_1 a_2 \otimes \dots \otimes a_{n+2} + \dots + (-1)^{n+1} a_0 \otimes \dots \otimes a_{n+1} a_{n+2}.$$

The module operation on P_n is

$$(a \otimes b)(a_0 \otimes \dots \otimes a_{n+1}) = aa_0 \otimes \dots \otimes a_{n+1} b.$$

The proof that the P_n are relatively projective is similar to our proof that $A \otimes X$ is projective. If $\mathfrak{Y}$ is a left Banach $\mathfrak{A} \hat{\otimes} \mathfrak{A}_e$ module, $\mathfrak{Z}$ is a quotient module with the quotient map q admissible and f is a continuous $\mathfrak{A} \hat{\otimes} \mathfrak{A}_e$ module homomorphism then we can find $g_0 \in \mathscr{L}(P_{n-2}, \mathfrak{Y})$ with $f(1 \otimes p \otimes 1) = qg_0(p)$ $(p \in P_{n-2})$ and we define $g(a \otimes p \otimes b) = (a \otimes b) g_0(p)$ $(a, b \in \mathfrak{A}$, $p \in P_{n-2})$. To show that the resolution is really exact and the maps ϕ are admissible we first see by calculation that $\phi_{n-1}\phi_n = 0$, and so $\operatorname{Ker} \phi_{n-1} \supseteq \operatorname{Im} \phi_n$. For each n define j_n; $P_n \to P_{n+1}$ by $j_n(p) = 1 \otimes p$. For $p \in \operatorname{Ker} \phi_{n-1}$ we have $\phi_n j_n(p) = p$, showing $\operatorname{Ker} \phi_{n-1} \subseteq \operatorname{Im} \phi_n$, so the sequence is exact. $\phi_n j_n$ maps P_n into $\operatorname{Im} \phi_n = \operatorname{Ker} \phi_{n-1}$ and is the identity on $\operatorname{Ker} \phi_{n-1}$. Thus $\phi_n j_n$ is a projection of P_n onto $\operatorname{Ker} \phi_{n-1}$, showing that the maps ϕ_n are admissible. Note that we do not assert that j_n is a module homomorphism.

Our isomorphism result is completed by establishing an isomorphism between $\mathscr{H}^n(\mathfrak{A}, \mathfrak{X})$ and $\operatorname{Hom}_{\mathfrak{A} \hat{\otimes} \mathfrak{A}_e}(P_n, \mathfrak{X})$. If $T \in \mathscr{H}^n(\mathfrak{A}, \mathfrak{X})$ then

$$\tilde{T}(a_0 \otimes \dots \otimes a_{n+1}) = a_0 T(a_1, \dots, a_n) a_{n+1}$$

is the required isomorphism. If $S \in \operatorname{Hom}_{\mathfrak{A} \hat{\otimes} \mathfrak{A}_e}(P_n, \mathfrak{X})$ then $T(a_1, \dots, a_n) = S(1 \otimes a_1 \dots \otimes a_n \otimes 1)$ gives an element T of $\mathscr{H}^n(\mathfrak{A}, \mathfrak{X})$ and $\tilde{T} = S$ because $a_0 S(1 \otimes a_1 \otimes \dots \otimes a_n \otimes 1) a_{n+1} = S(a_0 \otimes a_1 \dots \otimes a_n \otimes a_{n+1})$. We now have a diagram

$$\begin{array}{ccccccccc}
0 \longrightarrow \mathfrak{X} & \xrightarrow{\delta^1} & \mathscr{H}^1(\mathfrak{A}, \mathfrak{X}) & \xrightarrow{\delta^2} & \mathscr{H}^2(\mathfrak{A}, \mathfrak{X}) & \xrightarrow{\delta^3} & \mathscr{H}^3(\mathfrak{A}, \mathfrak{X}) & \xrightarrow{\delta^4} & \\
\downarrow & & \downarrow & & \downarrow & & \downarrow & & \\
0 \to \operatorname{Hom}(P_0, \mathfrak{X}) & \xrightarrow{\phi_0^*} & \operatorname{Hom}(P_1, \mathfrak{X}) & \xrightarrow{\phi_1^*} & \operatorname{Hom}(P_2, \mathfrak{X}) & \xrightarrow{\phi_2^*} & \operatorname{Hom}(P_3, \mathfrak{X}) & \xrightarrow{\phi_3^*} &
\end{array}$$

where $\operatorname{Hom} = \operatorname{Hom}_{\mathfrak{A} \hat{\otimes} \mathfrak{A}_e}$, and a calculation shows that it commutes. As the vertical arrows are bijections the cohomology groups constructed from the first row are isomorphic with those constructed from the second.

REFERENCES

The main reference for these lectures is [**5**].

1. S. Banach, *Théorie des opérations linéaires*, Chelsea, New York, 1955.
2. J. Dixmier, *Les algèbres d'opérateurs dans l'espace Hilbertien*, 2nd edition, Gauthier-Villars, Paris, 1969.
3. F. P. Greenleaf, *Invariant means on topological groups*, Van Nostrand, New York, 1969.
4. A. Ja. Helemskiĭ, On the homological dimension of normed modules over Banach

algebras, *Mat. Sb.* **81** (**123**), (1970), 430–444 (in Russian); *Math. USSR-Sb.* **10** (1970), 399–411 (English translation).

5. B. E. Johnson, Cohomology in Banach algebras, *Mem. Amer. Math. Soc.* **127**, (1972).
6. B. E. Johnson, The Wedderburn decomposition of Banach algebras with finite dimensional radical, *Amer. J. Math.* **90** (1968), 866–876.
7. B. E. Johnson, Approximate diagonals and cohomology of certain annihilator Banach algebras, *Amer. J. Math.* **94** (1972), 685–698.
8. I. Namioka and E. Asplund, A geometric proof of Ryll–Nardzewski's fixed point theorem, *Bull. Amer. Math. Soc.* **73** (1967), 443–445.

4. Operator Algebras

R. V. Kadison

Department of Mathematics
University of Pennsylvania, Philadelphia, U.S.A.

Introduction 101
1. Some C^*-algebra Basics 102
2. Von Neumann Algebra Basics 107
3. Algebraic Structure in von Neumann Algebras 112
4. Action of von Neumann Algebras on Spaces 116

INTRODUCTION

During this series of lectures, I want to outline for you some of the main results in the theory of von Neumann algebras. There are many subjects, of considerable importance, on which I will not touch. The subjects discussed are what many of us consider to be the core of the theory. These subjects could be classified under three headings: the Basics, Comparison Theory of Projections, and Unitary Equivalence. Under this last heading—and the main part of it—I include the theory of normal states.

It no longer makes very much sense to draw a sharp line between the results and methods of C^*-algebra theory and those of the theory of von Neumann algebras. Nonetheless there are areas of each of these subjects which are unambiguously identified with the one but not the other. For our purpose, we will want some of the tools of C^*-algebra theory. A description of these will provide us with an appropriate introduction.

As excellent general references, we cite the two books of J. Dixmier "Les algèbres d'opérateurs dans l'espace Hilbertien (Algèbres de von Neumann)" Cahiers Scientifiques Fasc. XXV: Gauthier-Villars, Paris, 1957, 2me éd. 1969 and "Les C^*-algèbres", Cahiers Scientifiques Fasc. XXIX, Gauthier-Villars, Paris, 1964, 2me éd. 1969, (especially pp. 1–55 of "Les C^*-algèbres"). In addition, S. Sakai's "C^*-algebras and W^*-algebras", Ergebnisse der Mathematik und Ihrer Grenzgebiete Bd. 60, Springer-Verlag, Berlin, 1971, gives an excellent account of fundamentals and recent work. The combined bibliography of the Dixmier–Sakai books is comprehensive.

1. SOME C^*-ALGEBRA BASICS

The Hilbert spaces with which we deal are complex (the field of scalars is $\mathbb{C}$). The inner product is denoted by $\langle x, y \rangle$ for a pair of vectors x, y in $\mathcal{H}$. The *length* or *norm* of x is denoted by $\| x \|$ $(= \langle x, x \rangle^{\frac{1}{2}})$. The *operators* on $\mathcal{H}$ are linear transformations of $\mathcal{H}$ into $\mathcal{H}$; and we assume that they are continuous unless otherwise stated. The *bound* or *norm* of an operator T is denoted by $\| T \|$ $(= \sup \{ \| Tx \| : \| x \| \leqslant 1 \})$, and we recall that the continuity of T is equivalent to its boundedness $(\| T \| < \infty)$. The set of all bounded operators on $\mathcal{H}$ will be denoted by $\mathcal{B}(\mathcal{H})$. It is an algebra under the usual operations of addition, multiplication by scalars, and multiplication (= iteration of transformations) (so $(A + B)(x) = Ax + Bx$, $(aA)x = a(Ax)$, and $(AB)x = A(Bx)$). The function $A \to \| A \|$ is a norm relative to which $\mathcal{B}(\mathcal{H})$ becomes a normed space. It is complete in this norm, so that it is a Banach space; and, indeed, $\| AB \| \leqslant \| A \| \, \| B \|$. Thus $\mathcal{B}(\mathcal{H})$ with the norm $A \to \| A \|$ is a Banach algebra. The metric topology on $\mathcal{B}(\mathcal{H})$ associated with the norm is called the *norm topology*.

The adjoint operation on $\mathcal{B}(\mathcal{H})$ provides an important piece of algebraic structure. Recall that, with A in $\mathcal{B}(\mathcal{H})$ there is associated an A^* in $\mathcal{B}(\mathcal{H})$, called the *adjoint* of A, characterized by the equality $\langle Ax, y \rangle = \langle x, A^*y \rangle$ for all x and y in $\mathcal{H}$. One verifies without difficulty that:

(1) $(aA + B)^* = \bar{a}A^* + B^*$

(2) $(AB)^* = B^*A^*$

(3) $(A^*)^* = A$

(4) $\| AA^* \| = \| A \| \, \| A^* \|$

(5) $\| A \| = \| A^* \|$.

An operator A such that $A = A^*$ is said to be *self-adjoint*. A subset $\mathcal{F}$ of $\mathcal{B}(\mathcal{H})$ such that $\mathcal{F}^* = \mathcal{F}$ (equivalently, $A^* \in \mathcal{F}$ if $A \in \mathcal{F}$) is said to be self-adjoint. A subalgebra $\mathfrak{A}$ of $\mathcal{B}(\mathcal{H})$ which is both norm closed and self-adjoint is called a *C^*-algebra*.

One of the key initial results of the theory—a slight generalization of a result of Gelfand and Neumark states:

THEOREM 1.1. *If $\mathcal{B}$ is a Banach algebra with an involution $A \to A^*$ satisfying* (1), (2), (3) *and* (4), *above, then there is a Hilbert space $\mathcal{H}$ and a C^*-algebra $\mathfrak{A}$*

acting on it such that $\mathscr{B}$ *is algebraically isomorphic to* $\mathfrak{A}$ *by means of an isomorphism* ϕ *for which* $\phi(B^*) = \phi(B)^*$.

In stating and proving the result it is usual to assume (5) as well as (1)–(4), and to assume that $\mathfrak{A}$ has a unit element. We will denote the unit element of $\mathscr{B}(\mathscr{H})$ by I (so that $Ix = x$) and refer to it as the identity operator. The theorem just noted establishes the "independent algebraic existence" of a C^*-algebra—independent of its action on a particular Hilbert space. It is often useful to think of the C^*-algebra in this way and to speak of its *representations* on a particular Hilbert space $\mathscr{H}$. A representation of the C^*-algebra $\mathfrak{A}$ on the Hilbert space $\mathscr{H}$ is a homomorphism ϕ of $\mathfrak{A}$ into $\mathscr{B}(\mathscr{H})$ such that $\phi(A^*) = \phi(A)^*$ for each A in $\mathfrak{A}$. It is a non-trivial fact that the image $\phi(\mathfrak{A})$ of $\mathfrak{A}$ under this mapping is norm closed—hence, a C^*-algebra. If ϕ is an isomorphism (the kernel of ϕ is (0)), we say that ϕ is a *faithful* representation of $\mathfrak{A}$. When the transforms $\phi(\mathfrak{A})x$ of a vector x in $\mathscr{H}$ by operators in $\phi(\mathfrak{A})$ lie dense in $\mathscr{H}$, we say that ϕ is a *cyclic representation* of $\mathfrak{A}$, and that x is a *cyclic vector* for $\phi(\mathfrak{A})$ (and for ϕ).

The technique of proof of Theorem 1.1, as developed by Segal was especially useful. It involved a construction of representations of $\mathfrak{A}$ based on a special type of linear functional on $\mathfrak{A}$. The functionals are called *states* and the procedure is known as the *GNS* (Gelfand–Neumark–Segal) *construction.* In order to describe this construction, we make use of another essential structure possessed by C^*-algebras—basic to their analysis—the *order structure.* If we think of $\mathfrak{A}$ as acting on $\mathscr{H}$, a *positive operator* in $\mathscr{B}(\mathscr{H})$ is an operator A such that $\langle Ax, x\rangle \geqslant 0$ for all x in $\mathscr{H}$; and the set of positive operators in $\mathfrak{A}$ forms a cone ($A + B \geqslant 0$ if A and B are positive; aA is positive if A is positive and $a \geqslant 0$; $A = 0$ if both A and $-A$ are positive). Relative to this cone, the real linear space of self-adjoint operators in $\mathfrak{A}$ is a partially-ordered vector space. We write "$A \geqslant B$" for "$A - B$ is positive". The unit element I of $\mathfrak{A}$ is an order unit: for each self-adjoint A there are constants a and b such that $aI \leqslant A \leqslant bI$. Moreover $-\|A\|I \leqslant A \leqslant \|A\|I$; and $\|A\|$ is the least non-negative constant for which this inequality is valid. A *state* of $\mathfrak{A}$ is a linear functional ρ on $\mathfrak{A}$ such that

(i) $\rho(A) \geqslant 0$ when $A \geqslant 0$

(ii) $\rho(I) = 1$.

The GNS construction proceeds as follows:

With ρ a state of $\mathfrak{A}$ define an inner product $\{,\}$ on $\mathfrak{A}$ by means of the formula $\{A, B\} = \rho(B^*A)$. As $\langle A^*Ax, x\rangle = \langle Ax, Ax\rangle = \|Ax\|^2 \geqslant 0$ for each A in $\mathfrak{A}$ (i.e. $A^*A \geqslant 0$ for each A in $\mathfrak{A}$) $\{,\}$ is a positive semi-definite inner product on $\mathfrak{A}$. This is enough in order that the Cauchy–Schwarz inequality should hold and

$$|\{A, B\}| = |\rho(B^*A)| \leqslant \{A, A\}^{\frac{1}{2}}\{B, B\}^{\frac{1}{2}} = \rho(A^*A)^{\frac{1}{2}}\rho(B^*B)^{\frac{1}{2}}.$$

It follows that $\rho(TA) = 0$ for all T in $\mathfrak{A}$ if $\rho(A^*A) = 0$ (of course, $\rho(A^*A) = 0$ if $\rho(TA) = 0$ for each T in $\mathfrak{A}$). The set of such A is a left ideal $\mathscr{K}$ in $\mathfrak{A}$ called the *left kernel* of ρ. It is the set of null vectors with respect to the inner product $\{\ ,\ \}$. If A is self-adjoint $-\|A\|I \leqslant A \leqslant \|A\|I$; so that $-\|A\| \leqslant \rho(A) \leqslant \|A\|$. Thus $|\rho(A)| \leqslant \|A\|$. In general,

$$|\rho(T)| = |\rho(IT)| \leqslant \rho(I)^{\frac{1}{2}}\rho(T^*T)^{\frac{1}{2}} \leqslant \|T^*T\|^{\frac{1}{2}} = \|T\|.$$

Thus states of $\mathfrak{A}$ are bounded linear functionals on $\mathfrak{A}$ of norm 1 (attaining their norm at I). The converse is also valid—functionals ρ on $\mathfrak{A}$ of norm 1 for which $\rho(I) = 1$ are states of $\mathfrak{A}$. This is not difficult to prove but requires some information we have not yet discussed.

The quotient Banach space $\mathfrak{A}/\mathscr{K}$ has a positive definite inner product,

$$\langle A + \mathscr{K}, B + \mathscr{K}\rangle = \rho(B^*A) = \{A, B\},$$

induced on it by $\{\ ,\ \}$. With $\phi_0(A)$ defined on $\mathfrak{A}/\mathscr{K}$ by:

$$\phi_0(A)(B + \mathscr{K}) = AB + \mathscr{K},$$

the resulting mapping is well-defined, since $\mathscr{K}$ is a left ideal and bounded relative to the norm on $\mathfrak{A}/\mathscr{K}$ associated with $\langle\ ,\ \rangle$ for

$$\begin{aligned}\|\phi_0(A)(B + \mathscr{K})\|^2 = \|AB + \mathscr{K}\|^2 = \rho(B^*A^*AB) &\leqslant \|A^*A\|\ \rho(B^*B)\\ &= \|A\|^2\ \|B + \mathscr{K}\|^2,\end{aligned}$$

where we have made use of the fact that $B^*HB \geqslant 0$ if $H \geqslant 0$ (since $\langle B^*HBx, x\rangle = \langle HBx, Bx\rangle \geqslant 0$) so that

$$\rho[B^*(\|A^*A\|I - A^*A)B] \geqslant 0.$$

It follows that $\|\phi_0(A)\| \leqslant \|A\|$ and that $\phi_0(A)$ can be extended to a bounded operator $\phi(A)$ on the completion $\mathscr{H}_\rho$ of $\mathfrak{A}/\mathscr{K}$ relative to the metric deduced from $\langle\ ,\ \rangle$. It is easy to check that ϕ is a homomorphism of $\mathfrak{A}$ into $\mathscr{B}(\mathscr{H}_\rho)$ (and from the preceding, $\|\phi(A)\| \leqslant \|A\|$). That ϕ preserves adjoints follows from:

$$\langle\phi(A)(B + \mathscr{K}), C + \mathscr{K}\rangle = \rho(C^*AB) = \langle B + \mathscr{K}, \phi(A^*)(C + \mathscr{K})\rangle.$$

Thus ϕ is a representation of $\mathfrak{A}$. We say that ϕ is the representation of $\mathfrak{A}$ *engendered* by ρ; and, when it is desirable to indicate the dependence of the representation on ρ, we denote it by π_ρ. The element $I + \mathscr{K}$ in $\mathscr{H}_\rho$, which we denote by x_ρ, for simplicity of notation, has special properties. To begin with

$\|x_\rho\|^2 = \rho(I) = 1$; so that x_ρ is a unit vector. In addition, $\phi(\mathfrak{A})x_\rho = \mathfrak{A}/\mathscr{K}$; so that, by construction, ϕ is a cyclic representation and x_ρ is a cyclic vector for ϕ. Finally, $\rho(A) = \langle\phi(A)x_\rho, x_\rho\rangle$. Note that the functional $\phi(A) \rightarrow \langle\phi(A)x_\rho, x_\rho\rangle$ is a state of $\phi(\mathfrak{A})$. We call such a state a *vector state* of $\phi(\mathfrak{A})$ and say that this vector state *represents* ρ.

In the description of the order structure on $\mathfrak{A}$, in particular, when defining positive operators, we assumed that $\mathfrak{A}$ acts on a Hilbert space. If $\mathfrak{A}$ is not so represented, a technique using the spectrum of elements in a Banach algebra allows us to define the order structure. I remind you that a complex number λ is said to lie in the *spectrum* of an element of $\mathfrak{A}$ (relative to $\mathfrak{A}$) when $A - \lambda I$ fails to have a two-sided inverse in $\mathfrak{A}$. The spectrum $\mathrm{sp}(A)$ is a non-empty, closed subset of $\mathbb{C}$ contained in the disc of radius $\|A\|$ (so that $\mathrm{sp}(A)$ is compact). The positive elements of $\mathfrak{A}$ are identified, now, as those self-adjoint elements A of $\mathfrak{A}$ for which $\mathrm{sp}(A)$ consists of non-negative real numbers.

It might be appropriate to pause, here, and note some specific examples of C^*-algebras.

(1) With $\mathscr{H}$ of dimension n, $\mathscr{B}(\mathscr{H})$ is a C^*-algebra, isomorphic to the algebra of $n \times n$ complex matrices when n is a finite cardinal.

(2) If X is a compact Hausdorff space and $C(X)$ is the algebra of complex-valued continuous functions on X (with pointwise operations) then $C(X)$ is a C^*-algebra—where complex conjugation of functions is taken as the involution. In this last case, the C^*-algebra is abelian. A specific example is had by choosing the interval $[0, 1]$ for X. It is worth noting that we have described all commutative C^*-algebras in this example (at least as far as their algebraic structure goes).

THEOREM 1.2. *If $\mathfrak{A}$ is a commutative C^*-algebra there is a compact Hausdorff space X such that $\mathfrak{A}$ is $*$-isomorphic to $C(X)$.*

This description of commutative C^*-algebras contains the algebraic content of the "spectral theorem". The set of all states of a C^*-algebra is a convex subset of the (continuous) dual space $\mathfrak{A}^\wedge$ of $\mathfrak{A}$. In the topology of convergence on elements of $\mathfrak{A}$, the w^*-topology, this convex set is compact (as a closed subset of the unit ball). The Krein–Milman theorem assures us that it is the closed convex hull of its extreme points—the *pure states* of $\mathfrak{A}$. The pure states of $\mathfrak{A}$ are those states ρ such that

$$\rho = a\rho_1 + (1 - a)\rho_2$$

with $0 < a < 1$ and ρ_1, ρ_2 states, only when $\rho_1 = \rho_2 = \rho$. The pure states of $C(X)$ are the functionals corresponding to evaluation of functions in $C(X)$ at a point of X. Theorem 1.2 can be proved by this technique: examine the pure states of a commutative C^*-algebra, show that they are multiplicative, linear functionals, and that they form a closed subset of the dual. In general the pure states of a C^*-algebra do not form a closed subset of the dual. The vector states of $\mathscr{B}(\mathscr{H})$ are among the pure states of $\mathscr{B}(\mathscr{H})$ but are not all pure states of $\mathscr{B}(\mathscr{H})$. All the others annihilate the compact operators.

If $\mathfrak{A}$ is a C^*-algebra and A is a self-adjoint operator in $\mathfrak{A}$, let $\mathfrak{A}(A)$ denote the C^*-subalgebra of $\mathfrak{A}$ generated by A and I. Since $\mathfrak{A}(A)$ (the norm closure of the polynomials in A) is commutative $\mathfrak{A}(A) \cong C(X)$, for some compact Hausdorff space X. With p a point of X, let ρ_0 be the state of $\mathfrak{A}(A)$ which assigns to each element the value of its corresponding function. The construction of the *-isomorphism of a commutative C^*-algebra with $C(X)$ carries with it the information that the isomorphism preserves order and norm so ρ_0 is a state. Applying the Hahn–Banach theorem, we extend ρ_0 to a functional ρ of norm 1 on $\mathfrak{A}$. Since $\rho_0(I) = 1$, ρ is a state of $\mathfrak{A}$. Now $\rho_0(A^2) = \rho_0(A)^2$ so that $\rho([A - \rho(A)I]^2) = 0$ and $A - \rho(A)I$ is in the left kernel of ρ. Thus $\rho(B(A - \rho(A)I)) = 0$ for each B in $\mathfrak{A}$. That is, $\rho(BA) = \rho(B)\rho(A)$ for each B in $\mathfrak{A}$. Symmetrically, $\rho(AB) = \rho(A)\rho(B)$. It follows that $A - \rho(A)I$ does not have an inverse in $\mathfrak{A}$ and that $\rho(A) \in \mathrm{sp}_{\mathfrak{A}}(A)$. From the outset, $\rho(A) = \rho_0(A) \in \mathrm{sp}_{\mathfrak{A}(A)}(A)$. It follows that $\mathrm{sp}_{\mathfrak{A}}(A)$ and $\mathrm{sp}_{\mathfrak{A}(A)}(A)$ coincide, for a self-adjoint A in $\mathfrak{A}$. What amounts to the same thing, $A - \lambda I$ has an inverse in $\mathfrak{A}$ if and only if it has an inverse in $\mathfrak{A}(A)$. For arbitrary T in $\mathfrak{A}$, if T lies in the C^*-subalgebra $\mathfrak{A}_0$ of $\mathfrak{A}$, T has an inverse in $\mathfrak{A}_0$ if and only if both T^*T and TT^* have inverses in $\mathfrak{A}_0$ (for then T has both a left and right inverse in $\mathfrak{A}_0$, hence a two-sided inverse). This last occurs if and only if T^*T and TT^* have inverses in $\mathfrak{A}$ which is the case if and only if T has an inverse in $\mathfrak{A}$.

Several useful facts emerge from this discussion:

(i) The spectrum of an element of a C^*-algebra is not dependent on the C^*-subalgebra containing it in which the spectrum is computed.

(ii) A state of a C^*-subalgebra of a C^*-algebra has an extension to the full algebra which is a state.

(iii) If ρ is a state of $\mathfrak{A}$ and A is a self-adjoint element of $\mathfrak{A}$ such that $\rho(A^2) = \rho(A)^2$, then

$$\rho(AB) = \rho(A)\rho(B) = \rho(BA).$$

(iv) If $\mathfrak{A}$ is a commutative C^*-algebra generated by the single self-adjoint element A then $\mathfrak{A} \cong C(\mathrm{sp}(A))$.

If $\mathfrak{A}$ acts on $\mathscr{H}$, and x is a unit vector in $\mathscr{H}$ such that $\langle A^2x, x\rangle = \langle Ax, x\rangle^2$ for some self-adjoint A in $\mathfrak{A}$, then, from the preceding:

$$\langle (A - \langle Ax, x\rangle I)^2 x, x\rangle = 0$$

so that $Ax = \langle Ax, x\rangle\, x$, and x is an eigenvector for A. Let $\mathscr{H}$ be $L_2([0,1])$ relative to Lebesgue measure and let $\mathfrak{A}$ be the C^*-algebra consisting of $\{M_f : f \text{ in } C([0,1])\}$ where $M_f(g) = f \cdot g$. We call M_f the *multiplication operator* corresponding to f. The state ρ_0 of $\mathfrak{A}$ defined by $\rho_0(M_f) = f(0)$

extends to a state ρ of $\mathscr{B}(\mathscr{H})$. Denoting by λ the identity function on $[0, 1]$, $\rho(M_\lambda)^2 = 0 = \rho(M_\lambda^2)$. If ρ were a vector state of $\mathscr{B}(\mathscr{H})$, that vector would be annihilated by M_λ. But no L_2-function on $[0, 1]$ other than 0 is annihilated by multiplication by λ. Thus ρ is not a vector state of $\mathscr{B}(\mathscr{H})$.

If ρ_0 is a state of the C^*-subalgebra $\mathfrak{A}_0$ of $\mathfrak{A}$ the set of all state extensions of ρ_0 to $\mathfrak{A}$ is a convex, w^*-compact set of states of $\mathfrak{A}$. If ρ is one of its extreme points and $\rho = a\rho_1 + (1 - a)\rho_2$, with $0 < a < 1$ and ρ_1, ρ_2 states of $\mathfrak{A}$, then this same relation persists on $\mathfrak{A}_0$. Since ρ_0 is pure, $\rho_0 = \rho_1 | \mathfrak{A}_0 = \rho_2 | \mathfrak{A}_0$; and ρ_1, ρ_2 are extensions of ρ_0. Since ρ is extreme in the set of such extensions, $\rho = \rho_1 = \rho_2$; and ρ is a pure state of $\mathfrak{A}$. Thus pure states of C^*-subalgebras have pure state extensions. In the case of the multiplication algebra, above, and the pure state ρ_0 of $\mathfrak{A}$ described there, if we take for ρ a pure state extension of ρ_0 to $\mathscr{B}(\mathscr{H})$, we have an example of a pure state of $\mathscr{B}(\mathscr{H})$ which is not a vector state.

If ϕ is a representation of the C^*-algebra $\mathfrak{A}$ on a Hilbert space $\mathscr{H}$, ϕ is said to be an *irreducible* representation of $\mathfrak{A}$ (equivalently, $\phi(\mathfrak{A})$ is said to act irreducibly on $\mathscr{H}$) when each non-zero vector in $\mathscr{H}$ is a cyclic vector for $\phi(\mathfrak{A})$. In this case no proper closed subspace of $\mathscr{H}$ is invariant under $\phi(\mathfrak{A})$. If $\mathscr{V}$ is a closed subspace of $\mathscr{H}$ the operator E which assigns to a vector its orthogonal projection on $\mathscr{V}$ is a projection (operator) with range $\mathscr{V}$. A check shows that E is self-adjoint and idempotent ($E^2 = E$) and that $\mathscr{V}$ is invariant under an operator A and its adjoint if and only if $AE = EA$. Thus ϕ is irreducible if and only if I and 0 are the only projections commuting with $\phi(\mathfrak{A})$. If ϕ is engendered by the state ρ, ϕ is irreducible if and only if ρ is pure. In effect, a commuting projection different from 0 or I provides a means for decomposing ρ.

If $\mathfrak{A}$ acts on $\mathscr{H}$ and x is a unit cyclic vector for $\mathfrak{A}$ the representation π_x corresponding to the vector state $A \to \langle Ax, x\rangle = \omega_x(A) = \rho(A)$ is unitarily equivalent to the action of $\mathfrak{A}$ on $\mathscr{H}$. The mapping $Ax \to \pi_x(A)x_\rho$ extends to an isomorphism (= unitary transformation) U of $\mathscr{H}$ onto $\mathscr{H}_\rho$ and $UAU^{-1} = \pi_x(A)$ for all A in $\mathfrak{A}$.

2. VON NEUMANN ALGEBRA BASICS

The *strong-operator topology* on $\mathscr{B}(\mathscr{H})$ is the topology for which the net (T_a) is convergent to T when $\|(T_a - T)x\| \to 0$ for each x in $\mathscr{H}$. The *weak-operator topology* on $\mathscr{B}(\mathscr{H})$ is that in which (T_a) converges to T when $\langle T_a x, y\rangle \to \langle Tx, y\rangle$ for each x and y in $\mathscr{H}$. The weak-operator topology is weaker (coarser) than the strong operator topology. Nevertheless

THEOREM 2.1. *The weak- and strong-operator closures of a convex subset of* $\mathscr{B}(\mathscr{H})$ *coincide.*

In essence, if $\mathcal{K}$ is convex its strong-operator closure is contained in its weak-operator closure. Suppose A is in its weak-operator closure but not in its strong-operator closure. Then there are vectors $x_1, \ldots, x_n$ such that $(Ax_1, \ldots, Ax_n)$ is not in the norm closure of $\{(Kx_1, \ldots, Kx_n)$: K in $\mathcal{K}\}$ in the direct sum $\mathcal{H} \oplus \ldots \oplus \mathcal{H}$ $(= \tilde{\mathcal{H}})$ of $\mathcal{H}$ with itself n times. The Separation Theorem tells us that there is a linear functional f on $\tilde{\mathcal{H}}$ and a scalar a such that $f(Ax_1, \ldots, Ax_n) > a$ and $f(Kx_1, \ldots, Kx_n) \leqslant a$ for each K in $\mathcal{K}$. But linear functionals on $\tilde{\mathcal{H}}$ arise from vectors; so that there is a vector $(y_1, \ldots, y_n)$ in $\tilde{\mathcal{H}}$ such that $\langle Ax_1, y_1 \rangle + \ldots + \langle Ax_n, y_n \rangle > a$ while $\langle Kx_1, y_1 \rangle + \ldots + \langle Kx_n, y_n \rangle \leqslant a$—which contradicts the choice of A in the weak-operator closure of $\mathcal{K}$.

It follows from this result that the strong- and weak-operator closures of a subalgebra of $\mathcal{B}(\mathcal{H})$ coincide. Those weak-operator closed subalgebras of $\mathcal{B}(\mathcal{H})$ stable under * are called *von Neumann algebras*. It follows from Theorem 2.1 that a linear functional on a von Neumann algebra $\mathcal{R}$ is weak-operator continuous on a convex subset $\mathcal{K}$ if and only if it is strong-operator continuous on $\mathcal{K}$. By choosing subbasic open sets appropriately in $\mathbb{C}$ it is enough to note that the linear functional has, as inverse image of a convex set, another convex set; which allows us to convert the condition on this inverse image of being strong-operator closed to one of being weak-operator closed. This works as well for a linear mapping η from one von Neumann algebra $\mathcal{R}_1$ into another $\mathcal{R}_2$. Here we assume that η is continuous on $\mathcal{K}$ in the strong-operator topology to $\mathcal{R}_2$ in the weak-operator topology, and conclude that it is continuous on $\mathcal{K}$ in the weak-operator topology to $\mathcal{R}_2$ in this same topology.

The change in closure assumption from norm closed for C^*-algebras to strong-operator closed for von Neumann algebras produces significant structural changes even though it seems like a fine technical distinction. For one thing, the von Neumann algebras have many projection operators while the C^*-algebras may have none. In a deeper sense, the passage from the C^*-algebras to the von Neumann algebras corresponds to the passage from the algebra of continuous functions to the algebra of bounded measurable functions. This correspondence can be made quite formal in the commutative case (Theorem 1.2 is part of the story).

A feature of the weak-operator topology is a certain compactness property it possesses.

THEOREM 2.2. *The unit ball* $(\mathcal{R})_1$ *in* $\mathcal{R}$ *is weak-operator compact, where* $\mathcal{R}$ *is a von Neumann algebra.*

The proof of this proceeds as does the proof that the unit ball in the dual of a normed space is compact—making use of the representation of bounded, conjugate bilinear functionals on $\mathscr{H}$ in terms of bounded operators and the definition of the weak-operator topology.

If $\{H_a\}$ is a monotone increasing net of self-adjoint operators on $\mathscr{H}$ then $\langle H_a x, x\rangle$ is monotone increasing for each x in $\mathscr{H}$. If $H_a \leqslant kI$, for all a, then $\langle H_a x, x\rangle$ converges for each x. By "polarization" $\langle H_a x, y\rangle$ converges for each x, y in $\mathscr{H}$. The resulting limit is a bounded conjugate bilinear functional on $\mathscr{H}$ and corresponds to a self-adjoint operator H on $\mathscr{H}$. Not only is H the weak-operator limit of $\{H_a\}$, but an argument with the Schwarz inequality shows that it is a strong-operator limit of $\{H_a\}$. Of course $H_a \leqslant H$ for all a and H is the least operator with this property. Thus H is characterized as the (unique) least upper bound of $\{H_a\}$. If all the H_a lie in a von Neumann algebra $\mathscr{R}$, then H lies in $\mathscr{R}$.

If $0 \leqslant A \leqslant I$, by passing to the function representation of $\mathfrak{A}(A)$, $(A^{1/n})$ can be seen to be a monotone increasing sequence bounded above by I. It has a least upper bound E which is its strong-operator limit. Then $(A^{2/n})$ has E^2 as its strong-operator limit. But $(A^{1/n}) = (A^{2/2n})$ is a subsequence of $(A^{2/n})$; so that $E = E^2$. One can show, now, that E is the projection on the closure of the range of A. We denote this *range projection* by $R(A)$. As $R(TT^*) = R(T)$ for each bounded T, we conclude that the range projection of each T in a von Neumann algebra $\mathscr{R}$ lies in $\mathscr{R}$. Thus von Neumann algebras have many projections. If $\{E_a\}$ is a family of projections their *union*, $\vee_a E_a$, and their *intersection*, $\wedge_a E_a$, are the projections on the subspace spanned by their ranges and on the intersection of their ranges, respectively. Since $R(E + F) = E \vee F$, we see that $E \vee F \in \mathscr{R}$ if the projections E and F lie in $\mathscr{R}$. If $\{E_a\}$ is a family of projections in the von Neumann algebra $\mathscr{R}$ then unions of finite subfamilies lie in $\mathscr{R}$ and form a monotone increasing net (bounded above by I) with least upper bound (strong-operator limit) $\vee_a E_a$. Thus $(E=)\vee_a E_a \in \mathscr{R}$. Since $E - \vee_a (E - E_a) = \wedge_a E_a$, $\wedge_a E_a \in \mathscr{R}$. Let P be the union of the range projections of all operators in the von Neumann algebra $\mathscr{R}$; then $PA = A$ for all A in $\mathscr{R}$ so that P is a unit for $\mathscr{R}$. For convenience, when we speak of von Neumann algebras, henceforth, we assume that they contain I.

The algebra $\mathscr{B}(\mathscr{H})$ is an example of a von Neumann algebra. Its centre consists of scalar multiples of I. Those von Neumann algebras with centre consisting of the scalar operators only are called *factors*. Another example is constructed from the algebra of multiplications on $L_2(S, \mu)$ (S a measure space with measure μ) by bounded measurable functions. This is an abelian von Neumann algebra. Recalling that an abelian C^*-algebra is *-isomorphic with some $C(X)$ and noting that each von Neumann algebra is a C^*-algebra, one naturally wonders about the special nature of X in the case of an abelian von Neumann algebra.

THEOREM 2.3. *If $\mathscr{A}$ is an abelian von Neumann algebra then $\mathscr{A} \cong C(X)$, with X a compact Hausdorff space in which the closure of each open set is open (as well as closed).*

We say that X is *extremely disconnected* in this case and call the sets which are both closed and open *clopen* sets.

Since each bounded monotone increasing net $\{A_a\}$ in $\mathscr{A}$ has a least upper bound A in $\mathscr{A}$ and since the isomorphism between $\mathscr{A}$ and $C(X)$ is order-preserving, the same is true for each such net $\{f_a\}$ of functions in $C(X)$. That is, there is an f in $C(X)$ which is a least upper bound for $\{f_a\}$. This condition will cause X to be extremely disconnected. From another viewpoint, we have seen that $\mathscr{A}$ has many projections. Each will correspond to an idempotent function in $C(X)$; and such a function is the characteristic function of a clopen set.

If A corresponds to f in $C(X)$ there is a largest clopen set O_λ on which f takes values not exceeding λ. A clopen set on which f takes values not exceeding λ has the closure of the set of points at which f takes values exceeding λ in its complement. This last set and its complement are clopen. The complement contains the first clopen set and is itself a clopen set on which f takes values not exceeding λ. It is O_λ. The characteristic function of O_λ is in $C(X)$ and corresponds to a projection E_λ in $\mathscr{A}$. The characterization of O_λ as the largest clopen set on which f takes values not exceeding λ allows us to conclude that

(1) $E_\lambda \leqslant E_\mu$ when $\lambda \leqslant \mu$,

(2) $\wedge_{\lambda > \lambda_0} E_\lambda = E_{\lambda_0}$,

(3) $\vee_\lambda E_\lambda = I$ and $\wedge_\lambda E_\lambda = 0$.

As a matter of fact, $E_\lambda = 0$ for $\lambda \leqslant -\|A\| - \varepsilon$ for each positive ε and $E_\lambda = I$ for $\lambda \geqslant \|\|A\|$. A family of projections $\{E_\lambda\}$ satisfying (1), (2) and (3) is called a *resolution of the identity*; and the particular one we constructed is called the *resolution of the identity for A*. If we assign to $\int_{-\infty}^{\infty} \lambda \, dE_\lambda$ the meaning of norm convergence of approximating Riemann sums, then

$$\int_{-\infty}^{\infty} \lambda \, dE_\lambda = \int_{-\|A\|-\varepsilon}^{\|A\|} \lambda \, dE_\lambda = A.$$

This last formula is the classical *Spectral Theorem*. We can read out of this discussion the fact that each self-adjoint operator is the norm limit of finite linear combinations of mutually orthogonal "spectral projections" for A with coefficients in sp(A).

There are two key approximation theorems at the base of the study of von Neumann algebras. If $\mathscr{F} \subseteq \mathscr{B}(\mathscr{H})$ we write

$$\mathscr{F}' = \{T : T \in \mathscr{B}(\mathscr{H}),\ TA = AT \text{ for all } A \text{ in } \mathscr{F}\}.$$

We call $\mathscr{F}'$ the *commutant* of $\mathscr{F}$.

THEOREM 2.4. (Double Commutant Theorem) *If $\mathscr{R}$ is a von Neumann algebra (containing I) then* $(\mathscr{R}')' = \mathscr{R}$.

Of course $\mathscr{R} \subseteq (\mathscr{R}')'$. Suppose A is in $(\mathscr{R}')'$. To show that A is in the strong-operator closure of $\mathscr{R}$ (hence in $\mathscr{R}$), we must show that given a finite set of vectors $x_1, \ldots, x_n$ there is a T in $\mathscr{R}$ such that $\|(T - A)x_j\|$ is small. For the idea of the argument, we do this for one vector x_0. Let E_0 be the projection with range $[\mathscr{R}x_0]$. Since the range of E_0 is stable under B and B^* for each B in $\mathscr{R}$, $E_0 \in \mathscr{R}'$. Thus A commutes with E_0 and $Ax_0 \in [\mathscr{R}x_0]$, so that there is a T in $\mathscr{R}$ with $\|(T - A)x_0\|$ small. The case of n vectors is handled by using $n \times n$ matrices with entries in $\mathscr{R}$ acting on $\mathscr{H} \oplus \ldots \oplus \mathscr{H}$ (n times) in this same fashion.

The second key approximation result is:

THEOREM 2.5. (Kaplansky Density Theorem) *If $\mathfrak{A}$ is a self-adjoint algebra of operators on a Hilbert space then each operator in the unit ball of the strong-operator closure, $\mathfrak{A}^-$, of $\mathfrak{A}$ is in the strong-operator closure of the unit ball of $\mathfrak{A}$. Moreover, self-adjoint operators in $(\mathfrak{A}^-)_1$ are approximable by self-adjoint operators in $(\mathfrak{A})_1$, positive operators by positive operators; and, if $\mathfrak{A}$ is norm-closed, unitary operators by unitary operators.*

The ingredients of the proof are the following. Suppose H is a self-adjoint operator in $(\mathfrak{A}^-)_1$. If (T_a) is a net of operators in $\mathfrak{A}$ tending to H in the weak-operator topology then $(\frac{1}{2}[T_a + T_a^*])$ tends to H in this topology. Since H is in the weak-operator closure of the set of self-adjoint operators in $\mathfrak{A}$ and this set is convex, H is in the strong-operator closure of this set. Let (H_a) be a net of self-adjoint operators in $\mathfrak{A}$ with strong-operator limit H. With the aid of the function representation of commutative C^*-algebras, we can apply continuous functions defined on the reals to self-adjoint operators. If f is such a function and $f(\lambda) = \lambda$ for λ in $[-1, 1]$ then $f(H) = H$. If, in addition, the range of f is in $[-1, 1]$, then $\|f(K)\| \leqslant 1$ for each self-adjoint K. Finally, if f defines a strong-operator continuous mapping on the self-adjoint operators, then $(f(H_a))$ has $f(H)$ $(=H)$ as strong-operator limit and $\|f(H_a)\| \leqslant 1$. Now $f(H_a)$ is in the norm closure of $\mathfrak{A}$ so that there is some self-adjoint operator in the unit ball of $\mathfrak{A}$ near $f(H_a)$ in norm, hence, strong-operator near $f(H_a)$ (and H).

What can be proved is that each continuous f which vanishes at ∞

defines a strong-operator continuous function on the self-adjoint operators in $\mathscr{B}(\mathscr{H})$. The fact that multiplication is strong-operator continuous on bounded sets yields the result that polynomials are strong-operator continuous on bounded sets of self-adjoint operators. With the Stone–Weierstrass theorem one concludes, now, that all continuous functions are strong-operator continuous on bounded sets. The Cayley Transform $H \to (H - iI)(H + iI)^{-1} = u(H)$ maps self-adjoint operators H into unitary operators and is strong-operator continuous—by inspection. Moreover, $u(H)$ does not have 1 in its spectrum. The function $-i(z+1)(z-1)^{-1}$ is an inverse to the Cayley Transform (where $|z| = 1$ and $z \neq 1$). If f is a continuous real-valued function on $\mathbb{R}$ vanishing at ∞, define $g(z)$ to be $f(-i(z+1)(z-1)^{-1})$ for z different from 1 and z of modulus 1. Then, letting $g(1)$ be 0, g is continuous on the unit circle (since f vanishes at ∞) and $g(u(H)) = f(H)$. This exhibits f as the composition of two strong-operator continuous mappings, the Cayley Transform and a continuous function g on the bounded set of unitary operators. For arbitrary operators T in $(\mathfrak{A}^-)_1$, we use $\mathfrak{A}_2^-$, the 2×2 matrices over $\mathfrak{A}^-$ acting on $\mathscr{H} \oplus \mathscr{H}$. The operator $\tilde{H}$ with 0 on the diagonal and T, T^* at the off-diagonal positions is self-adjoint, has norm 1 and is in $(\mathfrak{A}_2^-)_1$. It is a strong-operator limit of self-adjoint operators of norm 1 in $\mathfrak{A}_2$. Each entry has norm not exceeding 1 and tends to the corresponding entry of $\tilde{H}$. Thus T is the strong-operator limit of elements in $(\mathfrak{A})_1$.

3. ALGEBRAIC STRUCTURE IN VON NEUMANN ALGEBRAS

The first crude division of von Neumann algebras into distinct algebraic isomorphism classes can be effected in terms of minimal projections. The most forceful use of minimal projections occurs in connection with factors. A projection E in a von Neumann algebra $\mathscr{R}$ is said to be *minimal* (in $\mathscr{R}$) when $E \neq 0$ and $0 < F \leqslant E$ for a projection F in $\mathscr{R}$ only if $F = E$. Clearly the property of being minimal for a projection is preserved under *-isomorphisms. The one-dimensional projections in $\mathscr{B}(\mathscr{H})$ provide examples of minimal projections, and this situation is virtually general.

THEOREM 3.1. *If $\mathscr{M}$ is a factor with a minimal projection, then I is the sum of minimal projections in $\mathscr{M}$. The cardinal number n of all families of minimal projections in $\mathscr{M}$ with sum I is the same; and $\mathscr{M}$ is *-isomorphic to $\mathscr{B}(\mathscr{H})$, where $\mathscr{H}$ is n-dimensional.*

In the situation described in this theorem, $\mathscr{M}$ is said to be a *factor of type I_n*. For a factor with a minimal projection, the theorem stated constitutes a complete description of its algebraic structure (two factors with a minimal projection are *-isomorphic if and only if they have the same cardinal n).

Restricting this discussion to factors is not a serious limitation. Roughly speaking, each von Neumann algebra is a direct sum of factors. More precisely, when $\mathscr{H}$ is separable, a von Neumann algebra is a *direct integral* of factors. The indexing family for the "sum" is a measure space and, instead of summing, we must integrate the component "factors". In any event, the model of a von Neumann algebra as a direct sum of factors is an excellent guide to their structure. It places the proper emphasis on the rôle of factors in the theory.

For a finer analysis of the algebraic structure of von Neumann algebras, it is useful to develop a theory which compares the sizes of the ranges of projections in such an algebra, relative to that algebra.

If E and F are projections in a von Neumann algebra $\mathscr{R}$, we say that E is *equivalent* to F (modulo $\mathscr{R}$), and write $E \sim F$ (mod $\mathscr{R}$), when there is an operator V in $\mathscr{R}$ mapping the range of E isometrically onto that of F.

Replacing V by VE, we can require that V be "normalized" so that it annihilates the range of $I - E$. In this case, we say that V is a *partial isometry* with *initial projection* E and *final projection* F. A computation shows that $V^*V = E$ and $VV^* = F$. Conversely, if V satisifes these equations, it is a partial isometry with initial projection E and final projection F. The projection E is a partial isometry with initial and final projection E; so that $E \sim E$. If V is a partial isometry with initial and final projections E and F, respectively, then V^* has F and E as initial and final projections, respectively. Thus $F \sim E$ if $E \sim F$. If, in addition, W is a partial isometry with initial projection F and final projection G, then WV is a partial isometry with initial projection E and final projection G. Thus $E \sim G$, if $E \sim F$ and $F \sim G$. It follows that $\sim$ is an equivalence relation on the projections of $\mathscr{R}$. It determines when two projections have "the same size" as measured by operators in $\mathscr{R}$.

It may seem like a difficult project to find isometries in $\mathscr{R}$ comparing projections. Actually, arbitrary operators in $\mathscr{R}$ do almost as well. The key to this observation is the "polar decomposition" of operators. Noting that $\| Tx \|^2 = \| (T^*T)^{\frac{1}{2}}x \|^2$, we see that the operator V which maps $(T^*T)^{\frac{1}{2}}x$ onto Tx extends to an isometry of the closure of the range of $(T^*T)^{\frac{1}{2}}$ onto that of T. Extending V by defining it to be 0 on the orthogonal complement of the range of $(T^*T)^{\frac{1}{2}}$ produces a partial isometry with initial space the closure of the range of $(T^*T)^{\frac{1}{2}}$ (which is the closure of the range of T^*); and $T = V(T^*T)^{\frac{1}{2}}$. If $T \in \mathscr{R}$, then $(T^*T)^{\frac{1}{2}}$ is in the C^*-algebra generated by T, hence in $\mathscr{R}$. One shows without difficulty that V commutes with each self-adjoint operator commuting with T; so that $V \in (\mathscr{R}')' = \mathscr{R}$. It follows that $R(T) \sim R(T^*)$. In particular, if T maps some part of the range of E onto some part of the range of F, that is, if $FTE \neq 0$, $R(FTE) \sim R(ET^*F)$ so that E and F have equivalent non-zero subprojections. Now $\{TEx : x \text{ in } H,\ T \text{ in}$

$\mathscr{R}\}$ is stable under $\mathscr{R}$ and $\mathscr{R}'$, so that the projection Q on the subspace spanned by this set is in $\mathscr{R}'$ and $\mathscr{R}''$ $(=\mathscr{R})$. Thus Q is in the centre, $\mathscr{C}$, of $\mathscr{R}$. If $FTE = 0$ for all T in $\mathscr{R}$ then $FQ = 0$; and $F \leqslant I - Q$. In the situation where $E \leqslant Q$ and $F \leqslant I - Q$, no operator in $\mathscr{R}$ will map a non-zero vector in the range of E onto one in the range of F. We conclude, from this discussion, that E and F fail to have equivalent non-zero subprojections in $\mathscr{R}$ if and only if they are "separated" by a central projection ($E \leqslant Q$, $F \leqslant I - Q$).

Associated with the equivalence relation $\sim$, there is a partial ordering on the equivalence classes. We write $E \lesssim F$ when $E \sim E_0 \leqslant F$. (All the usual notational conventions related to a partial ordering will be used, e.g. $F \gtrsim E$ as well as $E \lesssim F$, etc.). There is no difficulty in showing that $E \lesssim E$ or that $E \lesssim G$ if $E \lesssim F$ and $F \lesssim G$. It is true that $E \sim F$ if $E \lesssim F$ and $F \lesssim E$; but this requires a Hilbert space analogue of the Cantor–Bernstein argument in set theory to establish it. The study of this partial ordering in a von Neumann algebra, is the *comparison theory of projections* in that algebra.

In a factor, there is no possibility of separating non-zero projections by a central projection. Such projections always have equivalent non-zero subprojections. Combining this with the fact that $E \sim F$ if $E = \sum E_a$, $F = \sum F_a$ and $E_a \sim F_a$ for all a, and an exhaustion argument, we have:

THEOREM 3.2. *If E and F are projections in a factor $\mathscr{M}$, then either $E \lesssim F$ or $F \lesssim E$.*

To parallel this general comparability in factors we have:

THEOREM 3.3. (The Comparison Theorem) *If E and F are projections in a von Neumann algebra $\mathscr{R}$, there is a central projection Q such that $QE \lesssim QF$ and $(I - Q)F \lesssim (I - Q)E$.*

By analogy with set theory, a projection E in $\mathscr{R}$ equivalent to a proper subprojection is said to be *infinite* (relative to $\mathscr{R}$), otherwise *finite*. A factor $\mathscr{M}$ with a non-zero finite projection but nominimal projection is said to be of *type II*; of *type II_1* if I is finite, of *type II_∞* if I is infinite. If $\mathscr{M}$ has no non-zero finite projection it is said to be of *type III*. Loosely speaking, a von Neumann algebra is of type I_n, II_1, II_∞, or III if all the factors appearing in its decomposition are of that type. It is possible to define von Neumann algebras of various types without reference to the factors appearing in the decomposition—that is, in "global" terms. Each von Neumann algebra is the direct sum of von Neumann of various "pure" types (some of the summands possibly not present).

It is apparent that type is preserved under *-isomorphism, and we have seen that all factors of type I_n (same n) are *-isomorphic. Are there factors not of type I? If G is a countable (discrete) group and $\mathscr{H}$ is $l_2(G)$, the square summable functions on G, and L_{g_0}, R_{g_0} are the operators defined on $\mathscr{H}$ by:

$$(L_{g_0}x)(g) = x(g_0^{-1}g), \qquad (R_{g_0}x)(g) = x(gg_0),$$

then L_g and $R_{g'}$ are commuting unitary operators. Let $\mathscr{L}_G$ and $\mathscr{R}_G$ be the von Neumann algebras generated by $\{L_g\colon g \text{ in } G\}$ and $\{R_g\colon g \text{ in } G\}$ respectively. Then $\mathscr{L}_G' = \mathscr{R}_G$ (so that $\mathscr{R}_G' = \mathscr{L}_G$). If each conjugacy class in G is infinite (with the exception of $\{e\}$) then $\mathscr{L}_G$ (and $\mathscr{R}_G$) is a factor of type II_1. Specific examples arise from the group Π of those permutations of the integers which move at most a finite number, and F_n the free (non-abelian) group on n generators ($n \geqslant 2$). It is known that $\mathscr{L}_\Pi$ and $\mathscr{L}_{F_n}$ are not isomorphic. Using weak-commutativity techniques that establish this, a non-denumerable collection of groups were constructed, recently, with associated II_1 factors pairwise non-isomorphic.

If $\mathscr{M}$ is a factor of type II_1, acting on $\mathscr{H}$ and $\tilde{\mathscr{M}}$ is the algebra of all those $\aleph_0 \times \aleph_0$ matrices with entries from $\mathscr{M}$ which, acting on $\mathscr{H}$, the $\aleph_0$-fold direct sum of $\mathscr{H}$ with itself, yield bounded operators, then $\mathscr{M}$ is a factor of type II_∞. Moreover, each factor of type II_∞ arises in this way from a factor of type II_1.

To exhibit factors of type III, we make use of the (unique) C^*-algebra $\mathfrak{A}$ which is generated as a C^*-algebra by an infinite sequence $(\mathscr{N}_j)$ of C^*-subalgebras $\mathscr{N}_j$, mutually commuting, each isomorphic to the algebra of complex 2×2 matrices. There is no difficulty in constructing $\mathfrak{A}$, explicitly, on $L_2(0, 1)$. Representing each element of $\mathscr{N}_j$ as a 2×2 matrix, let ρ_j^λ be the state of $\mathscr{N}_j$ which assigns $\lambda a + (1 - \lambda)b$ to A_j in $\mathscr{N}_j$, where a and b are the diagonal entries of A_j and λ is in $[0, 1]$. There is a state ρ_λ of $\mathfrak{A}$ with the property that

$$\rho_\lambda(A_{j_1} \ldots A_{j_n}) = \rho_{j_1}^\lambda(A_{j_1}) \ldots \rho_{j_n}^\lambda(A_{j_n}),$$

when $A_{j_1} \in \mathscr{N}_{j_1}, \ldots, A_{j_n} \in \mathscr{N}_{j_n}$ and $j_1, \ldots, j_n$ are distinct. Applying the GNS constructjon to ρ_λ we construct representations of π_λ of $\mathfrak{A}$ on $\mathscr{H}_\lambda$. Then π_0 and π_1 are irreducible; so that $\pi_0(\mathfrak{A})^- = \mathscr{B}(\mathscr{H}_0)$ and $\pi_1(\mathfrak{A})^- = \mathscr{B}(\mathscr{H}_1)$. In addition $\pi_{\frac{1}{2}}(\mathfrak{A})^-$ is a factor of type II_1 (*-isomorphic to $\mathscr{L}_\Pi$, curiously enough). With λ and λ' in $(0, \frac{1}{2})$, $\pi_\lambda(\mathfrak{A})^-$ and $\pi_{\lambda'}(\mathfrak{A})^-$ are not *-isomorphic and are factors of type III. Thus $\pi_\lambda(\mathfrak{A})^-$ is a factor of type III for $\lambda \neq 0, \frac{1}{2}, 1$; and this family contains a non-denumerable infnity of non-isomorphic factors of type III.

Recent results make deep inroads into the analysis of the structure of type III von Neumann algebras. In essence each such can be realized, in a canonical manner, as the von Neumann algebra generated by a one-parameter group of unitary operators, each inducing an automorphism of a von Neumann algebra $\mathscr{R}_0$ having no summand of type III, and $\mathscr{R}_0$.

4. ACTION OF VON NEUMANN ALGEBRAS ON SPACES

The problem of when two von Neumann algebras act in the same way on their underlying Hilbert spaces can be reduced to a comparison of their algebraic structure. The main result is:

THEOREM 4.1. (Unitary Implementation) *If $\mathscr{R}_1$ and $\mathscr{R}_2$ are von Neumann algebras acting on Hilbert spaces $\mathscr{H}_1$ and $\mathscr{H}_2$, respectively, x_1 and x_2 are unit vectors in $\mathscr{H}_1$ and $\mathscr{H}_2$ such that $[\mathscr{R}_1 x_1] = [\mathscr{R}'_1 x_1] = \mathscr{H}_1$ and $[\mathscr{R}_2 x_2] = [\mathscr{R}'_2 x_2] = \mathscr{H}_2$, and ϕ is a *-isomorphism of $\mathscr{R}_1$ onto $\mathscr{R}_2$, then there is a unitary transformation U of $\mathscr{H}_1$ onto $\mathscr{H}_2$ such that $A = U^{-1}\phi(A)U$.*

We say that U *implements* the isomorphism ϕ. Diligent use of this result and the comparison theory of projections reduces the question of action on the space to one of algebraic type for von Neumann algebras.

If the vector state $\omega_{x_1}|\mathscr{R}_1$ can be "transported" to a vector state $\omega_{y_1}|\mathscr{R}_2$ by means of ϕ (that is, if we can find a unit vector y_1 in $\mathscr{H}_2$ such that $\langle Ax_1, x_1\rangle = \langle\phi(A)y_1, y_1\rangle$ for all A in $\mathscr{R}_1$), and if y_1 can be chosen cyclic for $\mathscr{R}_2$ then the mapping $Ax_1 \to \phi(A)y_1$ is an isometric mapping of $\mathscr{R}_1 x_1$ onto $\mathscr{R}_2 y_1$ and extends to a unitary transformation U of $[\mathscr{R}_1 x_1]$ $(= \mathscr{H}_1)$ onto $[\mathscr{R}_2 y_2]$ $(= \mathscr{H}_2)$. There is no difficulty in showing that U implements ϕ.

In any event the functional ω defined by: $\omega(\phi(A)) = \langle Ax_1, x_1\rangle$, is a state of $\mathscr{R}_2$. It has certain continuity properties. The programme outlined above motivates the study of such states.

A state ω of a von Neumann algebra $\mathscr{R}$ is said to be *completely-additive* when $\omega(\sum_a E_a) = \sum_a \omega(E_a)$ for each orthogonal family $\{E_a\}$ of projections in $\mathscr{R}$.

A *-isomorphism of one von Neumann algebra $\mathscr{R}_1$ onto another $\mathscr{R}_2$ preserves order; so that $\sum_a \phi(E_a)$, the smallest projection larger than each $\phi(E_a)$, is the image $\phi(\sum_a E_a)$ of the projection $\sum_a E_a$, when $\{E_a\}$ is an orthogonal family of projections in $\mathscr{R}$. Thus $\omega \circ \phi$ is a completely additive state of $\mathscr{R}_1$ if ω is a completely-additive state of $\mathscr{R}_2$.

An ostensibly more stringent continuity requirement is that $\lim_a \omega(H_a) = \omega(H)$ for each monotone increasing net (H_a) of self-adjoint operators in $\mathscr{R}$. States which satisfy this condition are said to be *normal states* (of $\mathscr{R}$). Finally, there are the assumptions that ω is weak (or strong)-operator continuous on $(\mathscr{R})_1$. We saw that the equivalence of these two assumptions was a consequence of the fact that convex sets of operators have the same weak and strong-operator closures. The main result relating these conditions on ω is:

THEOREM 4.2. *If ω is a state of the von Neumann algebra $\mathscr{R}$ the following conditions are equivalent.*

(a) *There is a countable family $\{y_n\}$ of mutually orthogonal vectors such that $\sum \| y_n \|^2 = 1$ and $\omega = \sum_n \omega_{y_n} | \mathscr{R}$.*

(b)*There is a countable family of vectors* $\{x_n\}$ *such that* $\sum_n \| x_n \|^2 = 1$ *and* $\omega = \sum_n \omega_{x_n} | \mathscr{R}$.

(c) ω *is weak-operator continuous on* $(\mathscr{R})_1$.

(d) ω *is strong-operator continuous on* $(\mathscr{R})_1$.

(e) ω *is normal.*

(f) ω *is completely-additive.*

As for the possibility of realising a normal state ω as a vector state, the condition for this can be described, best, in terms of the *support* of ω. If $I - E$ is the union of all projections in $\mathscr{R}$ which are annihilated by ω, E is said to be the support of ω. It follows from the fact that ω is normal that $\omega(I - E) = 0$. It is not difficult to conclude that $\omega(H) > 0$ unless $R(H) \leqslant I - E$, for a positive H in $\mathscr{R}$.

THEOREM 4.3. *A normal state of a von Neumann algebra* $\mathscr{R}$ *acting on a Hilbert space* $\mathscr{H}$ *is a vector state if and only if its support E is a cyclic projection in* $\mathscr{R}$ *(that is, E has a range* $[\mathscr{R}'z]$ *for some vector z).*

In the circumstances of the Unitary Implementation Theorem, there is a vector x_2 such that $[\mathscr{R}'_2 x_2] = \mathscr{H}_2$. If E is any projection in $\mathscr{R}_2$, then $[\mathscr{R}'_2 E x_2] = [E\mathscr{R}'_2 x_2] = E(\mathscr{H}_2)$; so that the support of each normal state is a cyclic projection in $\mathscr{R}_2$, and each normal state of $\mathscr{R}_2$ is a vector state. An application of comparison theory allows us to choose the vector representing the state as a generator for $\mathscr{H}_2$ under $\mathscr{R}_2$ when the state is separating (since $[\mathscr{R}_2 x_2] = \mathscr{H}_2$), which supplies what is needed for the proof of Theorem 4.1.

5. Banach Algebras and Topology

J. L. TAYLOR

Department of Mathematics,
University of Utah, Salt Lake City, Utah, U.S.A.

Introduction ... 118
1. Axioms for Cohomology ... 120
2. Consequences of the Axioms ... 126
3. Čech Cohomology ... 131
4. Idempotents and Logarithms ... 138
5. The Functor K_1 ... 144
6. The Functor K_0 ... 148
7. Further Properties of K_0 ... 154
8. Relations between K_0 and K_1 ... 159
9. The Bott Periodicity Theorem ... 163
10. K-theory ... 169
11. Applications of Cohomology to Harmonic Analysis ... 175
12. K-theory and Harmonic Analysis ... 181
References ... 185

INTRODUCTION

Given the connection between commutative Banach algebras and compact Hausdorff spaces—that is, the cofunctor which assigns to a commutative Banach algebra A its maximal ideal space $\Delta(A)$—it is natural to ask what the various topological invariants of $\Delta(A)$ (cohomology groups, homotopy groups, etc.) mean to the structure of A. The first result along these lines was the Shilov Idempotent Theorem [**17**] relating the connectedness (0-dimensional Čech cohomology) of $\Delta(A)$ to the existence of idempotents in A. Then Arens [**2**] and Royden [**15**] identified the first Čech cohomology group of $\Delta(A)$ (with integral coefficients) as the group $A^{-1}/\exp(A)$ of invertible elements of A modulo its subgroup consisting of elements with logarithms.

Now each of these results has important implications in Banach algebra theory (for example, the author's interest in this subject stems from an application of the Arens–Royden theorem to the study of measure algebras which considerably simplifies the difficult problem of deciding which measures have convolution inverses, [**19**]). Hence, it is reasonable to assume that further information connecting the topology of $\Delta(A)$ and the structure of A will have important consequences.

However, we are not aware of much work in this direction in recent years. In particular, the relation between the structure of A and the higher Čech cohomology groups seems not very well understood (cf. [**20**] for more on this). The problem seems to be that the higher Čech groups are ill suited to simple characterization in terms of A. It turns out, however, that there is another cohomology theory (a generalized cohomology theory, in the terminology of these notes), called K-theory, which is eminently suited to such a characterization. The clue that leads to this fact is a generalization of the Arens–Royden theorem due to Arens [**3**]. The characterization is explicitly pointed out by Novadvorskii [**14**], and Forster [**9**].

The relation between K-theory and commutative Banach algebras seems to be known by a few abstract analysts and is touched on in some articles in the literature (cf. [**9**], [**13**], [**14**], [**21**]. However, the implications seem to be largely unexplored. Part of the reason for this may be the lack of an introductory treatment of the subject designed to be read by Banach algebra specialists. These lecture notes are an attempt to provide such a treatment.

The author is a novice at algebraic topology in general and k-theory in particular. However, he has for some time been curious about K-theory and its relationship to Banach algebra theory. The motivation to satisfy this curiosity was provided by the invitation to lecture at the Birmingham "Algebras in Analysis" conference. A 1972–73 graduate analysis course at the University of Utah provided the chance to try the subject first on a captive audience of students. The course was a disaster for the students, but the instructor learned a great deal.

The primary objective of these notes is to give a development of elementary K-theory for compact spaces and to describe the K-theory of the maximal ideal space of a commutative Banach algebra A directly in terms of the structure of A. We do this by defining groups $K_0(A)$ and $K_1(A)$ directly from A and then using the Arens theorem to show that these groups depend only on the maximal ideal space of A and agree with the groups of topological K-theory (as developed in [**4**]). Along the way, we prove a version of the Bott periodicity theorem in the context of Banach algebras and give a characterization of the second Čech cohomology group of the maximal ideal space of A due to Forster [**9**].

The material on K-theory comprises Sections 5–10 of the notes. It is preceded by an introductory discussion of generalized cohomology theories (Sections 1–3) and a discussion of the Shilov and Arens–Royden theorems

(Section 4). We do not prove these theorems since they are already well covered in the literature (in particular, see the treatment in Gamelin [**10**]), and since complete proofs are impossible without assuming a considerable background in several complex variables.

The K-theory sections are followed by two sections on applications in harmonic analysis. The first of these (Section 11) describes some work of the author in which the Čech cohomology groups of the maximal ideal space of a measure algebra are computed. In the zero and one-dimensional cases this information, together with the Shilov and Arens–Royden theorems, yields a proof of Cohen's idempotent theorem and a factorization theorem for invertible elements of a measure algebra. In Section 12 we give a brief discussion of applications of K-theory in harmonic analysis. Since the K-theory of the maximal ideal space of a measure algebra has not been computed precisely, this discussion is incomplete.

Our discussion of K-theory in Sections 5–10 is very limited and is not intended to prepare a student to use the theory effectively. We avoid vector bundles entirely. We do not mention the Thom isomorphism or the operations on K-theory. The ring structure of $K^*(X)$ and the Chern character are only mentioned in passing. The student who wants a working knowledge of K-theory should read Atiyah's notes [**4**]. We also recommend [**7**] and [**12**].

We can claim no originality for this material since the main results all appear in one form or another in the literature. Our list of references is minimal and is designed only to provide the student with examples of sources for background material, missing proofs, and different or further insights. Only occasionally do we attempt to give the original source of a theorem.

We express our appreciation to the organizers and supporters of the "Algebras in Analysis" conference and to the National Science Foundation for its support of our efforts under NSF Grant No. GP-32331.

Finally, we apologize to topologists in general for any violations of their mores and customs which may appear in these notes.

1. AXIOMS FOR COHOMOLOGY

After attempts to teach us the subject failed, we hoped, on leaving graduate school, never to encounter algebraic topology again. However, the analyst cannot escape topology. Too often he runs into problems where the solution its existence, uniqueness, or nature) depends on the topology of some underlying space. On the line every continuous function has a continuous antiderivative; not so on the circle. On the line (or any simply connected space) every continuous function has a continuous logarithm; again, this is not so on multiply connected spaces such as the circle. For a domain U in the plane, it is necessary and sufficient that U be simply connected in order that every analytic function on U be approximable by polynomials.

In general, "holes" in a topological space (such as the holes in the circle and the sphere) give rise to obstructions to the solutions of certain problems in both topology and analysis. The object of algebraic topology is to classify these holes and the obstructions they lead to in topology by assigning to a space certain algebraic invariants. The branch of algebraic topology that will concern us here is cohomology. A cohomology theory assigns to a space a sequence of abelian groups in such a way that certain axioms are satisifed.

We begin our discussion of cohomology by describing an axiom system for a (generalized) reduced cohomology theory for compact pointed spaces. We also discuss unreduced theories for compact pairs. However, for reasons of convenience most of our emphasis will be on the reduced theories.

1.1. Functors and cofunctors

Introducing a bit of economical language now (the language of category theory) will save much repetitive explanation later.

A *category* consists of:

(1) a class F of objects (e.g. compact spaces, abelian groups, Banach algebras);

(2) for each pair X, $Y \in F$ a set $\mathrm{Hom}(X, Y)$ of morphisms $f: X \to Y$ (e.g. continuous maps, homomorphisms, continuous homomorphisms);

(3) an assignment to each pair of morphisms $f: X \to Y$ and $g: Y \to Z$ a composite morphism $g \circ f: X \to Z$ such that the associative law $(f \circ g) \circ h = f \circ (g \circ h)$ holds when it makes sense;

(4) for each object X a morphism $1 \in \mathrm{Hom}(X, X)$ such that $1 \circ f = f$ and $g \circ 1 = g$ for all morphisms $f: Y \to X$ and $g: X \to Y$ (e.g. the identity map).

Given two categories F and G, a *functor* H from F to G is an assignment to each $X \in F$ an object $H(X) \in G$ and to each morphism $f: X \to Y$ $(X, Y \in F)$ a morphism $H(f)$: $H(X) \to H(Y)$ in such a way that $H(1) = 1$ and $H(f \circ g) = H(f) \circ H(g)$.

A *cofunctor* H (often called a contravariant functor) from F to G is the same kind of animal as a functor except that it reverses arrows and reverses composition. Thus, to $f: X \to Y$, H assigns a morphism $H(f)$: $H(Y) \to H(X)$ and $H(f \circ g) = H(g) \circ H(f)$.

An *isomorphism* $f: X \to Y$ is a morphism with an inverse g (a morphism g with $g \circ f = 1$ and $f \circ g = 1$).

A *commutative diagram* is a diagram of morphisms such as the square

$$\begin{array}{ccc} X & \xrightarrow{f} & Y \\ {\scriptstyle h}\downarrow & & \downarrow \\ U & \xrightarrow{g} & V \end{array}$$

in which any two sequences of morphism with the same starting and ending points have the same composition (in the example of the square this means $k \circ f = g \circ h$).

It follows immediately from the composition law and $H(1) = 1$ that:

PROPOSITION. *If H is a functor (or cofunctor) then H takes isomorphisms to isomorphisms and commutative diagrams to commutative diagrams.*

1.2. Locally compact spaces

We are ultimately interested in the category of topological spaces which arise as maximal ideal spaces of commutative Banach algebras—that is, the category of locally compact spaces. One must be careful here, however. The appropriate class of morphisms for this category is not the class of continuous maps. Rather, for X, Y locally compact a morphism $f: X \to Y$ will be a continuous, proper map $f: U \to Y$ from an open subset $U \subset X$ into Y (by a *proper map* we mean one such that the pre-image of each compact set is compact).

An example of a cofunctor is the correspondence $X \to C_0(X)$ which assigns to each locally compact space X the algebra $C_0(X)$ of complex valued continuous functions vanishing at infinity on X. If $\varphi: X \to Y$ is a morphism (with domain U) then we define an algebra homomorphism $C_0(\varphi): C_0(Y) \to C_0(X)$ by $C_0(\varphi)f(x) = f(\varphi(x))$ for $x \in U$ and $C_0(\varphi)f(x) = 0$ for $x \notin U$. That $C_0(\varphi)f$ is continuous and vanishes at infinity on X for $f \in C_0(Y)$ follows from the fact that $\varphi: U \to Y$ is a proper map. Obviously, $X \to C_0(X)$ and $\varphi \to C_0(\varphi)$ define a cofunctor from locally compact spaces to algebras.

It turns out that every algebra homomorphism $C_0(Y) \to C_0(X)$, for locally compact X and Y, has the form $C_0(\varphi)$ as above. Since we are ultimately interested in algebras such as $C_0(X)$, this justifies our choice of the morphisms for locally compact spaces.

1.3. Pointed compact spaces

Because of the way morphisms are defined, the above category is rather clumsy to deal with. Here we describe an equivalent category in which the morphisms are defined in a more natural way.

By a *pointed compact space* (X, x_0) we mean a compact space X with a distinguished element $x_0 \in X$ (the *base point*). A morphism $\varphi: (X, x_0) \to (Y, y_0)$ is a continuous map from X to Y with $\varphi(x_0) = y_0$.

In cases where the base point is understood, we shall usually write X rather than (X, x_0) for a pointed space.

Now if X is a locally compact space and X^+ is its one-point compactification, then we consider X^+ to be a pointed space with the point at infinity as base point. If $f: X \to Y$ is a morphism of locally compact spaces (with domain U) then we have a morphism $f^+: X^+ \to Y^+$ defined by $f^+(x) = f(x)$ for $x \in U$ and $f^+(x) = \infty \in Y^+$ if $x \in X^+\backslash U$. Thus, we have a functor $X \to X^+, f \to f^+$ from locally compact spaces to pointed compact spaces.

Conversely, if (X, x_0) is a pointed compact space, then $X\backslash\{x_0\}$ is locally compact and any morphism $f\colon (X, x_0) \to (Y, y_0)$, when restricted to $U = f^{-1}(Y\backslash\{y_0\})$, yields a morphism $X\backslash\{x_0\} \to Y\backslash\{y_0\}$ of locally compact spaces.

Now both compositions $X \to X^+ \to X^+\backslash\{\infty\}$ and $(X, x_0) \to X\backslash\{x_0\} \to (X\backslash\{x_0\})^+$ yield spaces homeomorphic to the initial space. With a similar statement for morphisms, we have what is meant by the following:

PROPOSITION. *The functor $X \to X^+$ is an equivalence between the category of locally compact spaces and the category of pointed compact spaces.*

If A is a commutative Banach algebra (such as $C_0(X)$) with maximal ideal space X, then X^+ is the maximal ideal space of the algebra A^+ obtained by adjoining an identity to A (for example, note that $C_0(X)^+ = C(X^+)$).

1.4. Smash product, suspension

For a pair $Y \subset X$ of compact spaces we can always construct a pointed compact space X/Y by identifying Y to a point, which is then the base point of X/Y. The resulting pointed space can obviously be identified with the one-point compactification $(X\backslash Y)^+$ of the complement of Y.

The cartesian product operation $(X, Y) \to X \times Y$ is an important one for locally compact spaces. For pointed compact spaces the corresponding operation is the *smash product* $(X, x_0) \wedge (Y, y_0)$. This is defined to be the pointed space $X \times Y/(X \times y_0) \cup (x_0 \times Y)$. Note that for locally compact spaces X, Y the space $X \times Y$ is the complement in $X^+ \times Y^+$ of $(X^+ \times \{\infty\}) \cup (\{\infty\} \times Y^+)$ and, hence, $X^+ \wedge Y^+ = (X \times Y)^+$. Thus, the functor $X \to X^+$ takes cartesian products to smash products.

Now since $\mathbb{R}^n$ is the n-fold cartesian product of $\mathbb{R}$, we have $(\mathbb{R}^n)^+ = \mathbb{R}^+ \wedge \ldots \wedge \mathbb{R}^+$ (n-times). However, $\mathbb{R}^+$ is the circle S^1 and $(\mathbb{R}^n)^+$ is the n-sphere S^n. Hence, $S^n = S^1 \wedge \ldots \wedge S^1$ (n-times).

Definition. For any pointed space X, the smash product $S^1 \wedge X$ is called the *reduced suspension* of X and is denoted SX. The space $S^n \wedge X = S^1 \wedge \ldots \wedge S^1 \wedge X$ is called the *n-fold reduced suspension of* X and is denoted S^nX.

Note that for X locally compact we have

$$S^n(X^+) = S^1 \wedge \ldots \wedge S^1 \wedge X^+ = (\mathbb{R}^n \times X)^+$$

so that nth suspension corresponds to cartesian product with $\mathbb{R}^n$.

1.5. Continuous cofunctors

A *cohomology theory* (for pointed compact spaces) is a sequence of cofunctors from pointed compact spaces to abelian groups which satisfies certain axioms. One of the axioms is that each cofunctor in the sequence should be continuous in the sense described below.

In this and later subsections $\tilde{H}$ will be a cofunctor from pointed compact spaces to abelian groups. For a morphism f, $\tilde{H}(f)$ will be denoted f^*.

For a pointed compact space X and a closed subspace Y (containing the base point), the inclusion $i\colon Y \to X$ is a morphism and, hence, induces $i^*\colon \tilde{H}(X) \to \tilde{H}(Y)$. We shall call this the restriction map from $\tilde{H}(X)$ to $\tilde{H}(Y)$.

Definition. The cofunctor $\tilde{H}$ is said to be *continuous* if, whenever $\{X_\alpha\}$ is a family of compact subspaces (containing the base point) of some compact pointed space, and the family is directed downward under inclusion with intersection X, then

(1) each element of $\tilde{H}(X)$ is the restriction of an element of some $\tilde{H}(X^\alpha)$; and

(2) if $u \in \tilde{H}(X_\alpha)$ restricts to zero in $\tilde{H}(X)$ then it restricts to zero in $\tilde{H}(X_\beta)$ for some $X_\beta \subset X_\alpha$.

In other words, $\tilde{H}$ is continuous if, with $\{X_\alpha\}$ and X as above, we have $\varinjlim \tilde{H}(X_\alpha) = \tilde{H}(X)$. This makes sense because $\{\tilde{H}(X_\alpha)\}$ is a direct system of groups with restriction defining the bonding maps.

1.6. Reduced generalized cohomology theories

A sequence

$$\ldots \to G_{n-1} \to G_n \to G_{n+1} \to \ldots$$

of group homomorphisms is said to be *exact* if, for each pair of consecutive maps, the image of the first is the kernel of the second.

By a compact pointed pair (X, Y) we mean a compact pointed space X together with a closed subspace Y containing the base point. For such a pair we let $i\colon Y \to X$ be the inclusion map and $\pi\colon X \to X/Y$ the map identifying Y to a point.

Definition. A *reduced generalized cohomology theory* is a sequence $\{\tilde{H}^p\}_{p=-\infty}^{\infty}$ of cofunctors from compact pointed spaces to abelian groups and a sequence $\{\delta^p\}$ of natural transformations $\delta^p\colon \tilde{H}^p(Y) \to \tilde{H}^p(X/Y)$, defined for compact pointed pairs (X, Y), which satisfy:

(a) Continuity: Each $\tilde{H}^p$ is continuous; and

(b) Exactness: for each compact pointed pair (X, Y), the sequence

$$\ldots \xrightarrow{\delta^{p-1}} \tilde{H}^p(X/Y) \xrightarrow{\pi^*} \tilde{H}^p(X) \xrightarrow{i^*} \tilde{H}^p(Y) \xrightarrow{\delta^p} \tilde{H}^{p+1}(X/Y) \xrightarrow{\pi^*} \ldots$$

is exact.

In this definition, the term "natural transformation" has a specific meaning. Given two cofunctors H and K from a category C to a category D, then a natural transformation $\alpha: H \to K$ is an assignment to each object $A \in C$ a morphism $\alpha: H(A) \to K(A)$ in such a way that the diagram

$$\begin{array}{ccc} H(B) & \xrightarrow{\alpha} & K(B) \\ \downarrow{\scriptstyle \varphi^*} & & \downarrow{\scriptstyle \varphi^*} \\ H(A) & \xrightarrow{\alpha} & K(A) \end{array}$$

is commutative for each morphism $\varphi: A \to B$. In the context of the definition, we think of $(X, Y) \to H^p(Y)$ and $(X, Y) \to H^{p+1}(X/Y)$ as cofunctors from compact pointed pairs to abelian groups. A morphism $\varphi: (X_1, Y_1) \to (X_2, Y_2)$ in the category of compact pointed pairs is a base point preserving continuous map $\varphi: X_1 \to X_2$ which maps Y_1 into Y_2. Hence, the statement "δ^p is a natural transformation" means that for such map $\varphi: (X_1, Y_1) \to (X_2, Y_2)$ the diagram

$$\begin{array}{ccc} \tilde{H}^p(Y_2) & \xrightarrow{\delta^p} & \tilde{H}^{p-1}(X_2/Y_2) \\ \downarrow{\scriptstyle \varphi^*} & & \downarrow{\scriptstyle \varphi^*} \\ \tilde{H}^p(Y_1) & \xrightarrow{\delta^p} & \tilde{H}^{p-1}(X_1/Y_1) \end{array}$$

is commutative.

1.7. The dimension axiom and cohomology

Ordinary (reduced) cohomology theories (such as reduced Čech cohomology) satisfy an additional axiom—the dimension axiom:

Definition. We shall say that a generalized reduced cohomology theory $\{\tilde{H}^n\}$ is a *reduced cohomology theory with coefficient group* G if $\tilde{H}^n(S^0) = 0$ for $n \neq 0$ and $\tilde{H}^0(S^0) = G$.

1.8. Cohomology for compact pairs

Consider the category of compact pairs (X, Y) $(Y \subset X)$ (no longer pointed spaces), where the morphisms $\varphi: (X_1, Y_1) \to (X_2, Y_2)$ are continuous maps of X_1 into X_2 which map Y_1 into Y_2.

Definition. By a generalized cohomology theory for the category of compact pairs we mean a sequence $\{H^p\}_{p=-\infty}^{\infty}$ of cofunctors from compact pairs to abelian groups such that the following conditions are satisfied:

(1) Continuity: If $\{X_\alpha, Y_\alpha\}$ is a directed system of compact pairs ($\beta > \alpha$ implies $X_\alpha \subset X_\beta$ and $Y_\alpha \subset Y_\beta$) with $X = \bigcap X_\alpha$, $Y = \bigcap Y_\alpha$, then $\varinjlim H^p(X_\alpha, Y_\alpha) = H^p(X, Y)$;

(2) Excision: If $X = K \cup L$ for compact subsets K and L, then $i: (K, K \cap L) \to (X, L)$ induces an isomorphism $H^p(X, L) \to H^p(K, K \cap L)$;

(3) Exactness: If we set $H^p(X) = H^p(X, \varnothing)$ for X compact, then there is a natural transformation $\delta: H^p(Y) \to H^{p+1}(X, Y)$, defined for pairs (X, Y), such that

$$\ldots \to H^p(X, Y) \xrightarrow{\pi^*} H^p(X) \xrightarrow{i^*} H^p(Y) \xrightarrow{\delta} H^{p+1}(X, Y) \to \ldots$$

is exact, where $i: (Y, \varnothing) \to (X, \varnothing)$ and $\pi: (X, \varnothing) \to (X, Y)$ are the inclusions.

We say $\{H^p\}$ is a *cohomology theory* (*with coefficient group* G) if we also have

(4) Dimension: $H^p(\text{pt}) = 0$ for $p \neq 0$ and $H^0(\text{pt}) = G$.

The four axioms above are the Eilenberg–Steenrod axioms except that continuity replaces the usual homotopy axiom (this stronger axiom is more appropriate for compact spaces, whereas homotopy suffices if one is dealing only with cell complexes).

1.9. THEOREM. *If $\{\tilde{H}^p\}$ is a generalized reduced cohomology theory, then we obtain a generalized cohomology theory $\{H^p\}$ by setting $H^p(X, Y) = \tilde{H}^p(X^+/Y^+)\,(=\tilde{H}^p(X/Y)$ if $Y \neq \varnothing)$.*

Proof. Let $\{\tilde{H}^p\}$ be a reduced theory and let $\{H^p\}$ be defined as above. That $\{H^p\}$ satisfies continuity follows immediately from the fact that $\{\tilde{H}^p\}$ does. Exactness for a pair (X, Y) follows by passing to the pointed pair (X^+, Y^+) and using the exactness of $\{\tilde{H}^p\}$.

To prove excision, note that we may as well assume $K \cap L \neq \varnothing$ for, otherwise, we may pass to $X^+ = K^+ \cup L^+$ where we let the additional point be common to K^+ and L^+. However, if $K \cap L \neq \varnothing$ then $H^p(X, L) = \tilde{H}^p(X/L)$, $H^p(K, K \cap L) = \tilde{H}^p(K/K \cap L)$ and $i: (K, K \cap L) \to (X, L)$ induces a homeomorphism $K/K \cap L \to X/L$. Excision follows.

Note that $H^p(\text{pt}) = \tilde{H}^p(\text{pt}^+) = \tilde{H}^p(S^0)$ and, hence, $\{H^p\}$ satisfies the dimension axiom if $\{\tilde{H}^p\}$ does. ∎

In Section 2 we shall prove the converse of this; that is, we shall prove that each generalized cohomology theory $\{H^p\}$ determines a generalized reduced cohomology theory $\{\tilde{H}^p\}$ which satisfies $\tilde{H}^p(X) = H^p(X, x_0)$ for a pointed space X with base point x_0 and $\tilde{H}^p(X/Y) = H^p(X, Y)$ for a compact pair (X, Y) with $Y \neq \varnothing$.

2. CONSEQUENCES OF THE AXIOMS

In this section we shall derive several consequences of the axioms for cohomology. We also discuss the question of uniqueness for a cohomology theory.

2.1. Homotopy

Let Λ be a compact space and X and Y pointed compact spaces. By a *parameterized family of maps* $\{\varphi_\lambda: X \to Y\}_{\lambda\in\Lambda}$ we mean a family of morphisms $\varphi_\lambda: X \to Y$ such that the map $(\lambda, x) \to \varphi_\lambda(x):\Lambda \times X \to Y$ is continuous.

PROPOSITION. *If $\{\varphi_\lambda:X \to Y\}_{\lambda\in\Lambda}$ is a parameterized family of maps and $\tilde{H}$ is a continuous cofunctor from pointed spaces to abelian groups, then for each $u \in \tilde{H}(Y)$ we have $\varphi_\lambda^* u \in \tilde{H}(X)$ is locally constant as a function of $\lambda \in \Lambda$.*

Proof. We adjoin a discrete base point to Λ obtaining Λ^+—a compact pointed space. The smash product $\Lambda^+ \wedge X$ can be identified as $\Lambda \times X/\Lambda \times \{x_0\}$, where x_0 is the base point of X.

Since each $\varphi_\lambda: X \to Y$ is base point preserving and $(\lambda, x) \to \varphi_\lambda(x): \Lambda \times X \to Y$ is continuous, the family $\{\varphi_\lambda\}$ defines a map $\varphi: \Lambda^+ \wedge X \to Y$ by $\varphi(\lambda, x) = \varphi_\lambda(x)$. Then $\varphi_\lambda = \varphi \circ j_\lambda$, where $j_\lambda: X \to \Lambda^+ \wedge X$ is defined by $j_\lambda(x) = (\lambda, x)$. Since $\varphi_\lambda^* = j_\lambda^* \circ \varphi^*$, the proof will be complete if we show that $j_\lambda^* u \in \tilde{H}(X)$ is locally constant in λ for $u \in \tilde{H}(\Lambda^+ \wedge X)$.

Let $\pi: \Lambda^+ \wedge X \to X$ be the projection map $(\lambda, x) \to x: \Lambda \times X/\Lambda \times \{x_0\} \to X$. Note that $\pi \circ j_\lambda = 1: X \to X$ and so $j_\lambda^* \circ \pi^* = 1$. Hence, if $w = \pi^* \circ j_\lambda^* u - u$ for $u \in \tilde{H}(X)$, then $j_\lambda^* w = j_\lambda^* \circ \pi^* \circ j_\lambda^* u - j_\lambda^* u = 0$. Now j_λ is just the homeomorphism $X \to \{\lambda\} \times X$ followed by the inclusion $i_\lambda: \{\lambda\} \times X \to \Lambda \times X/\Lambda \times \{x_0\}$. So we also have $i_\lambda^* w = 0$—that is, $w \in \tilde{H}(\Lambda^+ \wedge X)$ restricts to zero in $\tilde{H}(\{\lambda\} \times X)$. It follows from continuity that w restricts to zero on some compact set K containing $\{\lambda\} \times X$ in its interior. Then $\{\lambda_1\} \times X \subset K$ for λ_1 in some neighbourhood U of λ and w restricts to zero in $\tilde{H}(\{\lambda_1\} \times X)$ for $\lambda_1 \in U$. This implies that $i_{\lambda_1}^* w = 0$ and, hence, that $i_{\lambda_1}^* u = i_{\lambda_1}^* \circ \pi^* \circ i_\lambda^* u = i_\lambda^* u$. ∎

Note that the above proposition applies, in particular, to each $\tilde{H}^p$ in a generalized reduced cohomology theory.

Homotopic maps provide a particular example of the above situation. We say $\varphi: X \to Y$ and $\psi: X \to Y$ are homotopic if there exists a parameterized family $\{\varphi_t: X \to Y\}_{t\in[0,1]}$ with $\varphi_0 = \varphi$ and $\varphi_1 = \psi$. Since $[0, 1]$ is connected we obviously have:

2.2. COROLLARY. *If $\varphi, \psi: X \to Y$ are homotopic, then $\varphi^* = \psi^*: \tilde{H}(Y) \to \tilde{H}(X)$ for any continuous confunctor $\tilde{H}$.*

2.3. Contractible spaces, deformation-retracts

A pointed space (X, x_0) is *contractible* if there is a homotopy connecting the identity map $1: X \to X$ to the constant map $x \to x_0: X \to X$.

More generally, if (X, Y) is a compact pointed pair then we say Y is a *deformation retract* of X if there is a homotopy $\{\varphi_t : X \to X\}$ with $\varphi_1 = 1 : X \to X$ and φ_0 a map of X onto Y which is the identity on Y (a retract of X onto Y).

PROPOSITION. *If Y is a deformation retract of X, then $i: Y \to X$ induces an isomorphism $i^* : \tilde{H}(X) \to \tilde{H}(Y)$ for any continuous cofunctor $\tilde{H}$. In particular, $\tilde{H}^p(X) = \tilde{H}^p(\text{pt}) = 0$ if X is contractible and $\{\tilde{H}^p\}$ is a generalized reduced cohomology theory.*

Proof. If φ_t is a homotopy with $\varphi_0 = 1$ and $\varphi_2 = \varphi$ a retract of X onto Y, then $\varphi^* = 1 : \tilde{H}(X) \to \tilde{H}(X)$ by the homotopy property. Now $\varphi : X \to X$ is the composition of $X \xrightarrow{\psi} Y \xrightarrow{i} X$ where ψ is φ considered as a map of X to Y and i is the inclusion. Since $\varphi^* = 1$ we have $\psi^* \circ i^* = 1$. We also have $i^* \circ \psi^* = 1$ since $\psi \circ i = 1 : Y \to Y$. Thus i^* is an isomorphism.

A space X is contractible if its base point is a deformation retract of X. Hence, $\tilde{H}^p(X) = \tilde{H}^p(\text{pt}) = 0$ for X contractible and $\{\tilde{H}^p\}$ a generalized reduced cohomology theory. That $\tilde{H}^p(\text{pt}) = 0$ follows from the exactness axiom applied to the pair (pt, pt). ■

2.4. Inverse limits

As we have already seen, the continuity axiom is a powerful one. Another description of this axiom is as follows:

PROPOSITION. *A cofunctor $\tilde{H}$ is continuous if and only if* $\varinjlim \tilde{H}(X_\alpha) = \tilde{H}(X)$ *whenever* $\varprojlim X_\alpha = X$ *for an inverse limit system of compact pointed spaces X_α.*

Proof. Our definition of continuity is obviously a special case of the condition above. Hence, we must prove that continuity implies that $\varinjlim \tilde{H}(X_\alpha) = \tilde{H}(X)$ when $\varprojlim X_\alpha = X$.

We give a hint and then leave the proof as an exercise: Embed each X_α in a contractible space K_α (such as the cone CX_α described below). Let $K = \pi_\alpha K_\alpha$ and identify X with a subspace of K using the natural maps $X \to X_\alpha \subset K_\alpha$. Now express X as the intersection of a directed family $\{Y_\alpha \subset K\}$, such that each Y_α contains a copy of X_α which is the intersection of a directed family of deformation retracts of Y_α. ■

2.5. Cones and suspension

Note that the suspension SX of Section 1.4 can be described as the pointed space obtained from $X \times [0, 1]$ by identifying the subspace

$$(\{x_0\} \times [0, 1]) \cup (X \times \{0\}) \cup (X \times \{1\})$$

to a point, where X is a pointed space with x_0 as base point.

The *cone* CX over X is obtained from $X \times [0, 1]$ by identifying $(\{x_0\} \times [0, 1]) \cup (X \times \{0\})$ to a point. If we identify X with its image $X \times \{1\}$ in CX, then SX can be described as CX/X. The advantage of this description of SX is that CX is a contractible space. In fact, if $\varphi_s(x, t) = (x, st)$ for $s \in [0, 1]$ then $s \to \varphi_s$ provides a homotopy between $\varphi_1 = 1$ and the constant map φ_0 of CX onto its base point.

PROPOSITION. *If $\{\tilde{H}^p\}$ is a reduced generalized cohomology theory and X a compact pointed space, then there is an isomorphism $\delta^p: \tilde{H}^p(X) \to \tilde{H}^{p+1}(SX)$ for each p.*

Proof. Since we have identified X with a subspace $(X \times \{1\})$ of CX, the exactness axiom yields an exact sequence

$$\ldots \to \tilde{H}^p(CX) \xrightarrow{i^*} \tilde{H}^p(X) \xrightarrow{\delta^p} \tilde{H}^{p+1}(CX/X) \xrightarrow{\pi^*} \tilde{H}^{p+1}(CX) \to \ldots$$

However, since CX is contractible, we have $\tilde{H}^p(CX) = 0$ for all p. Hence, δ^p is an isomorphism. Since $SX = CX/X$, the proof is complete. ■

2.6. COROLLARY. *If $\{\tilde{H}^p\}$ is a reduced generalized cohomology theory, then $\tilde{H}^p(S^n) = \tilde{H}^{p-n}(S^0)$ for $n \geqslant 0$ and p any integer.*

Proof. We simply proceed by induction using the above proposition and the fact that $SS^k = S^{k+1}$. ■

2.7. COROLLARY. *If $\{\tilde{H}^p\}$ is a reduced cohomology theory with coefficient group G, then $\tilde{H}^p(S^n) = 0$ if $p \neq n$ and $\tilde{H}^n(S^n) = G$.*

2.8. Retracts

Let $\varphi: X \to Y$ be a retract of the pointed space X onto a subspace Y and let $i: Y \to X$ be the inclusion. If $\{\tilde{H}^p\}$ is a generalized reduced cohomology theory, then the induced maps $i^*: \tilde{H}^p(X) \to \tilde{H}^p(Y)$ and $\varphi^*: \tilde{H}^p(Y) \to \tilde{H}^p(X)$ satisfy $i^* \circ \varphi^* = 1$ (since $\varphi \circ i = 1$). It follows that i^* is onto and $\tilde{H}^p(X) \simeq \ker i^* \oplus \tilde{H}^p(Y)$.

Now from the exactness of the sequence

$$\ldots \to \tilde{H}^{p-1}(X) \xrightarrow{i^*} \tilde{H}^{p-1}(Y) \xrightarrow{\delta^*} \tilde{H}^p(X/Y) \xrightarrow{\pi^*} \tilde{H}^p(X) \xrightarrow{i^*} \tilde{H}^p(Y) \to \ldots$$

and the fact that i^* is onto for all p, we conclude that $\delta^* = 0$ and $\pi^*: \tilde{H}^p(X/Y) \to \tilde{H}^p(X)$ is an isomorphism of $\tilde{H}^p(X/Y)$ onto $\ker i^*$. Hence,

PROPOSITION. *If $Y \subset X$ is a retract of the pointed compact space X, then*

$\tilde{H}^p(X)$ is naturally isomorphic to $\tilde{H}^p(X/Y) \oplus \tilde{H}^p(Y)$ for any generalized reduced cohomology theory $\{H^p\}$.

2.9. Relation to unreduced theories

Now suppose $\{H^p\}$ is a generalized cohomology theory (unreduced). We define a reduced theory $\tilde{H}^p$ by setting $\tilde{H}^p(X) = H^p(X, \{x_0\})$ for a compact pointed space X with base point x_0.

PROPOSITION. *The sequence $\{\tilde{H}^p\}$ is a generalized reduced cohomology theory. Furthermore, $\{H^p\}$ can be recovered from $\{\tilde{H}^p\}$ since $H^p(X, Y) = \tilde{H}^p(X^+/Y^+)(= \tilde{H}^p(X/Y)$ if $Y \neq \varnothing)$.*

Proof. That each $\tilde{H}^p$ is continuous follows immediately from the continuity of H^p. Exactness for $\{\tilde{H}^p\}$ obviously follows from exactness for $\{H^p\}$ once we verify that $\tilde{H}^p(X/Y) = H^p(X, Y)$ when $Y \neq \varnothing$. The last statement of the proposition also follows from this, since we can replace X and Y by X^+ and Y^+ if $Y = \varnothing$.

Let W be the space obtained from $X \times [0, 1]$ by identifying $Y \times \{0\}$ to a point y_0. We identify X/Y with the subspace $(X \times \{0\})/(Y \times \{0\})$ of W and X with the subspace $X \times \{1\}$ of W. The subspace $(Y \times [0, 1])/(Y \times \{0\})$ we label CY. Now the pair (W, CY) can be continuously deformed to the pair $(X \cup CY, CY)$. It can also be continuously deformed to the pair $(X/Y, \{y_0\})$. Furthermore, the continuity axiom for unreduced theories yields (as in Proposition 2.2) a version of the homotopy property for unreduced theories. It follows that

$$\tilde{H}^p(X/Y) = H^p(X/Y, \{y_0\}) \simeq H^p(X \cup CY, CY).$$

Then excision (with $K = X, L = CY$) yields

$$H^p(X \cup CY, CY) \simeq H^p(X, Y).$$

Hence,

$$\tilde{H}^p(X/Y) \simeq H^p(X, Y)$$

and the proof is complete. ∎

The above proposition, together with Proposition 1.9, yields a one to one correspondence between generalized cohomology theories and generalized reduced cohomology theories.

2.10. PROPOSITION. *Let $\{H^p\}$ be a generalized cohomology theory and $\{\tilde{H}^p\}$ the corresponding reduced theory. If X is a compact space and $x_0 \in X$, then there is a natural decomposition $H^p(X) = \tilde{H}^p(X) \oplus H^p(\text{pt})$, where X is* *considered a pointed space with base point x_0 in computing $\tilde{H}^p(X)$. If X is connected, then this decomposition is independent of x_0.*

Proof. Since $\{x_0\}$ is a retract of X we have (by the unreduced analogue of Proposition 2.8) that

$$H^p(X) = H^p(X, \{x_0\}) \oplus H^p(\{x_0\}) = \tilde{H}^p(X) \oplus H^p(\text{pt}).$$

That the decomposition is independent of $x_0 \in X$ if X is connected follows from continuity. ■

Note that if $\{H^p\}$ is a cohomology theory, then $H^p(\text{pt}) = 0$ for $p \neq 0$ and, hence, $H^p(X)$ and $\tilde{H}^p(X)$ differ only when $p = 0$.

2.11. Uniqueness

It turns out that for a given coefficient group G there is only one cohomology theory in the sense of 1.8. For cell complexes, the equivalence of any two theories is a standard construction, versions of which appear in [**8**] and [**18**]. For the general result, one simply expresses a given compact space as an inverse limit of cell complexes and uses the continuity axiom.

The situation for generalized cohomology theories is considerably more complicated. However, one can prove that if $\{H^p(., G)\}$ denotes the unique cohomology theory with coefficients G (i.e., the Čech theory of the next section) and if $\{h^p\}$ is a generalized cohomology theory with $h^p(X)$ a vector space over the rationals for each X, then

$$h^n(X) \simeq \bigoplus_{p+q=n} H^p(X, h^q(\text{pt}))$$

(cf. [7]). In other words, such a generalized theory is always constructed from the cohomology theory $\{H^p\}$ by mixing dimensions. The significance of the condition that $h^p(X)$ be a rational vector space for all X is this: Two generalized theories may differ in an essential way in how they see torsion in a space; a theory with values in a rational vector space ignores torsion altogether.

3. ČECH COHOMOLOGY

We mentioned in the previous section that any two cohomology theories with a given coefficient group are equivalent. However, one might ask whether one exists at all. In fact, Čech cohomology is such a theory. In this section we define Čech cohomology theory and indicate how the axioms can be proved. The details are left as exercises.

We also show how the Čech groups $H^0(X, \mathbb{Z})$ and $H^1(X, \mathbb{Z})$ relate to the structure of the Banach algebra $C(X)$.

3.1. Čech cochains

Let X be a compact space and $\mathcal{U} = \{U_1, \ldots, U_n\}$ a finite open cover of $\mathcal{U}$. If $\sigma = (s_0, \ldots, s_p)$ $(p \geqslant 0)$ is a $p+1$-tuple of integers in $\{1, \ldots, n\}$, we let $|\sigma| = U_{s_0} \cap \ldots \cap U_{s_p}$ and call $|\sigma|$ the *support* of σ. If $|\sigma| \neq \varnothing$ then σ is called a *p-simplex*. The set of all p-simplices is denoted $N^p(\mathcal{U})$ and the union, $N(\mathcal{U}) = \bigcup_{p \geqslant 0} N^p(\mathcal{U})$ is called the *nerve* of $\mathcal{U}$.

Definition. A *Čech p-cochain* $(p \geqslant 0)$ for the cover $\mathcal{U}$ and the coefficient group G is a function $f: N^p(\mathcal{U}) \to G$. The group (under pointwise addition) of all p-cochains is denoted $C^p(\mathcal{U}, G)$. We set $C^p(\mathcal{U}, G) = 0$ for $p < 0$.

If $Y \subset X$ is a closed subspace, then the *group* $C^p(\mathcal{U}, Y, G)$ *of relative p-cochains* is defined to be the subgroup of $C^p(\mathcal{U}, G)$ consisting of cochains f such that $f(\sigma) = 0$ when $|\sigma| \cap Y \neq \varnothing$.

Note that $C^p(\mathcal{U}, \varnothing, G) = C^p(\mathcal{U}, G)$ since the condition $|\sigma| \cap Y \neq \varnothing$ is vacuous if $Y = \varnothing$.

3.2. The coboundary operator

We define a group homomorphism $\delta^p: C^p(\mathcal{U}, G) \to C^{p+1}(\mathcal{U}, G)$ by $\delta^p = 0$ for $p < 0$ and, for $p \geqslant 0$,

$$\delta^p f(s_0, \ldots, s_{p+1}) = \sum_{j=0}^{p+1} (-1)^j f(s_0, \ldots, \hat{s}_j, \ldots, s_{p+1})$$

where $(s_0, \ldots, \hat{s}_j, \ldots, s_{p+1}) \in N^p(\mathcal{U})$ is obtained from $(s_0, \ldots, s_{p+1}) \in N^{p+1}(\mathcal{U})$ by deleting the jth entry. Note that

$$\delta^p \circ \delta^{p-1} f(s_0, \ldots, s_{p+1}) = \sum_{j=0}^{p+1} (-1)^j \delta^{p-1} f(s_0, \ldots, \hat{s}_j, \ldots, s_{p+1})$$

$$= \sum_{j=0}^{p+1} \sum_{k<j} (-1)^{j+k} f(s_0, \ldots, \hat{s}_k, \ldots, \hat{s}_j, \ldots, s_{p+1})$$

$$+ \sum_{j=0}^{p+1} \sum_{j<k} (-1)^{j+k-1} f(s_0, \ldots, \hat{s}_j, \ldots, \hat{s}_k, \ldots, s_{p+1}) = 0$$

and, hence, $\operatorname{im} \delta^{p-1} \subset \ker \delta^p$.

If $Y \subset X$ is a closed subspace of X and $f \in C^p(\mathcal{U}, Y, G)$, then $\delta^p f \in C^{p+1}(\mathcal{U}, Y, G)$, since $|(s_0, \ldots, s_{p+1})| \cap Y \neq \varnothing$ implies $|(s_0, \ldots, \hat{s}_j, \ldots, s_{p+1})| \cap Y \neq \varnothing$ for each j. Thus, δ^p is also a homomorphism of $C^p(\mathcal{U}, Y, G)$ into $C^{p+1}(\mathcal{U}, Y, G)$.

Definition. We set

$$Z^p(\mathcal{U}, Y, G) = \ker \{\delta^p: C^p(\mathcal{U}, Y, G) \to C^{p+1}(\mathcal{U}, Y, G)\},$$

$$B^p(\mathscr{U}, Y, G) = \operatorname{im}\{\delta^{p-1}: C^{p-1}(\mathscr{U}, Y, G) \to C^p(\mathscr{U}, Y, G)\}, \text{ and}$$

$$H^p(\mathscr{U}, Y, G) = Z^p(\mathscr{U}, Y, G)/B^p(\mathscr{U}, Y, G).$$

Elements of $Z^p(\mathscr{U}, Y, G)$ are called *cocycles* while elements of $B^p(\mathscr{U}, Y, G)$ are called *coboundaries*.

Note that $H^p(\mathscr{U}, Y, G) = 0$ for $p < 0$ since $C^p(\mathscr{U}, Y, G) = 0$.

3.3. Refinement

Now suppose that we have two open covers $\mathscr{U} = \{U_1, \ldots, U_n\}$ and $\mathscr{V} = \{V_1, \ldots, V_m\}$ with $\mathscr{V}$ a refinement of $\mathscr{U}$—that is, each V_i is a subset of some U_j. We shall define a homomorphism from $H^p(\mathscr{U}, Y, G)$ to $H^p(\mathscr{V}, Y, G)$.

Let $r: \{1, \ldots, m\} \to \{1, \ldots, n\}$ be a function such that $V_1 \subset U_{r(i)}$ for each i. Note that if $\sigma = (s_0, \ldots, s_p) \in N^p(\mathscr{V})$, then $r\sigma = (r(s_0), \ldots, r(s_p)) \in N^p(\mathscr{U})$ and, in fact, $|\sigma| \subset |r\sigma|$. Hence, if $|\sigma| \cap Y \neq \varnothing$, also $|r\sigma| \cap Y \neq \varnothing$. It follows that $\tilde{r}f(\sigma) = f(r\sigma)$ defines a homomorphism $\tilde{r}: C^p(\mathscr{U}, Y, G) \to C^p(\mathscr{V}, Y, G)$.

Now it is easy to see that $\tilde{r}\delta^p = \delta^p\tilde{r}$ and, hence, that $\tilde{r}$ maps $Z^p(\mathscr{U}, Y, G)$ into $Z^p(\mathscr{V}, Y, G)$ and $B^p(\mathscr{U}, Y, G)$ into $B^p(\mathscr{V}, Y, G)$. We conclude that $\tilde{r}$ induces a homomorphism

$$r^*: H^p(\mathscr{U}, Y, G) \to H^p(\mathscr{V}, Y, G).$$

PROPOSITION. *The refinement homomorphism r^* is independent of the choice of r.*

Proof. Suppose r_1 is another choice. We define a map $\lambda^p: C^{p+1}(\mathscr{U}, Y, G) \to C^p(\mathscr{V}, Y, G)$ for each $p \geq 0$ *by*

$$\lambda^p f(s_0, \ldots, s_p) = \sum_{j=0}^{p} (-1)^j f(r(s_0), \ldots, r(s_j), r_1(s_j), \ldots, r_1(s_p))$$

and for $p < 0$ by $\lambda^p = 0$. One easily computes that

$$\tilde{r}_1 - \tilde{r} = \lambda^p \circ \delta^p + \delta^{p-1} \circ \lambda^{p-1}$$

Hence, if $f \in Z^p(\mathscr{U}, Y, G)$, then $\delta^p f = 0$ and $(\tilde{r}_1 - \tilde{r})f = \delta^{p-1} \circ \lambda^{p-1} f$—that is, $(\tilde{r}_1 - \tilde{r})f \in B^p(\mathscr{V}, Y, G)$. It follows that $r_1^* - r^*$ is the zero map. ■

3.4. COROLLARY. *Suppose $\mathscr{U}$ and $\mathscr{V}$ are refinements of one another. Then the map $r^*: H^p(\mathscr{U}, Y, G) \to H^p(\mathscr{V}, Y, G)$ is an isomorphism.*

Proof. Let r be as before and choose $t: \{1, \ldots, n\} \to \{1, \ldots, m\}$ such that $U_i \subset V_{t(i)}$ for each i. Then $r \circ t: \{1, \ldots, n\} \to \{1, \ldots, n\}$ expresses $\mathscr{U}$ as a refinement of itself. Hence, by Proposition 3.3 $r^* \circ t^* = (t \circ r)^* = 1^* = 1$. Similarly, $t^* \circ r^* = 1$ and, hence, r^* is an isomorphism. ■

3.5. The Čech groups

The collection of finite open covers of X is directed by refinement. With this as index set, the groups $H^p(\mathscr{U}, Y, G)$ form a direct system of groups (with the maps r^* as bonding maps). Hence, we may pass to the direct limit and define

$$H^p(X, Y, G) = \varinjlim H^p(\mathscr{U}, Y, G).$$

This is the pth *Čech cohomology group* of the pair (X, Y) with coefficients in G.

We would like $H^p(\cdot, \cdot, G)$ to be a cofunctor on compact pairs. Thus, for a map $\varphi\colon (X, Y) \to (X', Y')$ we must define an induced map

$$\varphi^*\colon H^p(X', Y', G) \to H^p(X, Y, G).$$

Let $\mathscr{U} = \{U_1, \ldots, U_n\}$ be an open cover of X' and note that $\varphi^{-1}\mathscr{U} = \{\varphi^{-1}U_1, \ldots, \varphi^{-1}N_n\}$ is an open cover of X. Also $U_{s_0} \cap \ldots \cap U_{s_p} \neq \varnothing$ whenever $\varphi^{-1}U_{s_0} \cap \ldots \cap \varphi^{-1}U_{s_p} \neq \varnothing$. Hence, $N^p(\varphi^{-1}(\mathscr{U}))$ is a subset of $N^p(\mathscr{U})$ and the restriction of $f \in C^p(\mathscr{U}, G)$ to $N^p(\varphi^{-1}(\mathscr{U}))$ defines an element $\tilde{\varphi} f \in C^p(\varphi^{-1}(\mathscr{U}), G)$. Since $\varphi^{-1}U_{s_0} \cap \ldots \cap \varphi^{-1}U_{s_p} \cap Y \neq \varnothing$ implies $U_{s_0} \cap \ldots \cap U_{s_p} \cap Y' \neq \varnothing$ we have that $\tilde{\varphi} f \in C^p(\varphi^{-1}(\mathscr{U}), Y, G)$ whenever $f \in C^p(\mathscr{U}, Y', G)$. Thus, $\tilde{\varphi}$ defines a homomorphism of $C^p(\mathscr{U}, Y', G)$ into $C^p(\varphi^{-1}(\mathscr{U}), Y, G)$.

It is easy to see that φ commutes with δ and, hence, defines a homomorphism φ^* of $H^p(\mathscr{U}, Y', G)$ into $H^p(\varphi^{-1}(\mathscr{U}), Y, G)$. Furthermore if $\mathscr{V}$ is a refinement of $\mathscr{U}$ then $\varphi^{-1}(\mathscr{V})$ is a refinement of $\varphi^{-1}(\mathscr{U})$ and the diagram

$$\begin{array}{ccc} H^p(\mathscr{U}, Y', G) & \xrightarrow{\varphi^*} & H^p(\varphi^{-1}(\mathscr{U}), Y, G) \\ \downarrow{\scriptstyle r^*} & & \downarrow{\scriptstyle r^*} \\ H^p(\mathscr{U}, Y', G) & \xrightarrow{\varphi^*} & H^p(\varphi^{-1}(\mathscr{V}), Y, G) \end{array}$$

commutes. That is, φ^* defines a homomorphism of directed systems of groups. and, on passing to the limit, a homomorphism $\varphi^*\colon H^p(X', Y', G) \to H^p(X, Y, G)$.

Obviously $1^* = 1$ and $\varphi \to \varphi^*$ behaves properly under composition. Hence, $(X, Y) \to H^p(X, Y, G)$ and $\varphi \to \varphi^*$ define a cofunctor from compact pairs to abelian groups.

3.6. The axioms

We leave most of the work of verifying that Čech cohomology satisfies the axioms of a cohomology theory for compact pairs as exercises. However, an outline of the proof of exactness will be given in the next subsection and comments (to be regarded as hints) on the other axioms will be given below (for complete proofs see [**8**] or [**18**]).

For an open cover $\mathcal{U} = \{U_1, \ldots, U_n\}$ of X and a subset $Y \subset X$, we let $\mathcal{U} \cap Y = \{U_1 \cap Y, \ldots, U_n \cap Y\}$ be the corresponding open cover of Y.

The following lemma has an elementary proof and is useful in verifying the axioms:

LEMMA. (a) *If* (X, Y) *is a compact pair and* $\mathcal{U}$ *is an open cover of* X, *then for each refinement* $\mathcal{W}$ *of* $\mathcal{U} \cap Y$ *there is a refinement* $\mathcal{V}$ *of* $\mathcal{U}$ *such that* $\mathcal{V} \cap Y$ *refines* $\mathcal{W}$;

(b) *for each open cover* $\mathcal{U}$ *of* X, *there is a refinement* $\mathcal{W}$ *of* $\mathcal{U}$ *and a neighbourhood* V *of* Y *such that* $\mathcal{W} \cap V$ *and* $\mathcal{W} \cap Y$ *have the same nerve.*

The continuity axiom follows very quickly from (a) and (b) of the above lemma.

One proves exision by first proving a weaker version (that $H^p(X, L) = H^p(X/U, L/U)$ when $\mathcal{U}$ is open and $\mathcal{U} \subset \bar{\mathcal{U}} \subset \operatorname{int} L$) and then applying continuity.

The dimension axiom (that $H^p(\text{pt}, G) = G$ for $p = 0$ and 0 otherwise) follows from Corollary 3.4, which implies that $H^p(\text{pt}, G) = H^p(\mathcal{U}, G)$ for the open cover $\mathcal{U}$ consisting of the single open set $\{\text{pt}\}$.

3.7. Exactness

The exactness axiom is proved by first deriving a version for the groups $H^p(\mathcal{U}, Y, G)$ and then passing to the limit. This procedure works because direct limits of exact sequences are exact. The following proposition takes care of the first part of this program:

PROPOSITION. *For a compact pair* (X, Y) *and an open cover* $\mathcal{U}$ *of* X, *there is an exact sequence*

$$\ldots \to H^p(\mathcal{U}, Y, G) \xrightarrow{\pi^*} H^p(\mathcal{U}, G) \xrightarrow{i^*} H^p(\mathcal{U} \cap Y, G) \xrightarrow{\delta^*} H^{p+1}(\mathcal{U}, Y, G) \to \ldots$$

Outline of Proof: We obtain an onto map of $i: C^p(\mathcal{U}, G) \to C^p(\mathcal{U} \cap Y, G)$ by letting of $(\sigma) = f(\sigma)$ on those $\sigma \in N^p(\mathcal{U})$ which are also in $N^p(\mathcal{U} \cap Y)$—that is, those for which $|\sigma| \cap Y \neq \varnothing$. Note that, by definition, $C^p(\mathcal{U}, Y, G)$ is precisely the kernel of i. Hence, if $\pi: C^p(\mathcal{U}, Y, G) \to C^p(\mathcal{U}, G)$ is the inclusion, then

$$0 \to C^p(\mathcal{U}, Y, G) \xrightarrow{\pi} C^p(\mathcal{U}, G) \xrightarrow{i} C^p(\mathcal{U} \cap Y, G) \to 0$$

is exact.

Now both π and i commute with δ and, hence, we have a commutative diagram

$$
\begin{array}{ccccccc}
 & 0 & & 0 & & 0 & \\
 & \downarrow & & \downarrow & & \downarrow & \\
\cdots \to & C^{p-1}(\mathcal{U}, Y, G) & \xrightarrow{\delta} & C^{p}(\mathcal{U}, Y, G) & \xrightarrow{\delta} & C^{p+1}(\mathcal{U}, Y, G) & \longrightarrow \cdots \\
 & \downarrow \pi & & \downarrow \pi & & \downarrow \pi & \\
\cdots \to & C^{p-1}(\mathcal{U}, G) & \xrightarrow{\delta} & C^{p}(\mathcal{U}, G) & \xrightarrow{\delta} & C^{p+1}(\mathcal{U}, G) & \longrightarrow \cdots \\
 & \downarrow i & & \downarrow i & & \downarrow i & \\
\cdots \to & C^{p-1}(\mathcal{U} \cap Y, G) & \xrightarrow{\delta} & C^{p}(\mathcal{U} \cap Y, G) & \xrightarrow{\delta} & C^{p+1}(\mathcal{U} \cap Y, G) & \longrightarrow \cdots \\
 & \downarrow & & \downarrow & & \downarrow & \\
 & 0 & & 0 & & 0 &
\end{array}
$$

with exact columns. To define δ^* one starts with $f \in Z^p(\mathcal{U} \cap Y, G)$, backs up to $g \in C^p(\mathcal{U}, G)$ with $ig = f$, hits g with δ and notes $i\delta g = \delta ig = \delta f = 0$, backs up to $h \in C^{p+1}(\mathcal{U}, Y, G)$ with $\pi ih = \delta g$, notes $\delta h = 0$, and lets $f \to h$ define the map $\delta^*: H^p(U \cap Y, G) \to H^{p+1}(U, Y, G)$. Similar chasing around the diagram proves that δ^* is well defined and makes the sequence of the proposition exact. ■

3.8. Description of $H^0(X, \mathbb{Z})$

Let X be a compact space and $f: X \to \mathbb{Z}$ a continuous integer valued function. Then f takes on only finitely many values $j_1, \ldots, j_n$. We set $U_i = f^{-1}(j_i)$ and note that $\mathcal{U} = (U_1, \ldots, U_n)$ is a disjoint open cover of X. Since f assigns a fixed integer j_i to each set U_i, it determines an element $c_f \in C^0(\mathcal{U}, \mathbb{Z})$. Since $U_i \cap U_j = \varnothing$ for $i \neq j$, we have $\delta(C^0(\mathcal{U}, \mathbb{Z})) = 0$ and, hence, $c_f \in Z^0(\mathcal{U}, \mathbb{Z}) = H^0(\mathcal{U}, \mathbb{Z})$. We let u_f denote the class of c_f in $H^0(X, \mathbb{Z})$.

PROPOSITION. *The map $f \to u_f$ is an isomorphism of the group of continuous integer valued functions on X onto the group $H^0(X, \mathbb{Z})$.*

Proof. The map $f \to u_f$ is certainly one to one since $f \neq 0$ implies $c_f \neq 0$ and, hence, $u_f \neq 0$ (the refinement maps are all one to one in dimension zero).

To prove that $f \to u_f$ is onto, let u be an element of $H^0(X, \mathbb{Z})$ and $c \in Z^0(\mathcal{V}, \mathbb{Z})$ a representative of u for some cover $\mathcal{V}$. Then $\delta c(i, j) = c(j) - c(i) = 0$ whenever $(i, j) \in N^1(\mathcal{V})$—that is, whenever $V_i \cap V_j \neq \varnothing$. It follows that, if we set $f(x) = c(i)$ for $x \in V_i$, then f is a well defined integer valued function on X. Furthermore, $\mathcal{V}$ refines the cover $\mathcal{U}$ defined as above by f and c is the image under refinement of c_f. Hence, $u = u_f$ and $f \to u_f$ is onto. ■

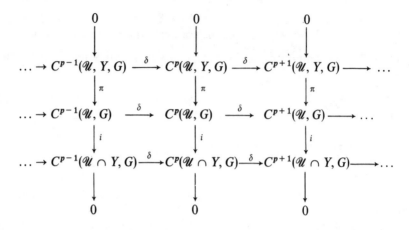

3.9. Description of $H^1(X, \mathbb{Z})$

Let $C(X)^{-1}$ denote the multiplicative group of invertible elements of $C(X)$—that is, the group of all non-vanishing continuous functions. Let $\exp(C(X))$ denote the subgroup of $C(X)^{-1}$ consisting of functions with continuous logarithms—that is, the range of the exponential map $f \to e^f : C(X) \to C(X)^{-1}$.

PROPOSITION. *There is a natural isomorphism from* $C(X)^{-1}/\exp(C(X))$ *onto* $H^1(X, \mathbb{Z})$.

Proof. If $f \in C(X)^{-1}$ we can choose a finite open cover $\mathscr{U} = \{U_1, \ldots, U_n\}$ such that, for each i, $f(U_i)$ is contained in a disc in $\mathbb{C}$ which does not contain zero. Then on each U_i one can construct a logarithm g_i for f by composing $f|U_i$ with some branch of the log function. For $(i, j) \in N^1(\mathscr{U})$ we have $e^{g_i} = f = e^{g_j}$ on $U_i \cap U_j$. It follows that $g_j - g_i = 2\pi i\, c(i, j)$ for some integer $c(i, j)$. Hence, we have defined an element $c \in C^1(U, \mathbb{Z})$. Note that

$$\delta c(i, j, k) = c(j, k) - c(i, k) + c(i, j)$$

$$= \frac{1}{2\pi i}[g_k - g_j) - (g_k - g_i) + (g_j - g_i)] = 0$$

on $U_i \cap U_j \cap U_k$ for any $(i, j, k) \in N^2(\mathscr{U})$. Hence, $c \in Z^1(\mathscr{U}, \mathbb{Z})$ and c determines an element $u_f \in H^1(X, \mathbb{Z})$.

We must prove that $f \to u_f$ is well defined. Thus, suppose $\mathscr{U}'$ is another cover and $\{g_i' \in C(U_i')\}$ another set of logarithms for f. Now we may as well assume $\mathscr{U} = \mathscr{U}'$ since, otherwise, we may pass to a common refinement of $\mathscr{U}$ and $\mathscr{U}'$. Thus, we assume $e^{g_i} = e^{g_i'} = f$ on $\mathscr{U}_i$ for $i = 1, \ldots, n$. Then $g_i - g_i' = 2\pi i k(i)$ for $k \in C^0(\mathscr{U}, \mathbb{Z})$. If $c' \in Z^1(\mathscr{U}, \mathbb{Z})$ is constructed from $\{g_i'\}$ as c was for $\{g_i\}$, then

$$c(i, j) - c'(i, j) = \frac{1}{2\pi i}[(g_j - g_i) - (g_j' - g_i')] = k(j) - k(i) = \delta k(i, j).$$

Hence, $c - c' \in B^1(\mathscr{U}, \mathbb{Z})$ and c and c' determine the same element of $H^1(X, \mathbb{Z})$.

Note that if f has a global logarithm g, then we may choose $g_i = g|U_i$ for each i and, hence, $u_f = 0$. Since the kernel of $f \to u_f$ contains $\exp(C(X))$, it determines a well defined homomorphism from $C(X)^{-1}/\exp(C(X))$ into $H^1(X, \mathbb{Z})$. We complete the argument by proving that this map is one to one and onto.

For $\mathscr{U} = \{U_1, \ldots, U_n\}$ an open cover of X, let $\{\varphi_i\}$ be a partition of unity for $\mathscr{U}$. That is, each $\varphi_i : X \to [0, 1]$ is continuous and has support in U_i and $\sum_{i=1}^{n} \varphi_i = 1$. If $c \in Z^1(\mathscr{U}, \mathbb{Z})$ is a representative of an element $u \in H^1(X, \mathbb{Z})$, then we define continuous functions $g_i \in C(U_i)$ by

$$g_i(x) = 2\pi i \sum_j c(j, i)\varphi_j(x) \qquad (x \in U_i).$$

Note that if $U_i \cap U_j = \varnothing$ (so $c(j, i)$ is not defined), then $x \notin U_j$, $\varphi_j(x) = 0$, and we interpret $c(j, i)\varphi_j(x)$ to mean zero.

For $(i, j) \in N^1(U)$ and $x \in U_i \cap U_j$ we have

$$g_j(x) - g_i(x) = 2\pi i \sum_k \varphi_k(x)\,[c(k, j) - c(k, i)] = 2\pi i \sum_k \varphi_k(x)c(i, j) = 2\pi i\, c(i, j),$$

since $\delta c = 0$. It follows that the element $u \in H^1(X, \mathbb{Z})$ determined by c is u_f for the function $f \in C(X)^{-1}$ defined by $f(x) = e^{g_i(x)}$ for $x \in U_i$. Hence $f \to u_f$ is onto.

Now suppose that $u_f = 0$ for some $f \in C(X)^{-1}$. By passing to a refinement, we may assume that this means that the element $c \in Z^1(\mathcal{U}, \mathbb{Z})$, constructed as above for f, satisfies $c = \delta k$ for some $k \in C^0(\mathcal{U}, \mathbb{Z})$. Thus, for $(i, j) \in N^1(\mathcal{U})$,

$$g_j - g_i = 2\pi i\, c(i, j) = 2\pi i (k(j) - k(i))$$

on $U_i \cap U_j$. Hence,

$$g_j - 2\pi i\, k(j) = g_i - 2\pi i\, k(i) \text{ on } U_i \cap U_j,$$

and we may define a global logarithm g for f by setting $g(x) = g_i(x) + 2\pi i\, k(i)$ for $x \in U_i$. Thus, $f \in \exp(C(X))$ and the proof is complete. ∎

3.10. The exponential map

Note that $f \in C(X)$ is integer valued if and only if $e^{2\pi i f} = 1$. If we observe that $f \to e^{2\pi i f}$ is a homomorphism from the additive group $C(X)$ to the multiplicative group $C(X)^{-1}$, then the results of the preceding two subsections may be interpreted as follows:

PROPOSITION. *The groups $H^0(X, \mathbb{Z})$ and $H^1(X, \mathbb{Z})$ are naturally isomorphic to the kernel and cokernel, respectively, of the exponential map $f \to e^{2\pi i f}: C(X) \to C(X)^{-1}$.*

4. IDEMPOTENTS AND LOGARITHMS

We now turn to the main question of these notes: for a commutative Banach algebra A with maximal ideal space Δ, how are the topological invariants of Δ (such as cohomology groups) related to the structure of A? For the low dimensional Čech cohomology groups $H^0(\Delta, \mathbb{Z})$ and $H^1(\Delta, \mathbb{Z})$ the answer is provided by theorems of Shilov, Arens, and Royden. In this section we describe these results and also lay the groundwork for the more general results on K-theory which will follow in later sections.

In this section A will denote a unital Banach algebra (a Banach algebra with identity). A unital homomorphism $\varphi: A \to B$ is one such that $\varphi(1) = 1$.

4.1. The exponential map

Let A^{-1} denote the group of invertible elements of A. For each $a \in A$ the element $e^a = \sum_{n=0}^{\infty} a^n/n!$ is in A^{-1} and has inverse e^{-a}. Thus, $a \to e^a$ is a map of A into A^{-1}.

Now an easy computation shows that $e^{a+b} = e^a e^b$ whenever a and b commute. In particular, if A is commutative then $a \to e^{2\pi i a}$ is a homomorphism from the group (under addition) A to the group (under multiplication) A^{-1}. In this case, the kernel $\{a: e^{2\pi i a} = 1\}$ and image of $a \to e^{2\pi i a}$ are subgroups of A and A^{-1}, respectively.

In general, whether A is commutative or not, we make the following definition:

Definition. We set $Q(A) = \{a \in A: e^{2\pi i a} = 1\}$ and let $\exp(A)$ be the subgroup of A^{-1} generated by elements of the form e^a for $a \in A$.

4.2. Idempotents and $Q(A)$

If $p \in A$ is idempotent ($p^2 = p$) then $p \in Q(A)$. In fact,

$$e^{2\pi i p} = 1 + \left(\sum_{1}^{\infty} (2\pi i)^n/n!\right) p = 1 + (e^{2\pi i} - 1)p = 1.$$

Moreover, we have:

PROPOSITION. (a) *Each sum $\sum n_i p_i$, with $p_1, \ldots, p_k$ a commuting set of idempotents and $n_1, \ldots, n_k$ a set of integers, belongs to $Q(A)$;*

(b) *if $a \in Q(A)$ then a has a unique representation as $a = \sum_{j=-\infty}^{\infty} j p_j$, where each p_j is idempotent, $p_j p_k = 0$ for $j \neq k$, and $\sum p_j = 1$. (Necessarily, only finitely many p_j's are non-zero). Furthermore, the p_j's depend continuously on $a \in Q(A)$ and commute with each element of A that commutes with a.*

Proof. Part (a) is evident since $e^{\hat{a}+b} = e^{\hat{a}} e^b$ if a and b commute.

To prove part (b) we use spectral theory and the functional calculus. For $a \in Q(A)$ let B be the closed subalgebra of A generated by a. Then B is a commutative Banach algebra containing a. If $\hat{a}$ is the Gelfand transform of a as an element of B then $\hat{a}$ satisfies $e^{2\pi i \hat{a}} = (e^{2\pi i a})\hat{} = 1$ and, hence, $\hat{a}$ is an integer valued function. It follows that the spectrum of a is a finite subset of the integers.

For each integer j we let D_j be the disc of radius $\frac{1}{4}$ centred at j and we set $D = \bigcup_{j \in \mathbb{Z}} D_j$. For any function f, analytic in a neighbourhood of D, the expression

$$f(a) = \frac{1}{2\pi i} \int_{\partial D} f(z)(z - a)^{-1}\, dz$$

makes sense (because only finitely many D_j's contain points of the spectrum of a) and defines a homomorphism $f \to f(a): \mathcal{O}(D) \to B$ mapping 1 to 1 and z to a. Here, $\mathcal{O}(D)$ is the algebra of functions analytic in a neighbourhood of D. Note that for fixed $f \in \mathcal{O}(D)$ the element $f(a) \in A$ depends continuously on $a \in Q(A)$.

We let q_j be an analytic function which is one in a neighbourhood of D_j and zero in a neighbourhood of $D \backslash D_j$. We set

$$p_j = q_j(a) = \frac{1}{2\pi i} \int_{\partial D} q_j(z)(z - a)^{-1}\, dz = \frac{1}{2\pi i} \int_{\partial D_j} (z - a)^{-1}\, dz.$$

Then, since $q_j q_k = \delta_{jk} q_j$ and $f \to f(a)$ is a homomorphism, $p_j p_k = \delta_{jk} p_j$. Similarly, $\sum q_j = 1$ in a neighbourhood of D and so $\sum p_j = 1$ in B.

Now if $f = \sum j q_j$ then $f(a) = \sum j p_j$. Furthermore, $z - f = 0$ on the integers and, hence, on the spectrum of a. It follows that $(z - f)(a) = a - f(a)$ is in the radical of B (since the identity $f(\hat{a}) = f \circ \hat{a} = \hat{a}$ holds for Gelfand transforms in the subalgebra B). Since a and $f(a)$ are commuting elements of $Q(A)$, we have $a - f(a) \in Q(A)$. That $a = f(a) = \sum j p_j$ now follows from the next lemma.

If $a = \sum j p'_j$ is any decomposition of a with $p'_j p'_k = \delta_{jk} p'_j$ and $\sum p'_j = 1$, then $(z - a)^{-1} = \sum (z - j)^{-1} p'_j$. It follows from the formula

$$p_j = \frac{1}{2\pi i} \int_{\partial D_j} (z - a)^{-1}\, dz,$$

defining p_j, that $p_j = p'_j$. Hence the decomposition is unique. That each p_j commutes with every element of A that commutes with a is also obvious from the defining formula. ■

Definition. A system $\{p_j\}$ of idempotents such that $p_j p_k = 0$ for $j \neq k$ and $\sum p_j = 1$ will be called a *partition of unity* in A. For $a \in Q(A)$ the unique representation $a = \sum j p_j$, with $\{p_j\}$ a partition of unity, will be called the *spectral decomposition of* a.

4.3. LEMMA. *If a is quasi-nilpotent (has spectral radius zero) and $a \in Q(A)$ then* $a = 0$.

Proof. If $a \in Q(A)$ then $e^b = 1$ where $b = 2\pi i a$. Hence,

$$\sum_1^\infty b^n/n! = e^b - 1 = 0$$

and $b(1 + c) = 0$ where

$$c = \sum_2^\infty b^{n-1}/n .$$

Now if a is quasi-nilpotent then it is in the radical of the commutative subalgebra B that it generates; but then b and c are also. This implies that $1 + c$ is invertible and, hence, that $b(1 + c) = 0$ can occur only if $b = a = 0$. ∎

4.4. The Shilov idempotent theorem

We now turn out attention to the case where A is a unital commutative Banach algebra with maximal ideal space Δ. Using the analytic functional calculus in several variables, Shilov [**17**] proved that for each open and closed subset K of Δ there is an idempotent $p \in A$ such that $\hat{p} = 1$ on K and $\hat{p} = 0$ on $\Delta \backslash K$. We won't prove this theorem here; we refer the reader to [**10**] for an excellent development of both the functional calculus and the Shilov idempotent theorem.

Now $Q(C(\Delta))$ consists exactly of those continuous functions f for which $e^{2\pi i f} = 1$, that is, those continuous functions which are integer valued. Such a function f has the form $f = \sum j\chi_{\Delta_j}$ where $\Delta_j = f^{-1}(\{j\})$ is open and closed in Δ and χ_{Δ_j} is its characteristic function. By Shilov's theorem, $\chi_{\Delta_j} = \hat{p}_j$ for an idempotent $p_j \in A$. It follows that $f = \hat{a}$ for the element $a = \sum jp_j \in Q(A)$.

Now since A is commutative, $Q(A)$ is closed under addition and, in fact, is a subgroup of the additive group of A. Note that if $\varphi: A \to B$ is any unital homomorphism of commutative unital Banach algebras, then φ maps $Q(A)$ homomorphically into $Q(B)$. In the particular case of the Gelfand transform $a \to \hat{a}: A \to C(\Delta)$ we have just seen that this map carries $Q(A)$ onto $Q(C(\Delta))$. By Lemma 4.3, it is one to one as well. In view of this and Proposition 3.8 we have the following interpretation of the Shilov idempotent theorem:

THEOREM. *The Gelfand transform yields an isomorphism of $Q(A)$ onto $Q(C(\Delta))$ and, hence, there is a natural isomorphism between $Q(A)$ and $H^0(\Delta, \mathbb{Z})$.*

This was the first result connecting the topology of Δ with the structure of A.

4.5. Direct sum decompositions

The main application of Shilov's idempotent theorem is in the decomposition of an algebra A into a direct sum.

If $A_1, \ldots, A_n$ are Banach algebras, then the direct sum $\overset{n}{\underset{1}{\oplus}} A_i$ is the algebra of n-tuples $(a_1, \ldots, a_n)$ $(a_i \varepsilon A_i)$ with coordinate-wise operations and norm defined by $\|(a_1, \ldots, a_n)\| = \max \|a_i\|$. Note that each A_i is embedded in $\overset{n}{\underset{1}{\oplus}} A_i$ as the ideal consisting of all n-tuples which are zero except in the ith coordinate. If A_i has an identity then its image $p_i \in \overset{n}{\underset{1}{\oplus}} A_i$ is an idempotent such that $A_i = p_i(\overset{n}{\underset{1}{\oplus}} A_i)$. If each A_i has an identity, then the corresponding system of idempotents $\{p_i\}_{i=1}^n$ is a partition of unity in A.

Conversely, given a partition of unity $\{p_i\}_{i=1}^n$ in the commutative Banach algebra A, the set of ideals $\{p_iA\}_{i=1}^n$ have the property that $\overset{n}{\underset{i=1}{\oplus}} p_iA$ is isomorphic to A. In view of the Shilov idempotent theorem, there is a one to one correspondence between partitions of unity, as above, and partitions $\{\Delta_i\}_{i=1}^n$ of Δ as the union of a pairwise disjoint family of open-closed sets. Putting all this together, we have:

COROLLARY. *Let A be a unital commutative Banach algebra with maximal ideal space Δ. Then for each partition*

$$\Delta = \bigcup_{i=1}^{n} \Delta_i$$

of Δ into the union of a family of pairwise disjoint open-closed sets, there exists a direct sum decomposition $A = \overset{n}{\underset{1}{\oplus}} A_i$, where for each i, $A_i = \{a \in A : \hat{a}(x) = 0$ for $x \notin \Delta_i\}$.

4.6. Logarithms

We now look at the other end of the exponential map: its range. Recall that $\exp(A)$ is the subgroup of A^{-1} generated by elements of the form e^a—that is, the subgroup generated by those elements of A^{-1} which have logarithms in A. Note that this subgroup consists of all finite products $e^{a_1}e^{a_2}\ldots e^{a_n}$. In fact, the set of such elements is closed under multiplication and also inversion, with $e^{-a_n}\ldots e^{-a_2}e^{-a_1}$ the inverse of $e^{a_1}e^{a_2}\ldots e^{a_n}$. If A is commutative, then $\exp(A) = \{e^a : a \in A\}$.

Now A^{-1} has a natural topology—the norm topology of A. With this topology A^{-1} is a topological group—that is, multiplication $(a, b) \rightarrow$

$ab: A^{-1} \times A^{-1} \to A^{-1}$ and inversion $a \to a^{-1}: A^{-1} \to A^{-1}$ are continuous.

In any topological group G, the component of the identity (largest connected set containing the identity) is always a closed normal subgroup H. That H is a subgroup follows from the fact that $H \cdot H$ and H^{-1} are connected sets containing the identity and, hence, are contained in H. Similarly $xHx^{-1} \subset H$ for each $x \in G$ and H is normal.

For the group A^{-1} we have:

PROPOSITION. *The component of the identity in A^{-1} is open, arcwise connected, and equal to* $\exp(A)$.

Proof. The subgroup $\exp(A)$ is arcwise connected since each element $e^{a_1} \dots e^{a_n}$ is connected to 1 by the arc $\{e^{ta_1} \dots e^{ta_n}: t \in [0, 1]\}$. We shall prove that $\exp(A)$ is both open and closed in A^{-1}.

If $a \in A$ satisfies $\|1 - a\| < 1$ then the power series

$$-\sum_{1}^{\infty} \frac{1}{n}(1 - a)^n$$

converges to a logarithm for a. Hence, we have that there is a neighbourhood V of the identity in A^{-1} which belongs to $\exp(A)$. Since for each $b \in A^{-1}$ the map $a \to ab: A^{-1} \to A^{-1}$ is a homeomorphism (with inverse $a \to ab^{-1}$), we have that Vb is a neighbourhood of b which lies in $\exp(A)$ if $b \in \exp(A)$. Hence, $\exp(A)$ is open in A^{-1}. However, an open subgroup H of a topological group G is always closed as well (since the union of all left cosets of H which are distinct from H is open and is the complement of H).

Now since $\exp(A)$ is connected, it is contained in the component of the identity in A^{-1}. However, since it is both open and closed, it must contain this component. Hence, the two are equal. ∎

4.7. The Arens–Royden theorem

Since $\exp(A)$ is a normal subgroup of A^{-1}, we can pass to the quotient group $A^{-1}/\exp(A)$. Note that a unital homomorphism $A \to B$ maps A^{-1} into B^{-1} and $\exp(A)$ into $\exp(B)$. Hence, we always have an induced map $A^{-1}/\exp(A) \to B^{-1}/\exp(B)$.

Recall that if Δ is a compact Hausdorff space then $C(\Delta)^{-1}/\exp(C(\Delta))$ is naturally isomorphic to $H^1(\Delta, \mathbb{Z})$. (Proposition 3.9). If A is a unital commutative Banach algebra with maximal ideal space Δ, Arens [**2**] and Royden [**15**] used the several variable functional calculus and results from several complex variables to prove:

THEOREM. (Arens–Royden). *The Gelfand transform induces an isomorphism of* $A^{-1}/\exp(A)$ *onto* $C(\Delta)^{-1}/\exp(C(\Delta))$. *Hence,* $A^{-1}/\exp(A)$ *is naturally isomorphic to* $H^1(\Delta, \mathbb{Z})$.

This is the second main result relating the topology of Δ to the structure of A. A nice proof appears in [**10**].

4.8. Lifting invertible elements

We conclude this section by giving an example of one reason why the group $A^{-1}/\exp(A)$ is interesting to the Banach algebraist.

PROPOSITION. *Let A be a unital Banach algebra (not necessarily commutative). If $a \in A^{-1}$ then $a \in \exp(A)$ if and only if, for each onto unital Banach algebra homomorphism $B \to A$, a is the image of an invertible element of B.*

Proof. If $a = e^{a_1} \ldots e^{a_n} \in \exp(A)$ then we simply choose pre-images $b_1, \ldots, b_n \in B$ for $a_1, \ldots, a_n$ and note that a is the image of the invertible element $b = e^{b_1} \ldots e^{b_n} \in B$.

To prove the converse, we choose a special algebra homomorphism $B \to A$. We let B be the algebra CA of all continuous functions from $[0, 1]$ into A such that $f(0)$ is a multiple of the identity. This is a Banach algebra under pointwise operations and norm $\|f\| = \sup_t \|f(t)\|$. The map $f \to f(1)$ is an onto homomorphism from CA to A.

Now if $a \in A$ and $a = f(1)$ for some $f \in (CA)^{-1}$, then $\{f(t) : t \in [0, 1]\}$ is an arc in A^{-1} connecting a to a multiple of the identity. Hence, $a \in \exp(A)$. ■

4.9. COROLLARY. *If A is commutative then $H^1(\Delta, \mathbb{Z}) = 0$ if and only if every onto unital Banach algebra homomorphism $B \to A$ maps B^{-1} onto A^{-1}.*

5. THE FUNCTOR K_1

So far we have descriptions of the Čech cohomology groups $H^0(\Delta, \mathbb{Z})$ and $H^1(\Delta, \mathbb{Z})$ for the maximal ideal space of a commutative Banach algebra A. There is also a description of $H^2(\Delta, \mathbb{Z})$ due to Forster [**9**] that we shall discuss in Section 10. It turns out that descriptions, in terms of the structure of A, of the higher Čech cohomology groups of Δ are difficult to obtain directly (see [**20**] for a discussion of this problem). However, there is a generalized cohomology theory, K-theory, which has a very natural description in terms of A. One is led to this fact by a theorem of Arens [**3**] which generalizes the Arens–Royden theorem.

We begin our discussion of K-theory in this section with a description of Arens' theorem and a functor $A \to K_1(A)$ whose definition is motivated by this theorem.

5.1. The Arens theorem

Throughout most of this section A will be a unital commutative Banach algebra with maximal ideal space Δ. By $M_n(A)$ we shall mean the algebra of $n \times n$ matrices with entries from A. This can be made into a Banach algebra using any one of a variety of equivalent norms. One such norm is the operator norm obtained by considering $M_n(A)$ as an algebra of operators on the Banach space $\overset{n}{\oplus} A$ of n-tuples of elements of A with norm

$$\|(a_1, \ldots, a_n)\| = \|a_1\| + \ldots + \|a_n\|.$$

The group of invertible elements of $M_n(A)$ will be denoted $GL_n(A)$ while its component of the identity, $\exp(M_n(A))$, will be denoted $GL_n^0(A)$. We denote the factor group $GL_n(A)/GL_n^0(A)$ by $L_n(A)$.

Now if $\varphi: A \to B$ is a unital Banach algebra homomorphism then φ induces a homomorphism of $M_n(A)$ into $M_n(B)$ (just apply φ to each entry of a matrix) which maps $GL_n(A)$ into $GL_n(B)$ and $GL_n^0(A)$ into $GL_n^0(B)$. Hence, φ induces a homomorphism $\varphi^*: L_n(A) \to L_n(B)$. In other words, $A \to L_n(A)$ is a functor from unital commutative Banach algebras to groups.

THEOREM. (Arens [**2**]) *The Gelfand transform $a \to \hat{a}: A \to C(\Delta)$ induces an isomorphism of $L_n(A)$ onto $L_n(C(\Delta))$ for each n.*

The proof of this theorem uses the several variable analytic functional calculus and some deep results from several complex variables due to Grauert. Note that the case $n = 1$ is just the Arens–Royden theorem.

According to the above theorem, the groups $L_n(A)$ are invariants of the space Δ rather than the algebra A. They may, of course, be non-abelian groups. We shall fix this shortly, but first we develop some conditions under which elements of $GL_n(A)$ determine the same equivalence class in $L_n(A)$.

5.2 Elementary operations

We describe three kinds of elementary matrices which belong to $GL_n^0(A)$. Since the non-zero complex numbers are connected, a matrix of the form

$$\begin{pmatrix} 1 & 0 & . & . & . & 0 \\ 0 & 1 & & & & . \\ . & & . & & & . \\ . & & & z & & . \\ . & & & & . & . \\ 0 & . & . & . & . & 1 \end{pmatrix} \quad (z \in \mathbb{C}\backslash\{0\})$$

is connected to the identity by an arc in $GL_n(\mathbb{C})$ and, hence, is in $GL_n^0(\mathbb{C})$.

The 2×2 matrix $\begin{pmatrix} 0 & 1 \\ 1 & 0 \end{pmatrix}$ is connected to $\begin{pmatrix} -1 & 0 \\ 0 & 1 \end{pmatrix}$ by the arc

$$t \to \begin{pmatrix} -\cos t & \sin t \\ \sin t & \cos t \end{pmatrix} \qquad (0 \leqslant t \leqslant \pi/2)$$

which lies in $GL_2(\mathbb{C})$. Hence, $\begin{pmatrix} 0 & 1 \\ 1 & 0 \end{pmatrix} \in GL_2^0(\mathbb{C})$. Similarly, any elementary elementary interchange matrix in $GL_n(\mathbb{C})$ is in $GL_n^0(\mathbb{C})$, where by an elementary interchange matrix we mean one whose action on $\mathbb{C}^n$ simply interchanges two coordinates and leaves the others fixed.

An elementary shear matrix in $GL_n(A)$ is one of the form $I_n + ae_{ij}$ $(i \neq j)$, where $a \in A$ and e_{ij} is the matrix whose only non-zero entry is a one in the ij-th position. The arc $t \to I_n + tae_{ij}$ $(0 \leqslant t \leqslant 1)$ lies in $GL_n(A)$ and connects such a matrix to the identity. Hence, elementary shear matrices in $GL_n(A)$ belong to $GL_n^0(A)$.

PROPOSITION. *The equivalence class in $L_n(A)$ of an element of $GL_n(A)$ is left unchanged by*

(a) *multiplying a row or column by a non-zero scalar;*

(b) *interchanging two rows or columns;*

(c) *adding a multiple (by an element of A) of one row or column to another row or column.*

Proof. Each of these operations can be described as multiplying on the left or right by one of the elementary matrices in $GL_n^0(A)$ described above. ■

5.3. COROLLARY. *For each n, $L_n(\mathbb{C})$ is the trivial group.*

Proof. An invertible scalar matrix can always be transformed into the identity through a sequence of elementary operations as above. ■

5.4. The group $K_1(A)$

For matrices $a \in M_n(A)$ and $b \in M_k(A)$ we denote by $a \oplus b$ the matrix

$$\begin{pmatrix} a & 0 \\ 0 & b \end{pmatrix} \in M_{n+k}(A).$$

Note that $a \to a \oplus I_{m-n}$ yields a canonical homomorphism of $GL_n(A)$ into $GL_m(A)$ for each $m > n$ and hence, a homomorphism $L_n(A) \to L_m(A)$. For $a \in GL_n(A)$ we call $a \oplus I_{m-n}$ a *trivial extension* of a.

Now although $L_n(A)$ may not be abelian, its image in $L_m(A)$ is abelian for large enough m (in fact, for $m \geqslant 2n$). This is a result of the following proposition:

PROPOSITION. *If $a, b \in GL_n(A)$ then $ab \oplus I_n$, $ba \oplus I_n$, $a \oplus b$, and $b \oplus a$ are all equivalent mod $GL_{2n}^0(A)$.*

Proof. By interchanging rows and columns, we get $a \oplus b \sim b \oplus a$ for any pair $a, b \in GL_n(A)$. Hence, we have

$$\begin{aligned} ab \oplus I_n &= (a \oplus I_n)(b \oplus I_n) \sim (a \oplus I_n)(I_n \oplus b) \\ &= a \oplus b \sim b \oplus a = (b \oplus I_n)(I_n \oplus a) \\ &\sim (b \oplus I_n)(a \oplus I_n) = ba \oplus I_n. \end{aligned}$$ ■

Note that the canonical maps $L_n(A) \to L_m(A)$ $(n < m)$ make the sequence $\{L_n(A)\}$ into a direct limit system of groups. Furthermore, it follows from the above Proposition that the limit group $\varinjlim L_n(A)$ is abelian. This is the object we are interested in.

Definition. For each unital commutative Banach algebra A, we set $K_1(A) = \varinjlim L_n(A)$.

Note that if $\varphi: A \to B$ is a unital homomorphism, then the induced maps $GL_n(A) \to GL_n(B)$ commute with trivial extensions. Hence the maps $L_n(A) \to L_n(B)$ commute with the bonding maps for the direct limit systems and yield a homomorphism $\varphi^*: K_1(A) \to K_1(B)$. In other words, K_1 is a functor from unital commutative Banach algebras to abelian groups.

5.5. Non-unital algebras and K_1

Now in the category of all commutative Banach algebras there is no unital condition on algebras or homomorphisms. We shall extend K_1 to a functor $\tilde{K}_1$ on this larger category.

For any commutative Banach algebra A we let A^+ denote the algebra obtained by adjoining an identity 1. Then A is a maximal ideal of A^+ and the factor map $A \to A^+/A \cong \mathbb{C}$ determines a distinguished complex homomorphism φ_0 of A^+.

Since A is the kernel of $A^+ \xrightarrow{\varphi_0} \mathbb{C}$, it would be natural to let $\tilde{K}_1(A)$ be the kernel of $\varphi_0^*: K_1(A^+) \to K_1(\mathbb{C})$. However, $K_1(\mathbb{C}) = 0$ since $L_n(\mathbb{C}) = 0$ for all n. Thus, we simply set $\tilde{K}_1(A) = K_1(A^+)$.

Now a homomorphism $\varphi: A \to B$ of Banach algebras always determines a natural unital homomorphism $\varphi^+: A^+ \to B^+$ by $\varphi^+(a + \lambda) = \varphi(a) + \lambda$. Hence, $A \to A^+$ is a functor from commutative Banach algebras to unital commutative Banach algebras. Since $\tilde{K}_1$ is just the composition of this with K_1, we have that $\tilde{K}_1$ is a functor on commutative Banach algebras.

We next prove that $\tilde{K}_1$ extends K_1. Thus, suppose A already had an identity p. Then in A^+, p and $q = 1 - p$ are idempotents which determine an algebra direct sum decomposition $A^+ = A \oplus \mathbb{C}$. It follows easily that $K_1(A^+) = K_1(A) \oplus K_1(\mathbb{C})$. Since $K_1(\mathbb{C}) = 0$ we have $\tilde{K}_1(A) = K_1(A^+) = K_1(A)$.

5.6. Weak exactness

The functor $\tilde{K}_1$ satisfies the following exactness condition:

Proposition. *If I is a closed ideal of the commutative Banach algebra A, then the exact sequence $I \xrightarrow{i} A \xrightarrow{\pi} A/I$ induces an exact sequence $\tilde{K}_1(I) \xrightarrow{i^*} \tilde{K}_1(A) \xrightarrow{\pi} \tilde{K}_1(A/I)$.*

Proof. We extend i and π to unital maps $i: I^+ \to A^+$ and $\pi: A^+ \to (A/I)^+ = A^+/I$. The composition $\pi \circ i$ then maps I^+ to the multiples of the identity in A^+/I. Since $K_1(\mathbb{C}) = 0$ it follows that the composition of $i^*: K_1(I^+) \to K_1(A^+)$ and $\pi^*: K_1(A^+) \to K_1(A^+/I)$ is zero. Hence, $\operatorname{im} i^* \subset \ker \pi^*$ and we need only prove the reverse containment.

Thus, suppose $u \in K_1(A^+)$ and $\pi^* u = 0$. Then for some representative $a \in GL_n(A^+)$ of u we have $\pi a \in GL_n^0(A^+/I)$. That is, $\pi a = e^{c_1} \dots e^{c_m}$ for elements $c_1, \dots, c_m \in M_n(A^+/I)$. We choose pre-images $b_1, \dots, b_m \in M_n(A^+)$ for $c_1, \dots, c_m$ and observe that if $d = e^{-b_m} \dots e^{-b_1} a \in GL_n(A^+)$, then $\pi d = 1$. However, this means that d is in the image of I^+ in A^+. Since $a \sim d \bmod GL_n^0(A^+)$ we have that $u \in \operatorname{im} i^*$. ■

5.7. Remark

By the Arens Theorem we have that the Gelfand transform $A \to C(\Delta)$ induces an isomorphism $L_n(A) \to L_n(C(\Delta))$ for each n. Thus, we have:

Proposition. *The Gelfand transform $A \to C(\Delta)$ yields an isomorphism $K_1(A) \to K_1(C(\Delta))$ for unital algebras A and an isomorphism $\tilde{K}_1(A) \to \tilde{K}_1(C_0(\Delta))$ for all commutative Banach algebras A.*

This will allow us in Section 10 to define a cofunctor $\tilde{K}^1$ from compact pointed spaces to abelian groups such that $\tilde{K}^1(\Delta^+) = \tilde{K}_1(A)$ for any commutative Banach algebra A with maximal ideal space Δ. This turns out to be one of the cofunctors for a generalized cohomology theory (K-theory).

6. THE FUNCTOR K_0

The group $K_1(A)$ was constructed from the group $L_n(A)$ obtained from

$GL_n(A)$ by factoring out the subgroup generated by the image of the exponential map. In this section we use the kernel of the exponential map $M_n(A) \to GL_n(A)$ to construct a group $K_0(A)$.

Our construction and subsequent results are greatly facilitated by using some elementary facts about constructing a group from a semigroup.

6.1. The universal group of a semigroup

A semigroup is a set S together with a single associative operation $S \times S \to S$. If this operation is commutative then S is said to be *abelian*. Here S will denote an abelian semigroup with operation

$$(x, y) \to x + y: S \times S \to S.$$

There is a canonical way of constructing an abelian group $U(S)$ from S. We begin with pairs $(x, u) \in S \times S$ and introduce an equivalence relation $\sim$ by setting $(x, u) \sim (y, v)$ if there exists $r \in S$ such that

$$x + v + r = y + u + r.$$

This relation is obviously symmetric and reflexive. Furthermore, if $(x, u) \sim (y, v)$ and $(y, v) \sim (z, w)$ then we have

$$x + v + r = y + u + r$$

and

$$y + w + s = z + v + s$$

for some $r, s \in S$. It follows that

$$x + w + (v + r + s) = y + w + u + r + s = z + u + (v + r + s)$$

and, hence, $(x, u) \sim (z, w)$. We conclude that "$\sim$" is transitive and, hence, is an equivalence relation.

If we let $U(S)$ be the set of equivalence classes of pairs (x, u) under $\sim$, then $U(S)$ inherits an operation from S (coordinatewise addition) under which it is a group. In fact, all pairs (x, x) belong to a single equivalence class in $U(S)$ and this class is an identity for $U(S)$ since $(y + x, u + x) \sim (y, u)$. For a pair (x, u) the pair (u, x) obviously determines an inverse for the class of (x, u).

Note that in defining the relation $(x, u) \sim (y, v)$, the element r in the equation

$$x + v + r = y + u + r$$

is unnecessary if S is *cancellative*—that is, if

$$x + r = y + r$$

implies $x = y$ for all $x, y, r \in S$.

If $x \in S$ then elements $(x + r, r)$ belong to a single equivalence class. Hence, $x \to (x + r, r)$ determines a map of S into $U(S)$. Since

$$(x + r, r) + (y + s, s) = (x + y + r + s, r + s),$$

this map is a homomorphism of S into $U(S)$. If we call this homomorphism μ then the group $U(S)$ and homomorphism $\mu: S \to U(S)$ are unique (up to isomorphism) in that they satisfy the following:

PROPOSITION. *If $\varphi: S \to G$ is any homomorphism of S into a group, then there is a unique homomorphism $\tilde{\varphi}: U(S) \to G$ such that φ is the composition $S \xrightarrow{\mu} U(S) \xrightarrow{\tilde{\varphi}} G$.*

Proof. Given $\varphi: S \to G$ and a class $(x, u) \in U(S)$ we define

$$\tilde{\varphi}(x, u) = \varphi(x) - \varphi(u)$$

(we use additive notation for the operation in G). This is well defined. In fact, if $(x, u) \sim (y, v)$ then

$$x + v + r = y + u + r$$

for some r,

$$\varphi(x) + \varphi(v) + \varphi(r) = \varphi(y) + \varphi(u) + \varphi(r)$$

and, hence,

$$\varphi(x) - \varphi(u) = \varphi(y) - \varphi(v).$$

The map $\tilde{\varphi}$ is obviously a homomorphism and is defined so that $\tilde{\varphi} \circ \mu = \varphi$. In fact, this equation obviously defines $\tilde{\varphi}$ uniquely. ■

6.2. The semigroup $J(A)$

Now let A be a unital commutative Banach algebra. We let $Q_n(A)$ denote the set of all $a \in M_n(A)$ such that $e^{2\pi i a} = 1$. This is the set $Q(M_n(A))$ of Section 4.

Now since $M_n(A)$ is not commutative, the sets $Q_n(A)$ are not groups under addition. If $a \in Q_n(A)$ then $-a \in Q_n(A)$, but for $a, b \in Q_n(A)$ we are only sure that $a + b \in Q_n(A)$ when a and b commute. We shall construct a group $K_0(A)$ from the sets $Q_n(A)$ and in this group the operation will correspond to $(a, b) \to a + b$ whenever a and b commute. We first construct a semigroup $J(A)$.

For $a \in Q_n(A)$, $b \in Q_m(A)$ the element

$$a \oplus b = \begin{pmatrix} a & 0 \\ 0 & b \end{pmatrix}$$

is obviously in $Q_{n+m}(A)$ (since $e^{2\pi i(a \oplus b)} = e^{2\pi ia} \oplus e^{2\pi ib}$). By a trivial extension of $a \in Q_n(A)$ we mean an element $a \oplus 0_{m-n} \in Q_m(A)$ for some $m > n$.

Two matrices $a, b \in M_n(A)$ are similar if $a = ubu^{-1}$ for some $u \in GL_n(A)$. We set $a \sim b$ for $a \in Q_n(A)$ and $b \in Q_m(A)$ if a and b have trivial extensions in some $Q_k(A)$ which are similar. This is obviously an equivalence relation on $\bigcup_n Q_n(A)$.

Note that if $a \sim a'$ and $b \sim b'$ then $a \oplus b \sim a' \oplus b'$. It follows that direct sum yields a well defined operation on the set of equivalence classes of $\bigcup_n Q_n(A)$ under $\sim$. Since $a \oplus b$ and $b \oplus a$ are similar, we have that this operation is abelian.

Definition. We denote $J(A)$ the abelian semigroup consisting of equivalence classes, under $\sim$, of $\bigcup_n Q_n(A)$.

6.3. PROPOSITION. *If $a, b \in Q_n(A)$ are in the same connected component of $Q_n(A)$ then they are similar and, hence, determine the same element of $J(A)$. Conversely, if $a \sim b$ then a and b have trivial extensions in some $Q_m(A)$ which lie in the same connected component of $Q_m(A)$.*

Proof. Recall from 4.3 that each element $a \in Q_n(A)$ has a unique spectral decomposition $a = \sum jp_j$, where $\{p_j\}$ is a set of idempotents in $M_n(A)$ forming a partition of unity ($p_ip_j = 0$ for $i \neq j$ and $\sum p_j = 1$). Also recall that the p_j's depend continuously on a. Hence, for $a \in Q_n(A)$ there is a neighbourhood U of a in $Q_n(A)$ such that for $b = \sum jq_j \in U$ we have

$$\| p_jq_j + (1 - p_j)(1 - q_j) - 1 \| < 1 \text{ for each } j.$$

It follows that

$$u_j = p_jq_j + (1 - p_j)(1 - q_j)$$

is invertible. Furthermore,

$$p_ju_j = p_jq_j = u_jq_j$$

and, hence, $p_j = u_jq_ju_j^{-1}$. If we set

$$u = \sum p_ju_j = \sum u_jq_j,$$

then u is invertible since

$$(\sum u_j^{-1}p_j)(\sum p_ju_j) = \sum u_j^{-1}p_ju_j = \sum q_j = 1.$$

From

$$au = \sum jp_ju_j = \sum ju_jq_j = ub,$$

we conclude that a and b are similar.

Conversely, suppose a and b are similar and choose $u \in GL_n(A)$ such that $a = ubu^{-1}$. By Proposition 5.4, $u \oplus 1$ and $1 \oplus u$ can be connected by an arc in $GL_{2n}(A)$. It follows that

$$a \oplus 0 = (u \oplus 1)(b \oplus 0)(u \oplus 1)^{-1}$$

and

$$b \oplus 0 = (1 \oplus u)(b \oplus 0)(1 \oplus u)^{-1}$$

are connected by an arc in $Q_{2n}(A)$. ■

6.4. The group $K_0(A)$

Since $J(A)$ is an abelian semigroup, we may pass to the universal group $UJ(A)$. For $a \in Q_n(A)$ we denote the class of a in $UJ(A)$ by $\{a\}$. The group $UJ(A)$ is not the one we want, however, because we do not have that $\{a + b\} = \{a\} + \{b\}$ whenever a and $b \in Q_n(A)$ commute. This defect is easily fixed, however.

Definition. We let $K_0(A)$ denote the quotient group of $UJ(A)$ modulo the subgroup generated by elements of the form $\{a + b\} - \{a\} - \{b\}$ for a and b commuting elements of some $Q_n(A)$. For $a \in Q_n(A)$ its class in $K_0(A)$ will be denoted $[a]$.

Now this definition looks rather complicated since it occurs in several stages, passing from $\bigcup_n Q_n(A)$ to $J(A)$ to $UJ(A)$, and then to $K_0(A)$. It is not as bad as it looks, however, since we can easily determine the homomorphisms of $K_0(A)$ into another group G.

6.5. Proposition. *If $\varphi:\bigcup_n Q_n(A) \to G$ is a map of $\bigcup_n Q_n(A)$ into a group, then there is a group homomorphism $\tilde{\varphi}:K_0(A) \to G$ satisfying $\tilde{\varphi}[a] = \varphi(a)$ provided φ has the following properties:*

(1) *$\varphi(a) = \varphi(b)$ if a and b are similar;*
(2) *$\varphi(a) = \varphi(b)$ if b is a trivial extension of a; and*
(3) *$\varphi(a + b) = \varphi(a) + \varphi(b)$ whenever a and b commute.*

Proof. Since $a \oplus 0$ and $0 \oplus b$ commute in $Q_{n+m}(A)$ for $a \in Q_n(A)$, $b \in Q_m(A)$, and since $a \oplus b = (a \oplus 0) + (0 \oplus b)$, we have from (3) that $\varphi(a \oplus b) = \varphi(a) + \varphi(b)$. This, together with (1) and (2) implies that φ determines a homomorphism of the semigroup $J(A)$ into G and, hence, a homomorphism of $UJ(A)$ into G. Now (3) implies that this homomorphism kills the subgroup of $UJ(A)$ generated by elements $\{a + b\} - \{a\} - \{b\}$ and, hence, that it determines a homomorphism $\tilde{\varphi}:K_0(A) \to G$. ■

We should point out one more thing about the construction of $K_0(A)$: The map $a \to [a]: \bigcup_n Q_n(A) \to K_0(A)$ is actually onto. That is, every element of $K_0(A)$ has a representative in some $Q_n(A)$. In fact, the image of $\bigcup_n Q_n(A)$ in $K_0(A)$ is obviously a semigroup which generates $K_0(A)$ but since $[a] + [-a] = [a - a] = 0$, this semigroup is closed under inversion and, hence, is all of $K_0(A)$.

6.6. Non-unital algebras and $\tilde{K}_0$

For a unital algebra homomorphism $\varphi: A \to B$, the induced map $M_n(A) \to M_n(B)$ obviously carries $Q_n(A)$ into $Q_n(B)$ and induces a homomorphism $\varphi^*: K_0(A) \to K_0(B)$. Thus, K_0 is a functor from the category of unital commutative Banach algebras to the category of abelian groups.

We now define a corresponding reduced functor $\tilde{K}_0$ on the category of commutative Banach algebras.

If A is a commutative Banach algebra and A^+ is the algebra with identity adjoined, then $A^+ \to A^+/A \simeq \mathbb{C}$ is a complex homomorphism φ_0 of A^+. It induces a homomorphism $\varphi_0^*: K_0(A^+) \to K_0(\mathbb{C})$. Similarly, the map $\lambda \to \lambda \cdot 1: \mathbb{C} \to A^+$ determines a homomorphism $K_0(\mathbb{C}) \to K_0(A^+)$ such that the composition $K_0(\mathbb{C}) \to K_0(A^+) \overset{\varphi_0^*}{\to} K_0(\mathbb{C})$ is the identity. Hence, if we let $\tilde{K}_0(A)$ denote the kernel of φ_0^*, then we have a natural direct sum decomposition

$$K_0(A^+) \simeq \tilde{K}_0(A) \oplus K_0(\mathbb{C}).$$

For an algebra homomorphism $A \to B$ the diagram

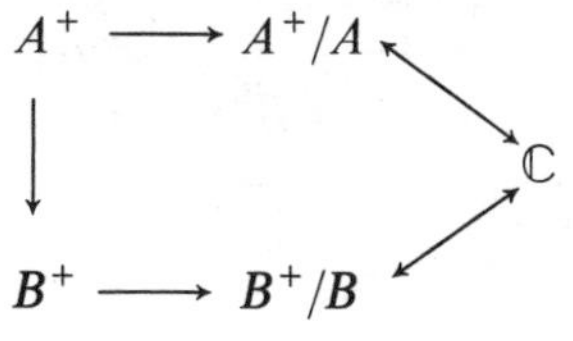

is commutative. Hence, so is

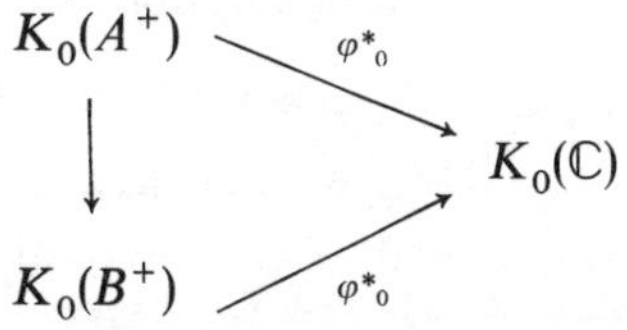

It follows that $K_0(A^+) \to K_0(B^+)$ maps $\tilde{K}_0(A)$ into $\tilde{K}_0(B)$. Hence, we have defined a functor $\tilde{K}_0$ on commutative Banach algebras.

One can easily prove that $K_0(A \oplus B) = K_0(A) \oplus K_0(B)$ whenever $A \oplus B$ is an algebra direct sum of unital algebras. Hence, if A is already unital then $K_0(A^+) = K_0(A) \oplus K_0(\mathbb{C})$—that is, $K_0(A)$ is the kernel of the map $K_0(A^+) \to K_0(\mathbb{C})$ and, hence, is equal to $\tilde{K}_0(A)$. In other words, $\tilde{K}_0$ extends K_0 to the category of all commutative Banach algebras.

6.7. Remark

As was true of $\tilde{K}_1$, it turns out that the Gelfand transform $A \to C_0(\Delta)$ induces an isomorphism $\tilde{K}_0(A) \to \tilde{K}_0(C_0(\Delta))$. This can be proved directly using results from several complex variables due to Grauert (cf. **[9]**, **[14]**). However, here we shall deduce it from the corresponding result for $\tilde{K}_1$ by using the version of the Bott periodicity theorem in Section 9.

Thus, as with $\tilde{K}_1$, we shall be able to show in Section 10 that $\tilde{K}_0(A) = \tilde{K}^0(\Delta^+)$ for an appropriate cofunctor $\tilde{K}_0$ from pointed compact spaces to abelian groups. With this and $\tilde{K}_1$ we can define the generalized cohomology theory called K-theory.

7. FURTHER PROPERTIES OF K_0

In this section we give a simpler description of K_0 and use it to determine some additional properties of K_0.

Since we made the exponential map the central object with which to begin our discussion of both K_1 and K_0, the object $\bigcup_n Q_n(A)$ was a natural one to use in constructing $K_0(A)$. However, we could have started with a much smaller set.

7.1. Idempotents

Each idempotent matrix $p \in M_n(A)$ is in $Q_n(A)$. We shall show that every element of $K_0(A)$ has the form $[p] - [q]$ for idempotents p and q and give a simple criterion for the equation $[p] - [q] = 0$ to hold in $K_0(A)$. To this end, we let $P_n(A)$ denote the set of idempotents in $M_n(A)$ and construct a group $T(A)$ from $\bigcup_n P_n(A)$.

Now $\bigcup_n P_n(A)$ is a subset of $\bigcup_n Q_n(A)$ which is closed under direct sum. Hence, $\bigcup_n P_n(A)$ determines a subsemigroup $S(A) \subset J(A)$. That is, $S(A)$ is the semigroup (under direct sum) of equivalence classes of elements of $\bigcup_n P_n$ under the equivalence relation $\sim$, where $p \sim q$ if p and q have similar trivial extensions. We let $T(A) = US(A)$ be the universal group of $S(A)$.

Now one must be careful here. Although $S(A) \subset J(A)$, one cannot directly conclude that $T(A) \subset UJ(A)$. The inclusion $i: S(A) \to J(A)$ does induce a map $\hat{i}: T(A) \to UJ(A)$—the map of $T(A)$ induced by the composition $S(A) \to J(A) \to UJ(A)$. However, one must work to prove that this map is one to one.

Actually, we shall prove that the map $i^*: T(A) \to K_0(A)$, obtained from $\tilde{i}$ by composing with the factor map $UJ(A) \to K_0(A)$, is an isomorphism. We do this by directly exhibiting an inverse.

7.2. The inverse map λ^*

For $a \in Q_n(A)$ we let $a = \sum jp_j$ be the unique spectral decomposition as in Proposition 4.2. We define an element $\lambda(a) \in T(A)$ by $\lambda(a) = \sum j\lambda(p_j)$, where $\lambda(p_j)$ is the class of the idempotent p_j in $T(A)$.

PROPOSITION. *The map* $\lambda: \bigcup_n Q_n(A) \to T(A)$ *satisfies the condition of Proposition 6.5 and, hence, defines a homomorphism*

$$\lambda^*: K_0(A) \to T(A).$$

Proof. If $a = \sum jp_j$ and $b = \sum jq_j$ are similar (say $a = ubu^{-1}$) then p_j and q_j are similar for each j. In fact, since

$$a = ubu^{-1} = \sum juq_ju^{-1},$$

the uniqueness of the spectral decomposition implies that $p_j = uq_ju^{-1}$ for each j. It follows from the definition of $T(A)$ that

$$\lambda(a) = \sum j\lambda(p_j) = \sum j\lambda(q_j) = \lambda(b).$$

If $a = \sum jp_j$ is a trivial extension of $b = \sum jq_j$, then obviously p_j is the corresponding trivial extension of q_j except for $j = 0$. Hence,

$$\lambda(a) = \sum j\lambda(p_j) = \sum j\lambda(q_j) = \lambda(b).$$

In order to verify the third condition of Proposition 6.5, we must derive the spectral decomposition of $a + b$ when a and b commute. Thus, let $a = \sum jp_j$ and $b = \sum jq_j$ and let a and b commute. Then each p_i commutes with each q_j and p_iq_j is an idempotent for each i, j. Since $\sum p_i = 1$ and $\sum q_j = 1$ we have

$$a = (\sum ip_i)(\sum q_j) = \sum_{i,j} ip_iq_j,$$

$$b = (\sum jq_j)(\sum p_i) = \sum_{i,j} jp_iq_j,$$

and

$$a + b = \sum_{i,j} (i + j)p_iq_j$$

It follows that if

$$r_k = \sum_{i+j=k} p_iq_j,$$

then $a + b = \sum kr_k$ is the spectral decomposition of $a + b$ (a simple computation shows that $\{r_k\}$ forms a partition of unity).

Now if p and q are disjoint idempotents in $P_n(A)$ ($pq = qp = 0$), then $p \oplus q$ is similar to a trivial extension of $p + q$. In fact, if $u \in GL_{2n}(A)$ is defined by $u = \begin{pmatrix} p & 1-p \\ 1-p & p \end{pmatrix}$ then $u^{-1} = u$ and

$$\begin{pmatrix} p & 1-p \\ 1-p & p \end{pmatrix}\begin{pmatrix} p+q & 0 \\ 0 & 0 \end{pmatrix}\begin{pmatrix} p & 1-p \\ 1-p & p \end{pmatrix} = \begin{pmatrix} p & 0 \\ 0 & q \end{pmatrix}.$$

That is, $p \oplus q = u[(p + q) \oplus 0]u^{-1}$.

It follows from the above that

$$\lambda(r_k) = \lambda(\sum_{i+j=k} p_i q_j) = \lambda(\bigoplus_{i+j=k} p_i q_j) = \sum_{i+j=k} \lambda(p_i q_j),$$

$$\lambda(p_i) = \lambda(\sum_j p_i q_j) = \lambda(\bigoplus_j p_i q_j) = \sum_j \lambda(p_i q_j),$$

and similarly

$$\lambda(p_j) = \sum_i \lambda(p_i q_j).$$

Hence, we have,

$$\lambda(a + b) = \sum k\lambda(r_k) = \sum_{i,j} (i + j)\lambda(p_i q_j)$$

$$= \sum_i i \sum_j \lambda(p_i q_j) + \sum_j j \sum_i \lambda(p_i q_j) = \sum i\lambda(p_i) + \sum j\lambda(q_j)$$

$$= \lambda(a) + \lambda(b).$$

This completes the proof. ■

7.3. PROPOSITION. *The natural map $i^*: T(A) \to K_0(A)$ is an isomorphism with inverse λ^*.*

Proof. We need only observe that $i^* \circ \lambda^* = 1$ and $\lambda^* \circ i^* = 1$. Furthermore, it suffices to verify these identities on generating subsets of $K_0(A)$ and $T(A)$, respectively. However, on elements determined by idempotents both identities obviously hold by the construction of λ. Since classes of idempotents generate both groups, the proof is complete. ■

We now forget about the group $T(A)$ and simply interpret what the above result says about $K_0(A)$.

7.4. COROLLARY. *Each element of $K_0(A)$ has the form $[p] - [q]$ for idempotents $p, q \in \bigcup_n P_n(A)$. Furthermore, given two idempotents p and q, $[p] -$*

$[q] = 0$ *if and only if there is an idempotent r such that* $p \oplus r$ *and* $q \oplus r$ *have similar trivial extensions.*

Proof. That $i^*: T(A) \to K_0(A)$ is onto says exactly that every element of $K_0(A)$ has the form $[p] - [q]$.

The condition $p \oplus r \sim q \oplus r$ for some r is exactly the condition that p and q determine the same element of the universal group $T(A)$ of $S(A)$.

Since $T(A) \to K_0(A)$ is one to one, we have that this is also the condition for $[p] = [q]$ in $K_0(A)$. ■

We can refine the above result somewhat by using the fact that if p is idempotent in $M_n(A)$ then so is $I_n - p$. In fact, for a class $[p] - [q] \in K_0(A)$ with $q \in P_n(A)$ we have

$$[p] - [q] = [p] - [q] + [q + I_n - q] - [I_n]$$
$$= [p] + [I_n - q] - [I_n] = [p \oplus (I_n - q)] - [I_n].$$

Since $[p \oplus (I_n - q)]$ is idempotent, we have that every element of $K_0(A)$ has the form $[p] - [I_n]$ for an idempotent p and an integer n.

Similarly, if $p \oplus r \sim q \oplus r$ for $r \in P_n(A)$ then $p \oplus I_n \sim p \oplus r \oplus (I_n - r) \sim q \oplus r \oplus (I_n - r) \sim q \oplus I_n$. Hence, we have:

7.5. COROLLARY. *Every element of* $K_0(A)$ *has the form* $[p] - [I_n]$ *for* $p \in \bigcup_n P_n(A)$ *and n an integer. Furthermore,* $[p] = [q]$ *in* $K_0(A)$ *for idempotents p and q if and only if* $p \oplus I_n$ *and* $q \oplus I_n$ *have similar trivial extensions for some integer n.*

7.6. The group $K_0(\mathbb{C})$

For any unital algebra A the group $K_0(A)$ always contains a canonical copy of the integers. In fact, the map $n \to [I_n]$ from the positive integers $\mathbb{Z}^+$ into $K_0(A)$ is a semigroup homomorphism and, hence, extends to the universal group $\mathbb{Z}$ of $\mathbb{Z}^+$. That this map is one to one follows from the previous corollary. In the one dimensional case ($A = \mathbb{C}$) this map is also onto.

PROPOSITION. *The canonical embedding* $\mathbb{Z} \to K_0(\mathbb{C})$ *is an isomorphism.*

Proof. If p is an idempotent complex $n \times n$ matrix then p is similar to a matrix $I_k \oplus 0_{n-k}$ and, hence, $[p] = [I_k]$. It follows from the previous corollary that $\mathbb{Z} \to K_0(\mathbb{C})$ is onto. ■

7.7. Description of $\tilde{K}_0(A)$

We determine when an element $[p] - [I_n] \in K_0(A^+)$ is in $\tilde{K}_0(A) = \ker(K_0(A^+) \to K_0(\mathbb{C}))$.

Since $K_0(A^+) \to K_0(\mathbb{C})$ is the map induced by the quotient map $\varphi_0 : A^+ \to A^+/A = \mathbb{C}$, we have that the image of $[p] - [I_n]$ is just $[\varphi_0(p)] - [I_n]$. Now $\varphi_0(p)$ is similar to $I_k \oplus 0$, where k is the rank of $\varphi_0(p)$. Hence:

PROPOSITION. *The element* $[p] - [I_n] \in K_0(A^+)$ *belongs to* $\tilde{K}_0(A)$ *if and only if the rank of the complex matrix* $\varphi_0(p)$ *is* n.

7.8. Weak exactness

We can now prove that $\tilde{K}_0$, like $\tilde{K}_1$, satisfies a portion of the exactness axiom.

PROPOSITION. *If* I *is a closed ideal of the commutative Banach algebra* A, *then the exact sequence* $I \xrightarrow{i^*} A \xrightarrow{\pi} A/I$ *induces an exact sequence*

$$\tilde{K}_0(I) \xrightarrow{i^*} \tilde{K}_0(A) \xrightarrow{\pi^*} \tilde{K}_0(A/I).$$

Proof. We pass to A^+ and consider the sequence

$$I^+ \xrightarrow{i} A^+ \xrightarrow{\pi} A^+/I.$$

The composition is just the complex homomorphism

$$I^+ \to I^+/I = \mathbb{C}$$

followed by the canonical embedding of $\mathbb{C}$ in A^+/I. It follows that the composition

$$i^* \circ \pi^* : K_0(I^+) \to K_0(A^+/I)$$

has $\tilde{K}_0(I)$ as kernel. Thus, the composition

$$\tilde{K}_0(I) \to \tilde{K}_0(A) \to \tilde{K}_0(A/I)$$

is the zero map.

Now suppose that $c \in \tilde{K}_0(A) \subset K_0(A^+)$ and that $\pi^* c = 0$. We represent c as $[p_1] - [I_k]$ for $p_1 \in \bigcup_n P_n(A^+)$ and note that $\pi p_1 \oplus I_n \sim I_k \oplus I_n = I_{k+n}$ for some n. Hence, if $p, q \in \bigcup_n P_n(A^+)$ are appropriate trivial extensions of $p_1 \oplus I_n$ and I_{k+n}, then $c = [p] - [q]$ and πp is similar to πq—say $\pi q = u(\pi p)u^{-1}$ for $u \in GL_m(A^+/I)$.

Now we may as well assume that $u \in GL_m^0(A^+/I)$ (the matrix $u \oplus u^{-1}$ is in $GL_{2m}^0(A^+/I)$ by Proposition 5.4, and

$$\pi p \oplus 0 = (u \oplus u^{-1})(\pi q \oplus 0)(u \oplus u^{-1})^{-1}).$$

However, if $u \in GL_m^0(A^+/I)$ then $u = \pi v$ for some $v \in GL_m^0(A^+)$ by Proposition 4.8. Hence, we have $[p] - [q] = [vpv^{-1}] - [q]$ and $\pi(vpv^{-1}) =$

$u(\pi p)u^{-1} = \pi q$. However, since q is just $I_{n+k} \oplus 0$, we have that $vpv^{-1} \in P_m(I^+)$ and modulo I has rank $n + k$. Hence, $c = [p] - [q]$ is the image of an element of $\tilde{K}_0(I)$. This completes the proof. ∎

8. RELATIONS BETWEEN K_0 AND K_1

In this section we begin to build a cohomology theory from K_0 and K_1. The first task is to construct, for an ideal $I \subset A$, a homomorphism $\delta^*: \tilde{K}_1(A/I) \to \tilde{K}_0(I)$ which pieces together the sequences of 5.6 and 7.8 and yields a six term exact sequence.

The construction of δ^* can be described intuitively as follows: An element a of $GL_k(A/I)$ can be used to "twist" a trivial idempotent $I_k \oplus 0 \in P_n(I^+)$ into a non-trivial idempotent $p \in P_n(I^+)$. The class of a is then mapped to the class $[p] - [I_k]$.

8.1. The construction of δ^*

We begin with a unital algebra A and construct $\delta^*; K_1(A/I) \to \tilde{K}_0(I)$. We then extend the definition of δ^* to the non-unital case.

Let $\pi: A \to A/I$ be the factor map. For $a \in GL_k(A/I)$ and some $n > k$ we may choose $b \in GL_{n-k}(A/I)$ so that $a \oplus b \in GL_n^0(A/I)$. In fact, it suffices to choose $n = 2k$ and $b = a^{-1}$, although the choice of b is not important.

Since $a \oplus b \in GL_n^0(A/I)$, it has a pre-image $u \in GL_n^0(A)$. We set $q = I_k \oplus 0_{n-k}$ and $p = uqu^{-1}$. Note that $\pi p = (a \oplus b)(I_k \oplus 0)(a \oplus b)^{-1} = I_k \oplus 0$ and, hence, $p \in P_n(I^+)$ (p is a scalar matrix modulo I). Since p has a rank k modulo I, we have that $[p] \in K_0(I^+)$ satisfies $[p] - [I_k] \in \tilde{K}_0(I)$.

Now replacing b by a trivial extension $b \oplus I$ and u by the corresponding trivial extension $u \oplus I$ simply increases n and replaces p by a trivial extension $p \oplus 0$. Moreover, we claim that the class $[p] \in K_0(I^+)$ is unaffected by the choices of b and u and, in fact, depends only on the class of a modulo $GL_k^0(A/I)$. To prove this, we suppose $c \sim a \bmod GL_k^0(A/I)$ and $d \in GL_{n-k}(A/I)$, $v \in GL_n(A)$ are chosen so that $\pi v = c \oplus d \in GL_n^0(A/I)$. We then have

$$(c \oplus d)^{-1}(a \oplus b) = c^{-1}a \oplus d^{-1}b \in GL_n^0(A/I).$$

Since $c^{-1}a \in GL_k^0(A/I)$ we conclude that $I_k \oplus d^{-1}b \in GL_n^0(A/I)$ and, hence, that d and b may be replaced by trivial extensions ($d \oplus I_k$ and $b \oplus I_k$) which are equivalent. That is, we may as well assume that $d^{-1}b \in GL_{n-k}^0(A/I)$.

We now choose pre-images $g \in GL_k(A)$ and $h \in GL_{n-k}(A)$ for $c^{-1}a$ and $d^{-1}b$. We then set $w = g \oplus h \in GL_n(A)$. Note that

$$\pi(vwu^{-1}) = (c \oplus d)(c^{-1}a \oplus d^{-1}b)(a^{-1} \oplus b^{-1}) = I_n$$

and, hence, $vwu^{-1} \in GL_n(I^+)$. Now with $p = uqu^{-1}$ ($q = I_k \oplus 0_{n-k}$) as before, we also have $p = u(g^{-1} \oplus h^{-1})(I_k \oplus 0)(g \oplus h)u^{-1} = uw^{-1}qwu^{-1}$.

Hence, $vqv^{-1} = vwu^{-1}\,p\,uw^{-1}v^{-1}$. Since $p = uqu^{-1}$ and vqv^{-1} are similar via an element of $GL_n(I^+)$, they have the same class in $K_0(I^+)$. This establishes our claim.

Now passing to a trivial extension $a \oplus I_j$ of a has the effect of replacing p by $p \oplus I_j$ and I_k by I_{k+j}. Hence, the class $[p] - [I_k]$ is unaffected.

We set $\delta(a) = [p] - [I_k]$ for $a \in GL_k(A/I)$. Obviously

$$\delta(a \oplus b) = \delta(a) + \delta(b).$$

Since $ab \oplus I \sim a \oplus b$ we have $\delta(ab) = \delta(a) + \delta(b)$. It follows from this and the above that setting $\delta^*([a]) = \delta(a)$ defines a homomorphism

$$\delta^*: K_1(A/I) \to \tilde{K}_0(I).$$

In the case where A is non-unital and $I \subset A$ is a closed ideal, we have that $A^+/I = (A/I)^+$ and, hence, that $\tilde{K}_1(A/I) = K_1((A/I)^+) = K_1(A^+/I)$. The homomorphism $\delta^*: K_1(A^+/I) \to \tilde{K}_0(I)$ is then our map

$$\delta^*: \tilde{K}_1(A/I) \to \tilde{K}_0(I).$$

8.2. Proposition. *If I is a closed ideal of A, then the sequence*

$$\tilde{K}_1(I) \xrightarrow{i^*} \tilde{K}_1(A) \xrightarrow{\pi^*} \tilde{K}_1(A/I) \xrightarrow{\delta^*} \tilde{K}_0(I) \xrightarrow{i^*} \tilde{K}_0(A) \xrightarrow{\pi^*} \tilde{K}_0(A/I)$$

is exact.

Proof. We must prove exactness at $\tilde{K}_1(A/I)$ and $\tilde{K}_0(I)$ (we already have it elsewhere by 5.6 and 7.8).

We first look at $\tilde{K}_0(I)$. Now for $a \in GL_k(A^+/I)$ the element

$$[p] - [I_k] = \delta^*([a]) \in \tilde{K}_0(I)$$

was chosen so that $p = uqu^{-1}$ was similar to $q = I_k \oplus 0$ in $M_n(A^+)$. It follows that $\operatorname{im}\delta^* \subseteq \ker i^*$. On the other hand if $[p] - [I_k]$ is any element of $\tilde{K}_0(I)$ in $\ker i^*$, then (on passing to trivial extensions) we have $p = uqu^{-1}$ for $q = I_k \oplus 0$ and some $u \in GL_n(A)$. Since $[p] - [I_k] \in \tilde{K}_0(I)$, πp is a scalar matrix of rank k which we may as well assume is $I_k \oplus 0$. Then

$$I_k \oplus 0 = (\pi u)(I_k \oplus 0)(\pi u)^{-1}$$

and, hence, πu commutes with $I_k \oplus 0$. This means πu must have the form $a \oplus b$ for $a \in GL_k(A^+/I)$, $b \in GL_{n-k}(A^+/I)$. By the construction of δ^*, $\delta^*([a]) = [p] - [I_k]$. Hence, $\operatorname{im}\delta^* = \ker i^*$.

We now prove exactness at $\tilde{K}_1(A/I)$. If $a \in GL_k(A^+/I)$ has a pre-image $v \in GL_k(A^+)$, then we may choose $u = v \oplus v^{-1}$ in the construction of $\delta^*([a])$. However, we then get $p = q = I_k \oplus O_k$ and, hence, $[p] - [I_k] = 0$. Thus, $\operatorname{im}\pi^* \subset \ker\delta^*$. Conversely, if $\delta^*([a]) = [p] - [I_k] = 0$, then (on passing to trivial extensions) we have $p = v^{-1}qv$ with $q = I_k \oplus 0$ and $v \in GL_n(I^+)$. Since $p = u^{-1}qu$ also, we have that $q = vu^{-1}quv^{-1}$. Hence, uv^{-1} commutes with q and, therefore, has the form $g \oplus h$ for $g \in GL_k(A^+)$ and $h \in GL_{n-k}(A^+)$. Since $\pi v = I_n$ we have

$$\pi g \oplus \pi h = \pi(uv^{-1}) = \pi u = a \oplus b.$$

Hence, a has a pre-image g and $[a] \in \operatorname{im} \pi^*$. This yields $\operatorname{im} \pi^* = \ker \delta^*$ and the proof is complete. ∎

8.3. Suspension

We now give a key application of the above result.

Let A be a commutative Banach algebra (not necessarily unital). By SA we shall mean the Banach algebra of all continuous functions $f:[0, 1] \to A$ such that $f(0) = f(1) = 0$. This can also be described as the algebra of continuous A-valued functions on $\mathbb{R}$ which vanish at infinity.

The unital algebra $(SA)^+$ can be described as the algebra of all continuous A^+-valued functions on $[0, 1]$ such that $f(0) = f(1) \in \mathbb{C}$.

The operation $A \to SA$ is analogous to the suspension operation of 1.4. In fact, we have the following proposition:

PROPOSITION. *If Δ is the maximal ideal space of A, then $\mathbb{R} \times \Delta$ is the maximal ideal space of SA and $S\Delta^+$ is the maximal ideal space of $(SA)^+$.*

The proof is a simple exercise.

Our application of 8.2 is the following:

8.4. PROPOSITION. *There is a natural isomorphism $\alpha^*: \tilde{K}_1(A) \to \tilde{K}_0(SA)$.*

Proof. If we let CA be the algebra of continuous functions from $[0, 1]$ into A with $f(0) = 0$, then SA is a closed ideal in CA. Furthermore, the map $f \to f(1): CA \to A$ is an onto homomorphism with kernel SA. Hence, we may identify A with CA/SA and Proposition 8.2 says we have an exact sequence

$$\tilde{K}_1(SA) \to \tilde{K}_1(CA) \to \tilde{K}_1(A) \xrightarrow{\delta^*} \tilde{K}_0(SA) \to \tilde{K}_0(CA) \to \tilde{K}_0(A).$$

Now the algebra CA is contractible in the following sense: If we set $\varphi_s f(t) = f(st)$ for $f \in CA$, then $s \to \varphi_s$ $(0 \leqslant s \leqslant 1)$ is a continuously parameterized family of endomorphisms of CA with $\varphi_1 = 1$ and $\varphi_0 = 0$. The natural extension $\tilde{\varphi}_s$ to $(CA)^+ (\tilde{\varphi}_s(a + \lambda) = \varphi_s(a) + \lambda)$ can then be used to deform each matrix in $\bigcup_n P_n((CA)^+)$ to a scalar matrix and each matrix in $\bigcup_n GL_n((CA)^+)$ to a scalar matrix. It follows that $\tilde{K}_0(CA) = \tilde{K}_1(CA) = 0$.

We conclude from the diagram that $\delta^*; \tilde{K}_1(A) \to \tilde{K}_0(SA)$ is an isomorphism. This is our map α^*. ■

8.5. A non-trivial element of $K_0(C(S^2))$

By following the construction of the map δ^* in Proposition 8.2, one can construct an explicit idempotent p of rank one in $P_2(C(S^2))$ such that $[p] - [I_1] \neq 0$ in $K_0(C(S^2))$. This means that $K_0(C(S^2))$ contains more than just the canonical copy of the integers.

We choose for A the algebra $C(D)$ where D is the unit disc $\{z \in \mathbb{C}: |z| \leqslant 1\}$. For I we choose $\{f \in C(D): f(z) = 0 \text{ for } |z| = 1\}$. Then $A/I = C(\mathbb{T})$ where $\mathbb{T}$ is the unit circle. For the element a in the proof of Proposition 8.2 we choose the identity function $e^{i\phi} \in C(\mathbb{T})^{-1} = GL_1(A/I)$. A choice for the element $u \in GL_2(A)$ is the matrix valued function

$$u(z) = u(re^{i\phi}) = \begin{pmatrix} e^{i\phi} \sin \frac{\pi}{2} r & -\cos \frac{\pi}{2} r \\ \cos \frac{\pi}{2} r & e^{-i\phi} \sin \frac{\pi}{2} r \end{pmatrix}$$

which is invertible for every $z \in D$, is a scalar matrix for $r = 0$, and is $\begin{pmatrix} e^{i\phi} & 0 \\ 0 & e^{-i\phi} \end{pmatrix}$ for $r = 1$.

If we set

$$p(z) = u(z) \begin{pmatrix} 1 & 0 \\ 0 & 0 \end{pmatrix} u(z)^{-1}$$

then

$$p(z) = \tfrac{1}{2} \begin{pmatrix} 1 - \cos \pi r & e^{i\phi} \sin \pi r \\ e^{-i\phi} \sin \pi r & 1 + \cos \pi r \end{pmatrix}$$

This is an idempotent of rank one for each $z \in D$ and on the unit circle $\mathbb{T}$ it is the constant matrix $\begin{pmatrix} 1 & 0 \\ 0 & 0 \end{pmatrix}$. Since $D/T = S^2$ we have that

$$p \in P_2(I^+) = P_2(C(S^2)).$$

Now $e^{i\phi}$ determines a non-trivial element of $K_1(C(\mathbb{T}))$ (see the next section). Furthermore, it follows as in the proof in Proposition 8.4 that δ^* is an isomorphism. Hence, $[p] - [I_1] \neq 0$ in $K_0(C(S^2))$. That is, p has no trivial extension which is similar to a matrix $I_1 \oplus 0$.

8.6. The relation between K_0 and H^0 and between K_1 and H^1

In this section, let A be a unital algebra. The trace defines an additive function $\mathrm{tr}: M_n(A) \to A$ which satisfies $\mathrm{tr}(uau^{-1}) = \mathrm{tr}(a)$ for $u \in GL_n(A)$. Furthermore, $\mathrm{tr}(a \oplus 0) = \mathrm{tr}(a)$. If $a \in Q_n(A)$ then $\mathrm{tr}(a) \in Q(A)$. It follows that tr defines a map of $\bigcup_n Q_n(A)$ into $Q(A)$ which satisfies the conditions of Proposition 6.5. Hence, we have a homomorphism $\mathrm{tr}^*: K_0(A) \to Q(A)$.

Since $Q(A) = Q_1(A) \subset \bigcup_n Q_n(A)$, the injection $i: Q(A) \to \bigcup_n Q_n(A)$ determines a homomorphism $i^* Q(A) \to K_0(A)$ such that $\mathrm{tr}^* \circ i^* = 1$. It follows that $Q(A)$ is a direct summand of $K_0(A)$. Since $Q(A) \simeq H^0(\Delta, \mathbb{Z})$, we have that $K_0(A)$ contains a copy of $H^0(\Delta, \mathbb{Z})$ as a direct summand.

Similarly, the determinant yields a map $\det: GL_n(A) \to A^{-1}$ which maps $GL_n^0(A)$ into $\exp(A)$. It follows that there is an induced map

$$\det^*: K_1(A) \to A^{-1}/\exp(A)$$

with the injection

$$i^*: A^{-1}/\exp(A) = L_1(A) \to \varinjlim L_n(A) = K_1(A)$$

as a right inverse. Hence, $A^{-1}/\exp(A) = H^1(\Delta, \mathbb{Z})$ is a direct summand of $K_1(A)$.

Now in the case where $A = C(\mathbb{T})$ we have that

$$C(\mathbb{T})^{-1}/\exp(C(\mathbb{T})) \simeq H^1(\mathbb{T}, \mathbb{Z}) = \mathbb{Z}.$$

In fact, the function $e^{i\phi} \in C(\mathbb{T})^{-1}$ is a generator for this group. It follows that $e^{i\phi}$ generates a subgroup of $K_1(C(\mathbb{T}))$ isomorphic to the integers. Actually, we shall prove in the next section that this subgroup is all of $K_1(C(\mathbb{T}))$ and, hence, $K_1(C(\mathbb{T})) = H^1(\mathbb{T}, \mathbb{Z}) = \mathbb{Z}$. On the other hand, note that the results of the previous subsection show that $K_0(C(S^2))$ contains more than $H^0(S^2, \mathbb{Z})$.

9. THE BOTT PERIODICITY THEOREM

In this section we prove the central theorem of K-theory, the Bott periodicity theorem, in the context of Banach algebras. In our setting, this theorem says that $\tilde{K}_0(S^2 A) \simeq \tilde{K}_0(A)$, where $S^2 A = SSA$ is the second suspension of A. We prove it by exhibiting an isomorphism

$$\beta^*: \tilde{K}_0(A) \to \tilde{K}(SA)$$

and then applying the map α^* of the previous section with A replaced by SA.

Our proof essentially follows that of Atiyah in [4].

9.1. The algebra ΩA

It will be convenient initially to work with a unital algebra A and the unital algebra ΩA of all continuous functions from $[0, 1]$ to A such that $f(0) = f(1) \in \mathbb{C}$.

The operation $A \to \Omega A$ for unital algebras is related to $A \to SA$ for non-unital algebras in the following way: If A is non-unital, then SA is the ideal in ΩA^+ consisting of functions with values in A, while $(SA)^+$ is the subalgebra of ΩA^+ consisting of functions which are constant modulo A.

For A unital, we shall call an element of $GL_n(\Omega A)$ a *loop*.

9.2. The exponential map

Let A be unital. If $a \in Q_n(A)$ ($a \in M_n(A)$ and $e^{2\pi i a} = 1$) then the function $t \to e^{2\pi i t a}$ defines a loop in $GL_n(\Omega A)$ which we shall call $\beta(a)$. Note that $\beta(a + b) = \beta(a)\beta(b)$ if a and b commute, $\beta(uau^{-1}) = u\beta(a)u^{-1}$ for $u \in GL_n(A)$, and $\beta(a \oplus 0) = \beta(a) \oplus I$.

If we set $\beta^*([a]) = [\beta(a)] \in K_1(\Omega A)$ for $[a] \in K_0(A)$, then it follows from Proposition 6.5 that β^* is a well defined homomorphism from $K_0(A)$ to $K_1(\Omega A)$. The periodicity theorem will follow easily from:

THEOREM. *The map β^*: $K_0(A) \to K^1(\Omega A)$ is an isomorphism.*
We shall prove this by constructing an auxiliary group G suvh that β^* factors as

$$K_0(A) \overset{\varphi^*}{\to} G \overset{\psi^*}{\to} K_1(\Omega A),$$

and then proving that φ^* and ψ^* are both isomorphisms.

9.3 Linear loops

If $p \in P_n(A) \subset Q_n(A)$ is an idempotent, then

$$\beta(p)(t) = e^{2\pi i t p} = 1 + (\sum_1^\infty (2\pi i t)^n/n!)p = 1 + (e^{2\pi i t} - 1)p = 1 - p + zp,$$

where z will henceforth denote the scalar loop $z(t) = e^{2\pi i t}$. Hence, $\beta(p)$ is a loop which is linear in z. We call such a loop a *linear loop*.

We construct a group G from the linear loops. Note that the direct sum $f \oplus g$ of two linear loops is also linear and, hence, a trivial extension of a linear loop is linear. We write $f \sim g$ for linear loops f and g if f and g have trivial extensions which are connected by an arc consisting of linear loops. The equivalence classes of linear loops under this relation form an abelian semigroup under $\oplus$. We let G denote the universal group of this semigroup.

The inclusion ψ (the map which forgets that a linear loop is linear and considers it just an element of $GL_n(\Omega A)$) obviously induces a homomorphism $\psi^*: G \to K_1(\Omega A)$. Furthermore, the map φ which takes an idempotent $p \in P_n(A)$ to the linear loop $(1 - p) + zp$ clearly induces a homomorphism

$\varphi^*\colon K_0(A) \to G$ (Corollary 7.5) and our map β^* is just the composition $\psi^* \circ \varphi^*$.

9.4. LEMMA. *There exists a homomorphism $\lambda^*\colon G \to K_0(A)$ such that $\lambda^* \circ \varphi^* = 1$ and $\psi^* = \psi^* \circ \varphi^* \circ \lambda^*$.*

Proof. Let g be a linear loop. Then $g(0) = g(1)$ is a scalar matrix and, hence, is deformable to the identity. It follows that g and $f = g(0)^{-1}g$ have the same class in G. Furthermore, $f(0) = f(1) = 1$.

Since f is linear, we have $f = a + zb$ for $a, b \in M_n(A)$. Since $f(1) = 1$, we have $a + b = 1$. It follows that f takes all its values in the commutative subalgebra of $M_n(A)$ generated by a. This allows us to compute a logarithm for $f(t)$ by a familiar method.

Note that $f(t)$ is differentiable as a function of t, with derivative $f'(t) = 2\pi i z b$. We set

$$l(f)(t) = \frac{1}{2\pi i}\int_0^t f'(s)f(s)^{-1}ds = \int_0^t zb(a + zb)^{-1}$$

Then $l(f)$ $(0) = 0$, $l(f)$ is a continuously differentiable A-valued function on $[0, 1]$, and

$$l(f)' = \frac{1}{2\pi i} f' f^{-1}.$$

It follows that the function $f^{-1}e^{2\pi i l(f)}$ has derivative zero and is 1 at $t = 0$. Hence $f(t) = e^{2\pi i l(f)(t)}$.

Now $l(f)(0) = 0$, but we may not have $l(f)(1) = 0$ (so $l(f)$ may not be in $M_n(\Omega A)$). We set $\lambda(g) = \lambda(f) = l(f)(1)$. Since $f(1) = 1$, $\lambda(g) \in Q_n(A)$. Obviously $\lambda(g)$ depends continuously on g and satisfies $\lambda(g_1 \oplus g_2) = \lambda(g_1) \oplus \lambda(g_2)$. It follows that λ defines a homomorphism $\lambda^*\colon G \to K_0(A)$.

If $p \in P_n$ is idempotent, then $\varphi(p) = (1 - p) + zp$ and

$$\varphi(p)^{-1} = (1 - p) + z^{-1}p.$$

Hence,

$$\lambda(\varphi(p)) = \int_0^1 zp[(1 - p) + z^{-1}p] = \int_0^1 p = p.$$

Since idempotents generate $K_0(A)$, we have $\lambda^* \circ \varphi^* = 1$.

If g is a linear loop and $f = g(0)^{-1}g$ then we set $h(t) = l(f)(t) - t\lambda(g)$ and note that $h(0) = h(1) = 0$ (since $\lambda(g) = l(f)(1)$). Hence, $h \in M_n(\Omega A)$ and $k = e^{2\pi i h} \in GL_n^0(\Omega A)$. Since $g = g(0)ke^{2\pi i t\lambda(g)} = g(0)k\varphi \circ \lambda(g)$, we have that g and $\varphi \circ \lambda(g)$ determine the same class in $K_1(\Omega A)$. Hence,

$$\psi^* = \psi^* \circ \varphi^* \circ \lambda^*. \qquad \blacksquare$$

9.5. Polynomial loops

According to the above Lemma, if we can prove ψ^* is an isomorphism then we will have $\varphi^* \circ \lambda^* = 1$ and, hence, φ^* is an isomorphism as well. We proceed to do this.

By a *polynomial loop* we mean an $f \in GL_n(\Omega A)$ of the form $f = \sum_{j=0}^{k} z^j a_j$ (with $z(t) = e^{2\pi i t}$ as before). Note that $f \oplus g$ is a polynomial loop if f and g are. As with linear loops, the homotopy classes of polynomial loops form an abelian semigroup under $\oplus$, where polynomial loops are identified with their trivial extensions.

LEMMA. *The universal group of the semigroup of homotopy classes of polynomial loops is isomorphic to* $K_1(\Omega A)$ *(via the inclusion map).*

Proof. It is an easy exercise to prove that elements of ΩA with finite Laurent series expansions $\sum_{-k}^{k} a_j z^j$ are dense in ΩA. One method is to use a partition of unity argument to prove that finite sums $\sum f_i a_i$ with $f_i \in C(\mathbb{T})$ and $a_i \in A$ are dense in ΩA, and then to use the Stone–Weierstrass Theorem to prove that the functions in $C(\mathbb{T})$ with finite Laurent expansions are dense.

Since $GL_n^0(\Omega A)$ is open in $M_n(\Omega A)$, it follows that for each loop $f \in GL_n(\Omega A)$ we can find a loop

$$g = \sum_{-k}^{k} z^j a_j \in GL_n(\Omega A)$$

which is equivalent to $f \bmod GL_n^0(\Omega A)$. For such a loop g (a Laurent loop) we have that $z^k g$ is a polynomial loop—that is, $g = z^{-k} h$ for a polynomial loop h and a non-negative integer k. Since z^{-k} is the inverse of the polynomial loop z^k, we have that the classes $[h] \in K_1(\Omega A)$, for h a polynomial loop, form a semigroup which generates $K_1(\Omega A)$.

Now suppose two polynomial loops f, g determine the same class in $K_1(\Omega A)$. Then (by taking trivial extensions) we may assume that f and g are connected by an arc in $GL_n(\Omega A)$. We may choose points $f = f_0, f_1, \ldots, f_m = g$ on this arc so close together that the line segments joining f_i and f_{i+1} still lie in $GL_n(\Omega A)$. Now by approximating the f_i's $(0 < i < m)$ by Laurent loops g_i and connecting these by line segments, we may construct a piecewise linear arc from f to g consisting entirely of Laurent loops. If we multiply this arc by z^k for large enough k we get an arc of polynomial loops connecting $z^k f$ to $z^k g$.

Since $z^k f \oplus I$ and $f \oplus z^k$ are connected by an arc of polynomial loops, as are $z^k g \oplus I$ and $g \oplus z^k$, we have that $[f] = [g]$ in $K_1(\Omega A)$ if and only if

$f \oplus h$ and $g \oplus h$ are homotopic in the set of polynomial loops for some polynomial loop h. Since this is precisely the relation we factor out on passing to the universal group of the semigroup of homotopy classes of polynomial loops, the proof is complete. ■

The next lemma completes the proof of Theorem 9.2:

9.6. LEMMA. *The map $\psi^*: G \to K_1(\Omega A)$ is an isomorphism.*

Proof. We use the above description of $K_1(\Omega A)$ and construct an inverse μ^* for ψ^*.

For a polynomial loop

$$f = \sum_0^k z^j a_j \in GL_n(\Omega A),$$

we let $\mu_k(f)$ be the $n(k+1) \times n(k+1)$ matrix

$$\mu_k(f) = \begin{pmatrix} a_0 & a_1 & \cdot & \cdot & \cdot & a_{k-1} & a_k \\ -zI_n & I_n & \cdot & \cdot & \cdot & 0 & 0 \\ \cdot & \cdot & & & & \cdot & \cdot \\ \cdot & \cdot & & & & \cdot & \cdot \\ \cdot & \cdot & & & & \cdot & \cdot \\ 0 & 0 & \cdot & \cdot & & -zI_n & I_n \end{pmatrix}$$

Obviously $\mu_k(f)$ is a linear loop. Furthermore, it is equivalent to $f \oplus I_{nk}$ modulo $GL^0_{n(k+1)}(\Omega A)$. In fact, we can construct $f \oplus I_{nk}$ from $\mu_k(f)$ by the following process: add z times the kth column to the $(k-1)$st column and then add $-a_k$ times the kth row to the first row. At this stage we have the matrix $b \oplus I_n$ where

$$b = \begin{pmatrix} a_0 & a_1 & \cdot & \cdot & \cdot & a_{k-2} & a_{k-1} & +za_k \\ -zI_n & I_n & \cdot & \cdot & \cdot & 0 & & 0 \\ \cdot & \cdot & & & & \cdot & & \cdot \\ \cdot & \cdot & & & & \cdot & & \cdot \\ \cdot & \cdot & & & & \cdot & & \cdot \\ 0 & 0 & \cdot & \cdot & \cdot & -zI_n & & I_n \end{pmatrix}$$

Now repeat the process. After k steps we arrive at $f \oplus I_{nk}$. It follows from Proposition 5.2 that $\mu_k(f)$ and $f \oplus I_{nk}$ are equivalent.

Now since $\mu_k(f)$ is a linear loop, it determines a class in G. We claim that this class is independent of k (for $k \geqslant \deg(f)$). In fact, if $a_k = 0$ then

$$\mu_k(f) = \begin{pmatrix} \mu_{k-1}(f) & & & & 0 \\ \cdot & & & & \cdot \\ \cdot & & & & \cdot \\ 0 & \cdot & \cdot & \cdot & -zI_n & I_n \end{pmatrix}$$

and

$$t \to \begin{pmatrix} \mu_{k-1}(f) & & & & 0 \\ \cdot & & & & \cdot \\ \cdot & & & & \cdot \\ \cdot & & & & \cdot \\ 0 & \cdot & \cdot & \cdot & -tzI_n & I_n \end{pmatrix}$$

is an arc connecting this to $\mu_{k-1}(f) \oplus I_n$.

It follows that if we let $\tilde{\mu}(f)$ be the class in G of $\mu_k(f)$ for $k \geqslant \deg(f)$, then $\tilde{\mu}$ is a well defined map from polynomial loops into G. Obviously, if f and g are connected by an arc of polynomial loops, then, for large k, μ_k applied to this arc yields an arc of linear loops connecting $\mu_k(f)$ and $\mu_k(g)$ and, hence, $\tilde{\mu}(f) = \tilde{\mu}(g)$. Clearly, $\tilde{\mu}(f \oplus g) = \tilde{\mu}(f) \oplus \tilde{\mu}(g)$. We conclude that $\tilde{\mu}$ defines a homorphism $\mu^*: K_1(\Omega A) \to G$.

Since $f \oplus I_n$ and $\mu_k(f)$ are connected by an arc of polynomial loops (for $k \geqslant \deg(f)$), we have that $\psi^* \circ \mu^* = 1$. On the other hand, if $g = a + bz$ is a linear loop, then

$$\mu_1(g) = \begin{pmatrix} a & b \\ -z & I_n \end{pmatrix} \text{ and } g \oplus I_n = \begin{pmatrix} a + zb & 0 \\ 0 & I_n \end{pmatrix}$$

are connected by an arc of linear loops. Hence, $\mu^* \circ \psi^* = 1$. This completes the proof. ∎

9.7. THEOREM. *For a non-unital algebra A, there is a natural isomorphism $\beta^*: \tilde{K}_0(A) \to \tilde{K}(SA)$.*

Proof. Recall that $K_0(A^+)$ factors as $\tilde{K}_0(A) \oplus \tilde{K}_0(\mathbb{C})$ where $\tilde{K}_0(A)$ consists of classes $[p] - [I_k]$ with k = rank of p modulo A. The copy of $\tilde{K}_0(\mathbb{C}) = \mathbb{Z}$ is generated by the classes $[I_k]$.

If we consider SA to be an ideal in ΩA^+, then there is a similar decomposition

$$K_1(\Omega A^+) = \tilde{K}_1(SA) \oplus K_1(C(\mathbb{T})),$$

where $\mathbb{T}$ is a circle. To see this, note that each $f \in GL_n(\Omega A^+)$ has a unique factorization $f = gh$ where g is constant mod A and $h \in GL_n(C(\mathbb{T}))$ has scalar entries. Thus $GL_n(\Omega A^+)$ is the product of $GL_n((SA)^+)$ and

$$GL_n(C(\mathbb{T})) = GL_n(\Omega\mathbb{C}).$$

It follows that $K_1(\Omega A^+)$ decomposes as above.

Now $e^{2\pi itp} \in GL_n(C(\mathbb{T}))$ if $p = I_k$ and so β^* maps $\tilde{K}_0(\mathbb{C})$ into $K_1(C(\mathbb{T}))$. Similarly, β^* maps $\tilde{K}_0(A)$ into $\tilde{K}_1(SA)$. Since $\beta^*: K_0(A^+) \to K_1(\Omega A^+)$ is an isomorphism, so is its restriction $\beta^*: \tilde{K}_0(A) \to \tilde{K}_1(SA)$.

9.8. Corollary. *The compositions*

$$\alpha^* \circ \beta^*: \tilde{K}_0(A) \to \tilde{K}_0(S^2A) \quad \textit{and} \quad \beta^* \circ \alpha^*: \tilde{K}_1(A) \to \tilde{K}_1(S^2A)$$

are isomorphisms.

10. K-THEORY

We are now prepared to use the functors K_0 and K_1 to define a cohomology theory (K-theory) for compact pairs, and to discuss some of its properties.

First we show that the periodicity theorem can be used to extend the six term exact sequence of 8.2.

10.1. A long exact sequence

Let I be a closed ideal in the commutative Banach algebra A and consider the diagram

$$\begin{array}{ccccccccccc}
\tilde{K}_0(I) & \to & \tilde{K}_0(A) & \to & \tilde{K}_0(A/I) & \dashrightarrow & \tilde{K}_1(I) & \to & \tilde{K}_1(A) & \to & \tilde{K}_1(A/I) \\
\downarrow\beta^* & & \downarrow\beta^* & & \downarrow\beta^* & & \downarrow\alpha^* & & \downarrow\alpha^* & & \downarrow\alpha^* \\
\tilde{K}_1(SI) & \to & \tilde{K}_1(SA) & \to & \tilde{K}_1(SA/SI) & \xrightarrow{\varepsilon^*} & \tilde{K}_0(SI) & \to & \tilde{K}_0(SA) & \to & \tilde{K}_0(SA/SI)
\end{array}$$

where α^* and β^* are the isomorphisms of 8.4 and 9.7 and the bottom sequence is the exact sequence of 8.2. It is important to note here that $SA/SI = S(A/I)$, so that $\beta^*: \tilde{K}_0(A/I) \to \tilde{K}_1(SA/SA)$ is defined. The fact that α^* and β^* are natural transformations means that the diagram commutes.

It follows from the diagram that we can fill in the dotted arrow with the map $\delta_0^* = (\alpha^*)^{-1} \circ \delta^* \circ \beta^*: \tilde{K}_0(A/I) \to \tilde{K}_1(I)$ and have the top row of the diagram exact as well. Together with 8.2, this yields:

Theorem. *For a closed ideal I in a commutative Banach algebra A there is a repeating exact sequence,*

$$\cdots \to \tilde{K}_1(A) \xrightarrow{\pi^*} \tilde{K}_1(A/I) \xrightarrow{\delta^*} \tilde{K}_0(I) \xrightarrow{i^*} \tilde{K}_0(A) \xrightarrow{\pi^*} \tilde{K}_0(A/I) \xrightarrow{\delta^*} \tilde{K}_1(I) \xrightarrow{i^*} \tilde{K}_1(A) \to \cdots$$

If one wants to go to the trouble of recalling the constructions of the maps α^*, β^*, and δ^*, he can verify that $\delta_0^* = (\alpha^*)^{-1} \circ \delta^* \circ \beta^*$ is induced in the following natural fashion from the exponential map: For an element $a \in Q_n((A/I)^+)$, we choose a pre-image $b \in M_n(A^+)$. Since $\pi(e^{2\pi ib}) = e^{2\pi ia} = 1$, we have $e^{2\pi ib} \in GL_n(I^+)$. If we let $\delta_0(a)$ be the class of $e^{2\pi ib}$ in $K_1(I^+) = \tilde{K}_1(I)$, then δ_0 induces the map δ_0^* above. Why not define δ_0^* this way and prove directly that it makes the sequence exact? We tried this and found ourselves reproving the periodicity theorem in the process. In fact, if one assumes the sequence is exact with δ_0^* as defined above by the exponential map, then the periodicity theorem follows immediately upon applying the sequence for the ideal $SA = \{f \in CA : f(1) = 0\}$, algebra CA, and factor algebra $CA/SA \simeq A$.

10.2. K-theory for compact pointed spaces

If X is a compact pointed space with base point x_0, we set

$$C_0(X) = \{f \in C(X) : f(x_0) = 0\}.$$

In other words, $C_0(X)$ is just the algebra $C_0(X \backslash \{x_0\})$ of functions which vanish at infinity on the locally compact space $X \backslash \{x_0\}$. Note that $C_0(X)^+ = C(X)$.

Definition. If X is a compact pointed space and p is an integer, we set

$$\tilde{K}^p(X) = \tilde{K}_0(C_0(X))$$

if p is even and

$$\tilde{K}^p(X) = \tilde{K}_1(C_0(X))$$

if p is odd.

Each $\tilde{K}^p$ is a cofunctor from compact pointed spaces to abelian groups. In fact, if $\varphi : X \to Y$ is a map of pointed spaces, then $f \to f \circ \varphi : C_0(Y) \to C_0(X)$ is an algebra homomorphism. This induces a map $\tilde{K}_i(C_0(Y)) \to \tilde{K}_i(C_0(X))$ ($i = 0, 1$) and, hence, a map $\varphi^* : \tilde{K}^p(Y) \to \tilde{K}^p(X)$ for each p. The results of the next two subsections show that $\{\tilde{K}^p\}$ forms a generalized reduced cohomology theory.

10.3. Exactness

If Y is a closed pointed subspace of the compact pointed space X, then $C_0(X/Y)$ may be identified with the ideal in $C_0(X)$ of all functions vanishing on Y. Furthermore, the factor algebra $C_0(X)/C_0(X/Y)$ is isomorphic to

$C_0(Y)$. Hence, with $I = C_0(X/Y)$, $A = C_0(X)$, and $A/I = C_0(Y)$, the exact sequence of 10.1 yields:

PROPOSITION. *The sequence of cofunctors* $\{\tilde{K}^p\}$ *satisfies the exactness axiom.*

10.4. Continuity

We let $\{X_\alpha\}$ be a directed family of pointed compact subsets of a compact pointed space Y and set $X = \bigcap X_\alpha$.

PROPOSITION. *We have* $\tilde{K}^p(X) = \varinjlim \tilde{K}^p(X_\alpha)$ *for each p and, hence,* $\{\tilde{K}^p\}$ *satisfies the continuity axiom.*

Proof. We carry out the proof for p odd. Hence $\tilde{K}^p(X) = \tilde{K}^1(X) = \tilde{K}_1(C_0(X)) = K_1(C(X))$.

If $f \in GL_n(C(X))$ then by the Tietze extension theorem we can extend f to a matrix $\tilde{f}$ in $M_n(C(Y))$. Since, on X, $\det \tilde{f} = \det f$ is non-vanishing, we must have $\det \tilde{f}$ non-vanishing in a neighbourhood of X and, hence, on some X_α. Thus $\tilde{f}_{|X_\alpha}$ determines a class in $\tilde{K}^1(X_\alpha) = K_1(C(X_\alpha))$ which restricts to $[f] \in \tilde{K}^1(X)$.

Now suppose $g \in GL_n(C(X_\alpha))$ is such that $g_{|X}$ determines the zero class in $\tilde{K}_1(X)$. By passing to trivial extensions, we may assume that $g_{|X} \in GL_n^0(C(X))$. Since restriction $M_n(C(Y)) \to M_n(C(X))$ is onto, it follows from 4.8 that $g_{|X}$ has an extension $h \in GL_n^0(C(Y))$. Since $h_{|X} = g_{|X}$, h and g will be close enough on some $X_\beta \subset X_\alpha$ that $(h_{|X_\beta})^{-1}(g_{|X_\beta})$ has a continuous logarithm on X_β. It follows that $g_{|X_\beta} \in GL_n^0(X_\beta)$ and, hence, that restriction $\tilde{K}^1(X_\alpha) \to \tilde{K}^1(X_\beta)$ maps $[g]$ to zero. This completes the proof that $\tilde{K}^1$ is continuous.

That $\tilde{K}^0 = \tilde{K}^p$ (p even) is continuous can be proved in a similar manner or by appealing to the isomorphism $\beta^*: \tilde{K}_0(A) \to \tilde{K}_1(SA)$. ∎

10.5. Periodicity

If X is a compact pointed space and SX its reduced suspension, then clearly $C_0(SX)$ is the suspension $SC_0(X)$ of $C_0(X)$. Hence, from 9.8 we have

THEOREM. *There are natural isomorphisms* $\alpha^*: \tilde{K}^1(X) \to \tilde{K}^0(SX)$ *and* $\beta^*: \tilde{K}^0(X) \to \tilde{K}^1(SX)$. *Hence,* $\tilde{K}^p(X) \simeq \tilde{K}^p(S^2X)$ *for all p.*

10.6. COROLLARY. *If* S^k $(k \geqslant 0)$ *is the k-sphere, then* $\tilde{K}^p(S^k) = \mathbb{Z}$ *if p and k have the same parity (are both even or both odd) and* $\tilde{K}^p(S^k) = 0$ *otherwise.*

Proof. By periodicity, it suffices to prove this for $k = 0$. However,

$$\tilde{K}^p(S^0) = \tilde{K}_0(\mathbb{C}) = H_0(\mathbb{C}) = \mathbb{Z}$$

if p is even and

$$\tilde{K}^p(S^0) = \tilde{K}_1(\mathbb{C}) = K_1(\mathbb{C}) = 0$$

if p is odd. ■

10.7. *K*-theory for compact pairs

According to 1.9, we get a generalized cohomology theory (K-theory for compact pairs) from our generalized reduced cohomology theory $\{\tilde{K}^p\}$ if we set

$$K^p(X, Y) = \tilde{K}^p(X/Y) \quad \text{for} \quad \varnothing \neq Y \subset X$$

and

$$K^p(X, \varnothing) = K^p(X) = \tilde{K}^p(X^+).$$

We show how $K^p(X, Y)$ and $K^p(X)$ relate to our functors K_0 and K_1.

If (X, Y) is a compact pair ($Y \neq \varnothing$) then

$$K^p(X, Y) = \tilde{K}^p(X/Y) = \tilde{K}_p(C_0(X/Y)) \qquad (p = 0, 1).$$

Similarly,

$$K^p(X) = \tilde{K}^p(X^+) = \tilde{K}_p(C_0(X^+)) = \tilde{K}_p(C(X)) = K_p(C(X))$$

for X compact.

Now suppose A is a commutative Banach algebra. In view of the Arens theorem we have that the Gelfand transform induces an isomorphism

$$\tilde{K}_1(A) \to \tilde{K}_1(C_0(\Delta)) = \tilde{K}^1(\Delta^+) \quad \text{(cf. 5.7)}.$$

From the fact that the maximal ideal space of $(SA)^+$ is $S\Delta^+$ we conclude that $\tilde{K}_1(SA) \to \tilde{K}^1(S\Delta^+)$ is also an isomorphism. On applying the isomorphism $\beta^*: \tilde{K}_0(A) \to \tilde{K}_1(SA)$, we conclude that $\tilde{K}_0(A) \to \tilde{K}^0(\Delta^+)$ is an isomorphism. If A happens to be unital we have $K_p(A) = \tilde{K}_p(A)$ and $K^p(\Delta) = \tilde{K}^p(\Delta^+)$ ($p = 0, 1$). Hence:

THEOREM. (Novadvorskii [**14**]). *The Gelfand transform $A \to C(\Delta)$ induces isomorphisms $K_p(A) \to K^p(\Delta)$ ($p = 0, 1$) for A unital and $\tilde{K}_p(A) \to \tilde{K}^p(\Delta^+)$ for A non-unital.*

10.8. Projective modules

In this subsection we give a more standard description of $K_0(A)$.

Let A be any unital algebra (not necessarily a Banach algebra). An A-*module* is then a vector space M together with a bilinear map $(a, m) \to am: A \times M \to M$ which is associative ($a(bm) = (ab)m$) and unital ($1m = m$). If M and N are A-modules, then a homomorphism $\varphi: M \to N$ is a linear map such that $\varphi(am) = a\varphi(m)$. A *submodule* N of an A-module M is a subspace such that $aN \subset N$ for each $a \in A$.

The direct sum $\bigoplus_\alpha M_\alpha$ of a family of A-modules is the vector space direct sum with the coordinatewise module operation $a(m_\alpha) = (am_\alpha)$.

Now A itself is an A-module under left multiplication. A module which is a direct sum of any number of copies of A is called a *free module*. That is, a module is free if it has a basis $\{e_\alpha\}$ such that each $a \in A$ can be written in a unique way as a finite sum $\sum a_\alpha e_\alpha$ with $a_\alpha \in A$. If A is a field then an A-module is just a vector space and, hence, always has a basis. However, in general A-modules need not be free if A is not a field.

If M is an A-module such that $M \oplus N$ is free for some A-module N, then M is called *projective*. Projective modules share many of the nice properties of free modules. The question of which modules are projective is central to the theory of homological algebra. The question of which projective (finitely generated) modules are free is the central question of algebraic K-theory (cf. [**5**]).

A module M is projective and finitely generated if $M \oplus N \simeq \overset{k}{\oplus} A$ for some module N and some integer k. In this case, the projection $(m, n) \to (m, 0)$ is an idempotent homomorphism of $\overset{k}{\oplus} A$ into itself. However, the homomorphisms of $\overset{k}{\oplus} A$ into itself are all determined by matrices in $M_k(A)$. Hence, the summand M of $\overset{k}{\oplus} A$ corresponds to an idempotent in $M_k(A)$. Conversely, if $p \in M_k(A)$ is idempotent, then $\overset{k}{\oplus} A = M \oplus N$ for the submodules $M = p(\overset{k}{\oplus} A)$ and $N = (1 - p)(\overset{k}{\oplus} A)$.

Proceeding in this way, it is not difficult to prove:

PROPOSITION. *The group $K_0(A)$ is isomorphic to the universal group of the semigroup (under $\oplus$) of isomorphism classes of finitely generated projective A-modules. Under this isomorphism, the classes $[I_k]$ correspond to the free modules $\overset{k}{\oplus} A$.*

If $A = C(X)$ then the isomorphism classes of finitely generated projective A-modules are in one to one correspondence with the isomorphism classes of complex vector bundles over X. This leads to the standard definition of K-theory in terms of complex vector bundles (cf. [**4**]).

10.9. The Chern character

If we set $K^*(X) = K^0(X) \oplus K^1(X)$ then $K^*(X)$ has a ring structure, where the multiplication is induced from the tensor product operation for $C(X)$-

modules (or vector bundles) (cf. [4]). Similarly, Čech cohomology has a ring structure (cup product) whenever the coefficient group is a ring (cf. [8]).

If $H^*(X, Q)$ is the infinite direct product of the Čech groups $H^p(X, Q)$ (rational coefficients), then $H^*(X, Q)$ is a ring. Furthermore, there exists a ring homomorphism (which we wouldn't attempt to describe here)

$$\mathrm{ch}: K^*(X) \to H^*(X, Q)$$

called the *Chern character* (cf. [7]). We list some of the properties of the Chern character without proof.

PROPOSITION. (a) *The Chern character maps $K^0(X)$ into the product of the even dimensional Čech groups and $K^1(X)$ into the product of the odd dimensional Čech groups;*

(b) *the kernel and cokernel of* ch *are torsion groups and, hence, $K^*(X) \otimes Q$ is isomorphic to $H^*(X, Q)$.*

The above is usually proved only for X a finite cell complex. However, the conclusions are preserved by passing to direct limits and, hence, they follow for general compact X by continuity.

10.10. Remark

Two facts which hold in the category of finite cell complexes, but which fail in the category of compact spaces are given by:

PROPOSITION. (cf. [7]) (a) *If $\varphi: X \to Y$ is a continuous map for which $\varphi^*: H^*(Y, \mathbb{Z}) \to H^*(X, \mathbb{Z})$ is an isomorphism, then $\varphi^*: K^*(Y) \to K^*(X)$ is also an isomorphism.*

(b) *If $H^*(X, \mathbb{Z})$ is torsion free, then so is $K^*(X)$ and, hence, the Chern character* $\mathrm{ch}: K^*(X) \to H^*(X, Q)$ *is one to one.*

We raised the question of whether or not this result extends to the category of compact spaces during the "Algebras in Analysis" conference. Several conference participants were kind enough to contact their colleagues among British topologists. Before the conference ended an answer came back from Andrew Casson and Frank Adams at Cambridge:

Based on Theorem 1.7 of Adams's paper [1], one can construct a sequence $\{\alpha_n: Y_n \to Y_{n-1}\}$ of maps between finite cell complexes such that each α_n induces an isomorphism on K-theory but induces the zero map on reduced cohomology. Then $Y = \varprojlim Y_n$ is a compact space with trivial cohomology but non-trivial K-theory. This supplies an immediate counter-example to both parts of the above proposition for compact spaces.

10.11. The group $H^2(\Delta, \mathbb{Z})$.

We end this section with a brief description of Forster's characterization of the second Čech group $H^2(\Delta, \mathbb{Z})$.

If P and Q are finitely generated projective A-modules, then $P \otimes Q$ is also, where $P \otimes Q$ denotes the A-module tensor product $P \otimes_A Q$. In fact, if $P = p(\overset{n}{\oplus} A)$ and $Q = q(\overset{m}{\oplus} A)$ for idempotent matrices p and q, respectively, then

$$P \otimes Q = (p \otimes q)((\overset{n}{\oplus} A) \otimes (\overset{m}{\oplus} A)) = (p \otimes q)(\overset{nm}{\oplus} A),$$

where $p \otimes q$ is the $nm \times nm$ matrix whose entries are all products of an entry from p and an entry from q.

The set of isomorphism classes of finitely generated projective A-modules forms an abelian semigroup S under $\otimes$ (just as it does under $\oplus$). The free module A on one generator determines an identity for S. Hence, the subset of S consisting of elements with inverses is a group in S. This is the *Picard group* for A and is denoted Pic(A) (cf. [**5**]).

It is not difficult to see that a finitely generated module P determines an element of Pic(A) if and only if it has rank 1 in the sense that whenever $P = p(\overset{n}{\oplus} A)$ for an idempotent matrix p, $\hat{p}(x)$ is a rank 1 complex matrix for each $x \in \Delta$. By assigning to each $x \in \Delta$ the vector space $\hat{p}(x)\mathbb{C}^n$, one constructs a complex line bundle (one dimensional vector bundle) for each such P. This determines a homomorphism from Pic(A) to the group (under tensor product of bundles) of isomorphism classes of complex line bundles on Δ. The latter group is canonically isomorphic to $H^2(\Delta, \mathbb{Z})$. Forster's theorem [**9**] is the following:

THEOREM. *The map* Pic$(A) \to H^2(\Delta, \mathbb{Z})$ *is an isomorphism.*

For a proof, see [**9**] or our paper [**20**].

11. APPLICATIONS OF COHOMOLOGY TO HARMONIC ANALYSIS

The Shilov and Arens–Royden theorems discussed in Section 4 characterize the zero and one dimensional Čech cohomology groups for the maximal ideal space of an algebra A. With these results, cohomology becomes a powerful tool for the study of certain kinds of questions about Banach algebras. In this section we illustrate this point by applying cohomology to the study of measure algebras on locally compact abelian groups.

If a Banach algebra A is sufficiently simple (for example, if $\hat{A}$ is dense in $C(\Delta)$) then the Shilov and Arens–Royden theorems for A can be obtained by elementary methods. Hence, for really deep applications of these results one must look at fairly complicated algebras. The measure algebra on a locally compact group is a very complicated algebra.

The first four subsections of this section are devoted to a quick review of elementary harmonic analysis. For details see [**16**].

11.1. L.c.a. groups

By a locally compact abelian group G (an l.c.a. group) we shall mean a topological group which is abelian as a group and locally compact as a topological space. Examples include $\mathbb{R}^n$ under addition, the circle group $\mathbb{T} = \{z \in \mathbb{C}: |z| = 1\}$ under multiplication, the torus $\mathbb{T}^n = \mathbb{T} \times \ldots \times \mathbb{T}$ (n times), and any abelian group with the discrete topology.

If G is an l.c.a. group then the dual group $\hat{G}$ is the group (under pointwise multiplication) of all continuous group homomorphisms $\gamma: G \to \mathbb{T}$. If $\hat{G}$ is given the compact-open topology then $\hat{G}$ is also an l.c.a. group. Furthermore, the dual $\hat{\hat{G}}$ of $\hat{G}$ is naturally topologically isomorphic to G. If G is compact then $\hat{G}$ is discrete and vice-versa.

It is easy to see that the dual group of $\mathbb{R}$ consists of the functions $t \to e^{ixt}$ for $x \in \mathbb{R}$ and, hence, that $\hat{\mathbb{R}}$ is isomorphic to $\mathbb{R}$. Similarly, $\hat{\mathbb{T}}$ consists of the functions $z \to z^n$ for $n \in \mathbb{Z}$ and, hence, is isomorphic to $\mathbb{Z}$.

11.2. Convolution

Let G be an l.c.a. group. By $M(G)$ we shall mean the Banach space of all finite regular Borel measures on G. That is, $M(G)$ is the dual space of the Banach space $C_0(G)$.

If $\mu, \nu \in M(G)$ then the equation

$$\int f \, d\mu * \nu = \int\int f(xy) \, d\mu(x) \, d\nu(y) \quad (f \in C_0(G))$$

defines a measure $\mu * \nu \in M(G)$. The operation $(\mu, \nu) \to \mu * \nu$ is associative and commutative and satisfies $\|\mu * \nu\| \leqslant \|\mu\| \, \|\nu\|$. Hence, $M(G)$ is a commutative Banach algebra under the operation $*$ (*convolution*).

If $x \in G$ then $\delta_x \in M(G)$ denotes the unit point mass at x. If e is the identity of G, then δ_e is an identity for $M(G)$. Hence, $M(G)$ is a unital algebra.

11.3. Group algebras

On each l.c.a. group G there exists a (generally non-finite) positive, inner regular measure m which is translation invariant ($m(xE) = m(E)$), finite on compact sets, and positive (or infinite) on open sets. This measure is unique up to a multiplicative constant and is called *Haar measure* for G.

If we denote by $L(G)$ the space of measures in $M(G)$ which are absolutely continuous with respect to Haar measure m, then $L(G)$ is a closed ideal in $M(G)$. The map which assigns to $\mu \in L(G)$ its Radon–Nikodym derivative with respect to m, is an isometry of $L(G)$ onto $L^1(G) = L^1(m)$. Under this map, the convolution operation for measures corresponds to the operation $(f, g) \to f*g$ on $L^1(G)$, where

$$f*g(x) = \int f(xy^{-1})g(y)\,dm(y).$$

We call the algebra $L(G)$ the *group algebra of* G. It is non-unital unless G is discrete, in which case m is counting measure and $M(G) = L(G)$.

In case $G = \mathbb{R}$, Haar measure is just Lebesgue measure and $L^1(\mathbb{R}) \simeq L(\mathbb{R})$ is just the usual space of Lebesgue integrable functions.

11.4. Maximal ideal space of $M(G)$

If G is an l.c.a. group, then each $\gamma \in \hat{G}$ determines a complex homomorphism $\mu \to \hat{\mu}(\gamma)$ of $M(G)$, where

$$\hat{\mu}(\gamma) = \int \gamma(x)\,d\mu(x)$$

In fact, $(\mu*\nu)\hat{}(\gamma) = \int \gamma\,d\mu*\nu = \int\int \gamma(xy)\,d\mu(x)\,d\nu(y)$

$$= \int\int \gamma(x)\,\gamma(y)\,d\mu(x)\,d\nu(y) = \hat{\mu}(\gamma)\,\hat{\nu}(\gamma).$$

Thus, $\hat{G}$ is embedded as a subset of the maximal ideal space Δ of $M(G)$. However, unless G is discrete Δ is very much larger than $\hat{G}$ (cf. [**16**] and [**19**]).

Now each complex homomorphism $\mu \to \hat{\mu}(\gamma)$ $(\gamma \in \hat{G})$ restricts to a non-zero complex homomorphism of the ideal $L(G)$. Actually, by using the fact that the dual space of $L(G)$ is $L^\infty(G) = L^\infty(m)$, one can easily prove that each complex homomorphism of $L(G)$ has this form. In fact, the maximal ideal space of the non-unital algebra $L(G)$ is precisely $\hat{G}$.

11.5. Other topologies on G

Let G be an l.c.a. group. By a refining topology on G we shall mean a topology which is at least as strong as the given topology and under which G remains an l.c.a. group. If τ is such a topology, then G_τ will represent the l.c.a. group which is G as a group, but which has the topology τ. Note that the discrete topology is always a refining topology.

If τ is a refining topology for G, then $M(G_\tau)$ may be identified as the subalgebra of $M(G)$ consisting of measures which remain regular for the τ-topology (the τ-topology may have fewer compact sets and so it is harder for a measure to be τ-regular). In particular, we may consider the group algebra $L(G_\tau)$ to be a subalgebra of $M(G)$.

If $L(G_\tau)^+$ is $L(G_\tau)$ with an identity adjoined, then we can map $L(G_\tau)^+$ into $M(G)$ by sending $\lambda + \mu$ ($\mu \in L(G_\tau)$, $\lambda \in \mathbb{C}$) to $\lambda\delta_e + \mu$. This is an isometry unless $L(G_\tau)$ already contains δ_e—which happens if and only if $\tau = d$ is the discrete topology.

11.6. Cohomology

For each refining topology τ on G, the canonical map $L(G_\tau)^+ \to M(G)$ induces a map $\Delta \to \hat{G}_\tau^+$ from the maximal ideal space of $M(G)$ to the maximal ideal space of $L(G_\tau)^+$. Hence, for each generalized cohomology theory $\{H^p\}$ we have induced maps $H^p(\hat{G}_\tau^+) \to H^p(\Delta)$. If we restrict each such map to $\tilde{H}^p(\hat{G}_\tau^+) \subset H^p(\hat{G}_\tau^+)$ and take the direct sum over all refining topologies τ, we have a homomorphism $\bigoplus_\tau \tilde{H}^p(\hat{G}_\tau^+) \to \mathrm{H}^p(\Delta)$ for each p. The following theorem shows that, although Δ is very complicated, its cohomology is rather simple:

THEOREM. (cf. [**9**]). *Let $\{H^p\}$ be any cohomology theory and let Δ be the maximal ideal space of $M(G)$. Then the map $\bigoplus_\tau \tilde{H}^p(\hat{G}_\tau^+) \to H^p(\Delta)$ is an isomorphism for each p, where the direct sum is over all refining topologies τ for G.*

The proof of this theorem is very complicated and we shall do no more than make a few comments regarding it. We refer the reader to [**19**] for a complete discussion. The proof breaks into two parts (Chapters 6 and 7 of [**19**]). In the first part we use the fact that Δ has natural semigroup structure, and the continuity, exactness, and dimension axioms, to prove that $H^p(\Delta)$ is the direct sum of subgroups of the form $\tilde{H}^p(\Gamma_h^+)$, where the Γ_h's are certain locally compact subgroups of the semigroup Δ. In the second part, we use some deep analysis in $M(G)$ to prove that each such subgroup Γ_h corresponds to a unique τ such that $\Delta \to \hat{G}_\tau^+$ maps Γ_h homeomorphically onto $\hat{G}_\tau$.

We have not yet been able to prove the analogous theorem for generalized cohomology theories (such as K-theory). The difficulty is that the dimension axiom enters into the proof in a certain induction argument. Without the dimension axiom there is no place to start the induction.

11.7. Cohomology of compact groups

If G is an l.c.a. group then G has an open subgroup of the form $\mathbb{R}^n \times K$ where K is a compact group (cf. [**16**]). Since $(\mathbb{R}^n \times K)^+ = S^n K^+$ is the nth suspension of K^+, we have $\tilde{H}^p((\mathbb{R}^n \times K)^+ = \tilde{H}^{p-n}(K^+) = H^{p-n}(K)$ by Proposition 2.5. Hence, in principle, to compute the groups $\tilde{H}^p(\hat{G}_\tau^+)$ of the previous theorem, we need only be able to compute the cohomology groups of compact groups. This is accomplished in [**11**]. For our purposes it is enough to state the result in the case $p = 1$, where it has an elementary proof.

Thus, let K be a compact abelian group. Recall from Section 3 that $H^1(K, \mathbb{Z}) \simeq C(K)^{-1}/(\exp(C(K))$. Now each $\gamma \in K$ is, in particular, an element of $C(K)^{-1}$. Hence, the inclusion $\hat{K} \to C(K)^{-1}$ determines a group homomorphism $\hat{K} \to C(K)^{-1}/\exp(C(K))$. In fact, we have:

PROPOSITION. *Let K be a compact abelian group. Then the map $\hat{K} \to C(K)^{-1}/\exp(C(K))$ is an isomorphism and, hence, $H^1(K, \mathbb{Z})$ is isomorphic to $\hat{K}$.*

An elementary proof of this result appears in [**19**].

11.8. Cohen's idempotent theorem

In the case where $p = 0$, Theorem 11.6 says that

$$H^0(\Delta, \mathbb{Z}) \simeq \bigoplus_\tau \tilde{H}^0(\hat{G}_\tau^+, \mathbb{Z}).$$

However, $H^0(\Delta, \mathbb{Z})$ and $\tilde{H}^0(G_\tau^+, \mathbb{Z})$ are, respectively, the additive groups generated by the idempotents in $M(G)$ and $L(G_\tau)$ (by Theorem 4.4). Now an elementary argument shows that in $L(G_\tau)$ this subgroup is generated by idempotents of the form $\gamma\mu$, where $\gamma \in \hat{G}$ and μ is Haar measure on some compact subgroup of G_τ (cf. [**16**]). Since compact subgroups of G_τ are also compact subgroups of G, we have the following interpretation of Theorem 11.6 in the case $p = 0$:

THEOREM. (Cohen). *The subgroup of $M(G)$ generated by idempotents in $M(G)$ agrees with the subgroup generated by idempotents $\gamma\mu$ for $\gamma \in \hat{G}$ and μ the Haar measure of some compact subgroup.*

This is a potent result. In particular, it can be used to determine the homomorphisms of $L(G)$ into $L(H)$ for l.c.a. groups G and H (cf. [**16**]). Cohen's original proof is quite different from ours; however, it is interesting to note that the theorem can be obtained as a special case of a general result on the cohomology of Δ.

11.9. A factorization theorem

We now turn to application of Theorem 11.6 in the case $p = 1$.

The map $L(G_\tau)^+ \to M(G)$ induces a homomorphism

$$(L(G_\tau)^+)^{-1}/\exp(L(G_\tau)^+) \to M(G)^{-1}/\exp(M(G)).$$

If we use the Arens–Royden theorem (Theorem 4.7) to identify these two groups with the Čech groups $H^1(\hat{G}_\tau^+, \mathbb{Z}) = \tilde{H}^1(\hat{G}_\tau^+, \mathbb{Z})$ and $H^1(\Delta, \mathbb{Z})$, respectively, then we can interpret 11.6 in the following way when $p = 1$:

THEOREM. *Each $\mu \in M(G)^{-1}$ has a factorization as $\mu = \nu_1 * \ldots * \nu_n * e^{\omega}$, with $\omega \in M(G)$ and each $\nu_i \in (L(G_{\tau_i})^+)^{-1}$ for some refining topology τ_i. Furthermore, each ν_i is unique modulo* $\exp(L(G_{\tau_i})^+)$.

This theorem has several implications. In particular, it leads to a considerable simplification of the problem of determining when a measure in $M(G)$ is invertible (cf. [**19**]).

11.10. The case $G = \mathbb{R}$

There are only two refining topologies on $\mathbb{R}$—the original topology and the discrete topology. Hence, by the above result, we have that each $\mu \in M(\mathbb{R})^{-1}$ has a factorization $\mu = \nu_1 * \nu_2 * e^{\omega}$ with $\nu_1 \in (L(\mathbb{R})^+)^{-1}$ and $\nu_2 \in L(\mathbb{R}_d)^{-1} = M(\mathbb{R}_d)^{-1}$, where $\mathbb{R}_d$ is $\mathbb{R}$ with the discrete topology.

Now the maximal ideal space of $L(\mathbb{R})^+$ is $\hat{\mathbb{R}}^+ = \mathbb{R}^+ = S^1$. Furthermore, $C(S^1)^{-1}/\exp(C(S^1)) \simeq H^1(S^1, \mathbb{Z})$ is isomorphic to the integers (cf. Section 4.7), where any function $f \in C(S^1)^{-1}$ of winding number one about zero can be chosen as the generator. Hence, we can choose ν_1 above to be of the form η^k for some $k \in \mathbb{Z}$ and η a measure in $L(\mathbb{R})^+$ such that $\hat{\eta}$ has winding number one.

Since $\hat{\mathbb{R}}_d$ is a compact group, Proposition 11.7 implies that each element of $C(\hat{\mathbb{R}}_d)^{-1}$ is equivalent mod $\exp(C(\hat{\mathbb{R}}_d))$ to a unique element $\gamma \in \hat{\hat{\mathbb{R}}}_d$. Each such γ has the form $\hat{\delta}_x$ for a unique $x \in \mathbb{R}_d$. Hence, the measure ν_2 above may be chosen to be δ_x for some $x \in \mathbb{R}$. Hence, we have:

THEOREM. *Let μ be any measure in $M(\mathbb{R})^{-1}$ and let $\eta \in L(\mathbb{R})^+$ be chosen so that $\hat{\eta}$ has winding number one. Then there exists a unique integer k and a unique real number x such that $\mu = \eta^k * \delta_x * e^{\omega}$ for some $\omega \in M(\mathbb{R})$.*

We call k and x the *discrete* and *continuous indices* of μ, respectively.

11.11. Wiener–Hopf operators

A key application of the above result is in the study of Wiener–Hopf operators with measure kernel (cf. [**6**]).

For $\mu \in M(\mathbb{R})$ and $f \in L^1(\mathbb{R})$ we write

$$C_\mu f(x) = \mu * f(x) = \int f(x - y)\, d\mu(y).$$

It follows that $C_\mu f \in L^1(\mathbb{R})$ and, in fact, $\mu \to C_\mu$ is an isomorphism of $M(\mathbb{R})$ onto the algebra of operators on $L^1(\mathbb{R})$ that commute with translation (cf. [**19**]. The operators C_μ are called *convolution operators.*

Let $L^1(\mathbb{R}^+)$ denote the subspace of $L^1(\mathbb{R})$ consisting of functions which vanish off $\mathbb{R}^+ = [0, \infty)$. Let $P: L^1(\mathbb{R}) \to L^1(\mathbb{R}^+)$ be the projection operator defined by $Pf(x) = f(x)$ for $x \in \mathbb{R}^+$ and $Pf(x) = 0$ for $x \notin \mathbb{R}^+$. (Note that in this section $\mathbb{R}^+$ has a meaning different from that of the previous section.) For $\mu \in M(\mathbb{R})$ the Weiner–Hopf operator W_μ on $L^1(\mathbb{R}^+)$ is defined by $W_\mu f = PC_\mu f$ for $f \in L^1(\mathbb{R}^+)$.

Now the map $\mu \to W_\mu$ is not multiplicative. However, it is easy to see that $W_{\mu * \nu} = W_\mu W_\nu$ provided either $\nu \in M(\mathbb{R}^+)$ or $\mu \in M(\mathbb{R}^-)$. In particular, if $\mu \in M(\mathbb{R}^+)^{-1}$ or $\mu \in M(\mathbb{R}^-)^{-1}$ then

$$W_\mu W_{\mu^{-1}} = W_{\mu^{-1}} W_\mu = W_{\mu * \mu^{-1}} = 1$$

and, hence, W_μ is invertible. Thus, if μ has a factorization as $\mu = \mu_1 * \mu_2$ with $\mu_1 \in M(\mathbb{R}^-)^{-1}$ and $\mu_2 \in M(\mathbb{R}^+)^-$ then $W_\mu = W_{\mu_1} W_{\mu_2}$ is invertible.

Suppose $\mu = e^\omega \in \exp(M(\mathbb{R}))$. Then $\mu = e^{\omega_1} * e^{\omega_2}$ with $\omega_1 = \omega_{|\mathbb{R}^+} \in M(\mathbb{R}^+)$ and $\omega_2 = \omega - \omega_1 \in M(\mathbb{R}^-)$. It follows that μ has a factorization as above and W_μ is invertible. Using this, Theorem 11.10, and an analysis of W_μ when $\mu = \delta_x$ or $\mu = \eta$, one can prove:

PROPOSITION. (cf. [6], [19]). *If $\mu \in M(\mathbb{R})$ then the Wiener–Hopf operator W_μ on $L^1(\mathbb{R}^+)$ is invertible if and only if $\mu \in M(\mathbb{R})^{-1}$ and the indices k and x of Theorem* 11.10 *are both zero. The operator W_μ is Fredholm (has finite dimensional kernel and cokernel) if and only if $\mu \in M(\mathbb{R})^{-1}$ and $x = 0$. In this case, the index* (dim ker W_μ − dim coker W_μ) *of the Fredholm operator W_μ is $-k$.*

11.12. Remark on $H^\infty(U)$

For another nice application of the Arens–Royden Theorem we refer the reader to [10], IV 7.7. Here the Banach algebras in question are algebras of the form $H^\infty(U)$, where U is a finitely connected domain in $\mathbb{C}$ (or an appropriate kind of Riemann surface) and $H^\infty(U)$ is the algebra of bounded analytic functions on U. If Δ is the maximal ideal space of such an algebra, then it turns out that $H^1(\Delta, \mathbb{Z}) = \mathbb{Z}^n \oplus V$, where n is the number of holes in U and V is a real vector space which arises from boundary effects in $H^\infty(U)$. Using this fact and the Arens–Royden theorem one can prove that, for an appropriate representing measure μ on ∂U, $H^\infty(U)$ is maximal among subalgebras of $L^\infty(\mu)$ for which μ remains a representing measure.

12. *K*-THEORY AND HARMONIC ANALYSIS

We had hoped in these notes to present a fairly detailed discussion of applications of K-theory to harmonic analysis. However, we ran into difficulties both with the mathematics and with lack of time. Hence, we are prepared to give no more than a few sketchy ideas on the subject and suggest that it may be a fruitful area for further research.

12.1. Multipliers

One of the primary reasons for being interested in the complicated algebra $M(G)$, rather than the much simpler algebra $L^1(G)$, is that $M(G)$ is the algebra of multipliers on $L^1(G)$ (cf. [19]).

In general, if A is a commutative Banach algebra, then a *multiplier* on A is bounded linear operator T on A such that $T(ab) = aT(b)$ for $a, b \in A$—that is, an operator that commutes with all multiplication operators. Obviously the set M of all multipliers on A is a Banach algebra with identity under the operator norm. Furthermore, the regular representation embeds A as an ideal of M (since $aT(b) = T(ab) = T(a)b$ for $T \in M$, $a, b \in A$). If A has a bounded approximate identity, then M is commutative and the regular representation $A \to M$ is an isomorphism of A onto a closed ideal of M. Of course, if A is unital then $A = M$ (since $T(a) = T(a)1$ for $T \in M$, $a \in A$).

Now if the non-unital algebra A has a direct sum decomposition $A = A_1 \oplus A_2$ into ideals A_1 and A_2, then the projection on either factor is a multiplier which is idempotent—that is, an idempotent in the algebra M. Thus, to study direct sum decompositions of a non-unital algebra we must study idempotents in M. Similarly, A-module automorphisms of A (invertible multipliers) are elements of M^{-1}.

In this way we are led to study

$$H^0(\Delta, \mathbb{Z}) = Q(M) \quad \text{and} \quad H^1(\Delta, \mathbb{Z}) = M^{-1}/\exp(M)$$

for the maximal ideal space Δ of the more complicated algebra M, even though our initial interest was with A. In the case where $A = L^1(G)$ and $M = M(G)$, this project was the topic of the preceding section.

12.2. Matrices of multipliers

Let A be a non-unital commutative Banach algebra and M its algebra of multipliers.

Consider the following question: what are the complemented submodules of the n-dimensional free A-module $\overset{n}{\oplus} A$? More specifically, can we describe the A-modules K such that for some other A-module L we have $K \oplus L \simeq \overset{n}{\oplus} A$? If $p: \overset{n}{\oplus} A \to \overset{n}{\oplus} A$ is the projection on K, for such a decomposition, then p is an idempotent operator on $\overset{n}{\oplus} A$ which commutes with multiplication by elements of A—that is, an idempotent endomorphism of the A-module $\overset{n}{\oplus} A$.

Now it is easy to see that each endomorphism of $\overset{n}{\oplus} A$ is given by an $n \times n$ matrix with entries from M—that is, an element of $M_n(M)$. Thus, direct sum decompositions of $\overset{n}{\oplus} A$ are determined by idempotents in $M_n(M)$ (elements of $P_n(M)$). Similarly, A-module automorphisms of $\overset{n}{\oplus} A$ are determined by invertible matrices in $M_n(M)$ (elements of $GL_n(M)$).

Returning to the case $A = L^1(G)$ and $M = M(G)$, if we are interested in module direct sum decompositions or automorphisms of $\overset{n}{\oplus} L^1(G)$, we must study idempotents and invertible elements of $M_n(M(G))$. This leads to the study of $K^0(\Delta) = K_0(M(G))$ and $K^1(\Delta) = K_1(M(G))$, where Δ is the maximal ideal space of $M(G)$.

12.3. *K*-theory for *M(G)*

Recall from Section 10.9 that there is a homomorphism $\mathrm{ch}: K^*(\Delta) \to H^*(\Delta, Q)$ for any compact space Δ, where $K^*(\Delta) = K^0(\Delta) \oplus K^1(\Delta)$ and $H^*(\Delta, Q)$ is the infinite direct product of the Čech groups $H^p(\Delta, Q)$. If we pick a base point for Δ then ch maps $\tilde{K}^*(\Delta)$ into $\tilde{H}^*(\Delta)$.

Now let τ be a refining topology for the l.c.a. group G. Then the embedding $L(G_\tau) \to M(G)$ induces a homomorphism

$$\tilde{K}_*(L(G_\tau)) \to K_*(M(G)).$$

Hence, we have a homomorphism

$$\varphi: \bigoplus_\tau \tilde{K}_*(L(G_\tau)) \to K_*(M(G)).$$

Here,

$$\tilde{K}_*(L(G_\tau)) = \tilde{K}_0(L(G_\tau)) \oplus \tilde{K}_1(L(G_\tau)) = \tilde{K}^*(\hat{G}_\tau^+)$$

and, similarly, $K_*(M(G)) = K^*(\Delta)$, where Δ is the maximal ideal space of $M(G)$.

If we apply the Chern character (which is a natural transformation), we obtain a commutative diagram

$$\begin{array}{ccc} \bigoplus_\tau \tilde{K}_*(L(G_\tau)) & \xrightarrow{\varphi} & K_*(M(G)) \\ \downarrow \mathrm{ch} & & \downarrow \mathrm{ch} \\ \oplus_\tau \tilde{H}^*(\hat{G}_\tau^+, Q) & \xrightarrow{\psi} & H^*(\Delta, Q), \end{array}$$

where the map ψ is the isomorphism of Theorem 11.6 in the case of Čech cohomology with rational coefficients.

Now we have not as yet been able to extend the methods of [19] to prove that φ is an isomorphism. If either part of Proposition 10.10 were true in the category of compact spaces, we could use the Chern character to prove that φ is an isomorphism. Since this is not so, however, the best we can do at this time is to prove:

PROPOSITION. *The map* $\varphi: \oplus \tilde{K}_*(L(G_\tau)) \to K_*(M(G))$ *is an isomorphism up to torsion ; that is*:

(a) *if* $u \in \ker \varphi$ *then* $ku = u + u + \cdots + u$ (*k-times*) *is zero for some integer* k; *and*

(b) *if* $v \in K_*(M(G))$ *then* $kv \in \operatorname{im} \varphi$ *for some integer* k.

Proof. This follows from the fact that if we extend ch to a map $\text{ch}: K^*(X) \otimes Q \to H^*(X, Q)$, then we obtain an isomorphism for every compact Hausdorff space X. This is proved in [7] for cell complexes X, but it holds for general compact X because each such X can be represented as $\varprojlim X_\alpha$ for an inverse limit system of cell complexes. The conclusion of the proposition is preserved by such limits. ■

12.4. Idempotents in $M_n(M(G))$

If we apply part (b) of the above proposition to $K_0(M(G))$ and remember what it means for the equality $[p] = [q]$ to hold in $K_0(A)$ for idempotents $p, q \in \bigcup P_n(A)$ (Corollary 7.5), we have:

THEOREM. *If $p \in M_n(M(G))$ is an idempotent matrix, then there are integers k, l, m such that $(\overset{k}{\oplus} p) \oplus I_l \oplus 0_m$ is similar to an idempotent matrix $q_1 \oplus q_2 \oplus \ldots \oplus q_j$, where each q_i has entries in $L(G_{\tau_i}^+)$ for refining topologies $\tau_1, \ldots, \tau_j$.*

If we knew that φ were actually an isomorphism in 12.3, then we would have $p \oplus I_l \oplus 0_m$ similar to a matrix $q_1 \oplus \ldots \oplus q_j$ as above.

12.5. Invertible matrices in $M_n(M(G))$

If we apply Proposition 12.3 to $K_1(M(G))$ we get:

THEOREM. *If $u \in GL_n(M(G))$ then there are integers k and l such that $u^k \oplus I_l = v_1 \ldots v_j e^{w_1} \ldots e^{w_m}$ for matrices $w_i \in M_n(M(G))$ and $v_i \in GL_n(L(G_{\tau_i})^+)$, where $\tau_1, \ldots, \tau_j$ are refining topologies for G.*

If we knew that φ were an isomorphism, we could draw the same conclusion for $u \oplus I_l$ (some l).

We conjecture that φ is an isomorphism and, hence, that the two previous theorems can be strengthened as indicated. If φ fails to be an isomorphism, then the maximal ideal space Δ of $M(G)$ would provide another counter-example to the problem of extending the result mentioned in 10.10 to the category of compact spaces. This would indicate rather strange new pathology in Δ and the existence of some rather strange direct summands of $\overset{n}{\oplus} L^1(G)$.

12.6. K-theory for $L(G)$

Even if φ is shown to be an isomorphism, to complete the study of K-theory for $M(G)$ one must be able to compute the K-theory for the algebras $L(G_\tau)$. That is, we want to compute $\tilde{K}_0(L(G)) = \tilde{K}^0(\hat{G}^+)$ and $\tilde{K}_1(L(G)) = \tilde{K}^1(\hat{G}^+)$ for any l.c.a. group G—preferably in a way which describes explicit representatives for each element.

Now the problem can easily be reduced to the case where $\hat{G}$ has the form $\mathbb{R}^n \times H$ for H compact. Then $\hat{G}^+ = S^n H^+$ and, hence, we can compute the K-theory of $\hat{G}^+$ from that of H by using the isomorphisms α^* and β^* of Sections 8 and 9. This reduces the problem to one of computing $K^*(H)$ for H a compact abelian group.

If $H = \mathbb{T}^n$ is an n-dimensional torus then the Künneth formula (cf. [**4**]) yields that $K^*(H)$ is an exterior algebra (over $\mathbb{Z}$) on n generators, where the elements of even degree are in $K^0(H)$ and those of odd degree are in $K^1(H)$. The generators of this exterior algebra correspond to the generators of the dual group $\mathbb{Z}^n$ of $\mathbb{T}^n$. A similar statement can probably be proved for general compact connected abelian groups using techniques like those of [**11**]. However, we have not had time to check into this.

12.7. The Picard group of $M(G)$

Recall that for a Banach algebra A, Pic(A) is the invertible group of the semigroup (under $\otimes$) of isomorphism classes of finitely generated projective A-modules and is naturally isomorphic to $H^2(\Delta, \mathbb{Z})$ (cf. Section 10.11).

If we apply the isomorphism of Theorem 11.6 in the case $p = 2$, we obtain an isomorphism

$$\bigoplus_{\tau} \operatorname{Pic}(L^1(G_\tau)^+) \to \operatorname{Pic}(M(G)).$$

Recall that Pic(A) can also be described as the group of isomorphism classes of rank 1 projective modules. Hence, we have:

THEOREM. *If P is a rank one projective $M(G)$ module, then there are refining topologies $\tau_1, \ldots, \tau_n$ of G and for each τ_i a rank one projective $L^1(G_{\tau_i})^+$ module Q_i, such that $P \simeq Q_1 \otimes \ldots \otimes Q_n$.*

This implies that each rank one projective $M(G)$-module can be represented as the image of an idempotent matrix $p: A^m \to A^m$ with entries from $\bigoplus_{\tau} L^1(G_{\tau_i})$.

REFERENCES

1. J. F. Adams, On the groups $J(X)$, *Topology* **5** (1966), 21–71.
2. R. Arens, The group of invertible elements of a commutative Banach algebra, *Studia Math. (Ser. Specjalna) Zeszyt* **1** (1963), 21–23.
3. R. Arens, To what extent does the space of maximal ideals determine the algebra?, *Function Algebras*, ed. F. T. Birtel; Scott-Foresman, Chicago, 1966.
4. M. F. Atiyah, *K-Theory*, Benjamin, New York, 1967.
5. H. Bass, *Algebraic K-Theory*, Benjamin, New York, 1968.
6. R. G. Douglas and J. L. Taylor, Wiener–Hopf operators with measure kernel, *Colloquia Math. Soc. János Bolai* **5**, *Hilbert space operators*, Tihany (Hungary), 1970.
7. E. Dyer, *Cohomology Theories*, Benjamin, New York, 1969.
8. S. Eilenberg and N. Steenrod, *Foundations of Algebraic Topology*, Princeton Univ. Press, Princeton, 1952.

9. O. Forster, Functionentheoretische Hilfsmittel im der Theorie der kommutativen Banach-Algebren, (in manuscript).
10. T. W. Gamelin, *Uniform Algebras*, Prentice-Hall, Englewood Cliffs, N. J., 1969.
11. K. H. Hofmann, Categories with convergence, exponential functors, and the cohomology of compact abelian groups, *Math. Z.* **104** (1968), 106–144.
12. D. Husemoller, *Fiber Bundles*, McGraw-Hill, New York, 1966.
13. M. Karoubi, *K-Théorie*, University of Montreal Press, Montreal, 1971.
14. M. E. Novodvorskii, Certain homotopical invariants of spaces of maximal ideals, *Mat. Zametki*, **1** (1967), 487–494.
15. H. L. Royden, Function algebras, *Bull. Amer. Math. Soc.* **69** (1963), 281–298.
16. W. Rudin, *Fourier Analysis on Groups*, Interscience, New York, 1962.
17. G. E. Shilov, On decomposition of a commutative normed ring in a direct sum of ideals, *Mat. Sb.* **32** (**74**) (1953), 353–364; English transl., *Amer. Math. Soc. Transl.* (2) **1** (1955), 37–48.
18. E. H. Spanier, *Algebraic Topology*, McGraw-Hill, New York, 1966.
19. J. L. Taylor, *Measure Algebras*, CBMS Regional Conference Series in Mathematics No. 16, Amer. Math. Soc., Providence, 1972.
20. J. L. Taylor, Topological invariants of the maximal ideal space of a Banach algebra, *Adv. in Math.*, to appear.
21. R. Wood, Banach algebras and Bott periodicity, *Topology* **4** (1966), 371–389.

6. The Holomorphic Functional Calculus and Non-Banach Algebras

L. WAELBROECK†

Free University of Brussels,
Brussels, Belgium

Introduction 187

Part I
1. Holomorphic Functional Calculus in Banach Algebras 189

Part II
2. Spaces and Algebras with Bounded Structures 197
3. Examples 200
4. Commutative *b*-algebras with Idempotent Boundedness 207
5. Regular Elements and Equiregular Sets 209
6. Continuous Inverse Algebras 215
7. Locally Pseudo-Convex Spaces and Algebras 217

Part III
8. Heuristic Considerations 221
9. Identities Involving Differential Forms 225
10. The Spectrum 234
11. The Holomorphic Functional Calculus and Non-compact Spectra ... 241
12. Applications 244
References 249

INTRODUCTION

I should say at the beginning a little about what will be in the course. There will be three parts:

I. Holomorphic functional calculus in Banach algebras.
II. Algebras with bounded structures.

† Notes elaborated with the collaboration of Peter Ludvik.

III. Holomorphic functional calculus in algebras where spectra are not compact.

These will occupy approximately two lectures, five lectures and five lectures, respectively: the twelve sections of the published version correspond approximately (although not exactly) to this division.

The first part is intended to remind you of what the holomorphic functional calculus is about, in the simplest and now classical case.

In the second part we look at a rather direct generalization of Banach algebra theory: here the joint spectra turn out to be compact. Our algebras are unions or limits of Banach algebras and we solve problems by reducing them to the Banach algebra case.

In the third part we deal with genuine non-Banach algebra theory. Now joint spectra are not necessarily compact, and the whole theory has a rather different appearance.

The first reasonable question about the title of this series of lectures is: what is a "non-Banach algebra"? This question has an obvious and obviously silly answer: an algebra is non-Banach if it is not a Banach algebra.

One of the first theorems in non-Banach algebra theory would then be the following. Let A be an algebra and $a \in A$. If $\operatorname{sp}(a) = \varnothing$ then A is non-Banach. The proof of this theorem was given by Barry Johnson in his elementary course on Banach algebra theory.

The 'holomorphic functional calculus' is also part of the title. Precisely what that object is will be described soon. The idea in this series of talks will be to show how Banach algebra theory can be extended to contain algebras that are not Banach. Several generalizations are possible. The theorems become more complicated when the hypotheses are weakened.

There would be good reason to survey the generalizations of Banach algebra theory, keeping as touchstone the holomorphic functional calculus. But life is made up of compromises, and one should not strive to the ideal, at least not in an instructional conference. At least part of what I say should be understood by at least part of the audience.

So, I shall try to show what the operational calculus is in Banach algebras. I shall describe some classes of non-Banach algebras, and show what theorems can be proved involving these. I am afraid that I shall not give many application of the theory developed. It is not so much that these do not exist. But I shall prove many technical theorems. The applications of these theorems will tend to come after the last lecture of this series.

Peter Ludvik's help during the elaboration of these notes has been greatly appreciated. It is also a pleasure to thank Professor Williamson and the organizing committee for all they have done to make the Birmingham meeting a success.

PART I

1. THE HOLOMORPHIC FUNCTIONAL CALCULUS IN BANACH ALGEBRAS

1.1. So, what is the holomorphic functional calculus? One might say loosely that a holomorphic functional calculus is any device that allows us to find the value of a holomorphic function at an element of a Banach algebra (on a system of elements of a Banach algebra if we are dealing with holomorphic functions in n variables). The origins of the holomorphic functional calculus go back to, I presume, Fantappiè, Dunford, Gelfand and Lorch. The idea is that you have an element a, say, of a Banach algebra A, which is unital and commutative. The unit in A will be denoted by 1 and we will regard the field of complex numbers, $\mathbb{C}$, as a subfield of A (via $s \leftrightarrow 1s$, $s \in \mathbb{C}$).

We now take the spectrum of a, sp(a). You all know what that is, it is a compact non-empty subset of $\mathbb{C}$. We take an open neighbourhood $U \supset \mathrm{sp}(a)$. Now take another neighbourhood V of sp(a) such that $\mathrm{sp}(a) \subset V \subset U$, V is relatively compact in U and with smooth boundary ∂V and such that all points of sp(a) have winding number 1 with respect to ∂V.

Now consider

$$\frac{1}{2\pi i}\int_{\partial V} f(z)(z-a)^{-1}\,dz \tag{1}$$

If f is a holomorphic function on U and a is a complex number, i.e. $a \in \mathbb{C}$, then (1) gives the value of f at a i.e. $f(a)$. It turns out that (1) makes sense for $a \in A$ as well, and then one will write $f[a]$ for (1). I use square brackets, because strictly speaking $f[a]$ is not the value of f at a (since $a \in A$ and f is only defined on $U \subset \mathbb{C}$) but is some element of A which we obtain when we substitute a in some formal expression involving f. And it is classical—I shall not prove it—that the mapping $f \leftrightarrow f[a]$ is a continuous homomorphism of $\mathcal{O}(\mathrm{sp}(a) \to A$ ($\mathcal{O}(\mathrm{sp}(a)$ is topological in the usual way), which maps z onto a (z is the function $z \leftrightarrow z$) and the constant function $z \leftrightarrow 1$ onto the unit of A.

All of this and more is proved in Barry Johnson's introductory course on Banach algebras so it serves no useful purpose to expound it here. Moreover, I will not be using the integral formula (1) because for what follows I only need power series representations of holomorphic functions.

1.2. Assume now that f is holomorphic in $D_r = \{z \in \mathbb{C} : |z| \leqslant r\}$ and that $a \in A$ has $\|a\| < r$. Then f can be expressed as $\sum_{n=0}^{\infty} c_n z^n$ where the c_n are obtained by Cauchy's evaluation. Then if we write

$$f[a] = \sum_{n=0}^{\infty} c_n a^n,$$

it is fairly obvious, once we can remember the Cauchy product of power series and the fact that $\sum c_n z^n$ converges absolutely and uniformly in all the ways we want, that the mapping $f \leftrightarrow f[a]$ from the algebra of functions holomorphic on D_r into A is a continuous homomorphism.

Next we take $\{a_1, \ldots, a_n\} \in A$ and we take $r_1, r_2, \ldots, r_n$ so that $\|a_i\| < r_i$ $\forall\, i$ and let $D = D_{r_1} \times D_{r_2} \ldots D_{r_n} \subset \mathbb{C}^n$, i.e. D is a polydisc with a polyradius $(r_1, \ldots, r_n)$. Then if $f \in \mathcal{O}(D)$ we can write f as

$$\sum c_{k_1 \ldots k_n} z_1^{k_1} \ldots z_n^{k_n}. \tag{2}$$

Now we can substitute a_i for z_i in (2) and define

$$f[a_1, \ldots, a_n] = \sum c_{k_1 \ldots k_n} a_1^{k_1} \ldots a_n^{k_n}.$$

So up to now all we have done is trivial.

1.3. In what follows to avoid profusion of indexes we shall write z for $(z_1, \ldots, z_n)$ and $a = (a_1, \ldots, a_n)$ if the dimension n is fixed and understood.

Take now a finite number of polynomials $P_1, P_2, \ldots, P_N$ in z and consider $P_k(a)$ as the element of A which is obtained by putting $z = a$. Also choose N numbers $R_k > 0$ such that $R_k > \|P_k(a)\|$. Then $a_1, a_2, \ldots, a_n, b_1, \ldots, b_N$, where $b_k = P_k(a)$, form a system of $n + N$ elements of A; take further $D' = D_{r_1} \times \ldots \times D_{r_n} \times D_{R_1} \times \ldots \times D_{R_N}$ $f \in \mathcal{O}(D')$ and define $f[a, b] = f[a, P(a)]$. This is not really any more complicated than before, only the notations are.

Once we are this far we look at the mapping

$$f \to f[a, P(a)] \; (f \text{ is a function of } (z, y) = (z_1, \ldots, z_n, y_1, \ldots, y_N)).$$

This maps $\mathcal{O}(D')$ into A in such a way that it maps $(z, y) \leftrightarrow z_i$ onto a_i, $(z, y) \leftrightarrow y_k$ onto $P_k(a)$ and $z \leftrightarrow 1$ onto 1 of A. We notice that this mapping is a continuous homomorphism of $\mathcal{O}(D')$ into A and that it vanishes on the ideal in $\mathcal{O}(D)$ generated by the functions $(z, y) \leftrightarrow y_k - P_k(z)$, $k = 1, \ldots, N$. Let us call this ideal α. Then $\mathcal{O}(D')/\alpha$ will be mapped into A.

And now we use a miracle—well, it isn't really a miracle, it is a theorem. And if you look for this theorem in the first chapter of Gunning and Rossi's book **[9]** you will not find it exactly, but if you carry the analysis of Sections D, E, F of Chapter 1 a little further you get the theorem I will use, namely

THEOREM 1: *$\mathcal{O}(D')/\alpha$ is isomorphic to $\mathcal{O}(\Delta)$, where*

$$\Delta = \{z \in \mathbb{C}^n \mid z \in D_{r_1} \times \ldots \times D_{r_n} \text{ and } |P_k(z)| < R_k\},$$

i.e.
$$\Delta = \{z \in \mathbb{C}^n \mid (z, P(z)) \in D'\}.$$

In fact Gunning and Rossi describe a circular proof that climbs one storey after each loop. If one continues their argument for a couple of loops more, one has Theorem 1. Consider the map of $\mathcal{O}(D')$ into $\mathcal{O}(\Delta)$ given by $f(z) \mapsto f(z, P(z))$, it is quite clear that this map is a continuous homomorphism of $\mathcal{O}(D')$ into $\mathcal{O}(\Delta)$ and the two theorems which Gunning and Rossi do not prove are†

(i) this homomorphism is surjective.

(ii) the kernel of the homomorphism is α.

Furthermore, since we are functional analysts and $\mathcal{O}(D')$ and $\mathcal{O}(\Delta)$ are both Fréchet spaces we know that this homomorphism is open. So $\mathcal{O}(\Delta)$ is topologically exactly $\mathcal{O}(D')/\alpha$.

This is what we need to define a mapping from $\mathcal{O}(\Delta)$ into A. We can define $f \mapsto f[a]$ as follows:

$$\forall\, f \in \mathcal{O}(\Delta)\ \exists\, F \in \mathcal{O}(D) \text{ such that } f(z) = F(z, P(z))$$

and now put $f[a] = F[a, P(a)]$.

1.4. We must now try and organize the sets Δ that we have obtained a little better. First I would like to say that through some sort of coquetry I am going to try to avoid the application of the axiom of choice as far as I can in my treatment of the joint spectrum. One can't get all the theorems without it, but why use it immediately when you can use it later? So I can't speak of the spectrum of the Banach algebra A the way Barry Johnson has.

You know that the spectrum of $a \in A$ can be defined as the set values of $\hat{a}(m)$, where $m \in \mathfrak{M}_A$ (maximal ideal space of A) where $\hat{a}(m)$ is the value of the image of a under the Gelfand homomorphism at m. It is reasonable to define the joint spectrum of $a_1, \ldots, a_n$ as

$$\operatorname{sp}(a_1, \ldots, a_n) = \{(\hat{a}_1(m), \hat{a}_2(m), \ldots, \hat{a}_n(m)) \subset \mathbb{C}^n : m \in \mathfrak{M}_A\}.$$

Now how can it be defined without mentioning the maximal ideals? There are two cases:

(i) the ideal generated by $s_i - a_i$, $s_i \in \mathbb{C}$, $i = 1, \ldots, n$ which we denote $\operatorname{idl}(s_i - a_i \mid i = 1, \ldots, n)$, is a proper ideal, (i.e. does not contain 1): then it is contained in some maximal ideal m and so we have $\hat{a}_i(m) = s_i\ \forall\, i$.

† The statement (i) comes trivially out of Lemma 7, Section F. As for statement (ii), it is a trivial corollary of theorems proved elsewhere in the book, e.g. Chapter IV, Section C, but Chapter I is so elementary compared to the sequel that the non specialist reader will find it easier to prove the result out of Chapter I than apply the later theorems.

(ii) the ideal $\mathrm{idl}(s_i - a_i | i = 1, \ldots, n)$ can be improper in which case for no single ideal m we can have $\hat{a}_i(m) = s_i \, \forall \, i$.

So the joint spectrum of $a_1, a_2, \ldots, a_n$ could be defined as

$$\{(s_1, \ldots, s_n) \subset \mathbb{C}^n \,|\, 1 \notin \mathrm{idl}(s_i - a_i | i = 1, 2, \ldots, n)\}.$$

However, we needed for this the notion of maximal ideal space, or at least of maximal ideals. So let us try something else. I define

$$\tilde{\mathrm{sp}}(a_1, \ldots, a_n) = \{(s_1, \ldots, s_n) \in \mathbb{C}^n \,|\, |P(s)| < \| P(a) \| \quad \forall \text{ polynomials in } (z_1, \ldots, z_n)\}.$$

This is not in general $\mathrm{sp}(a_1, \ldots, a_n)$ so that the tilde $\sim$ is a completely justified notation. It turns out that $\tilde{\mathrm{sp}}(a_i)$ is the polynomially convex hull of $\mathrm{sp}(a_i)$.

For those who do not know what a polynomially convex hull of a set $K \subset \mathbb{C}^n$, K compact, is, I define it as follows:

Let $P(\mathbb{C}^n)$ be the collection of all polynomials in $(z_1, \ldots, z_n)$, and $F \subset P(\mathbb{C}^n)$ a *finite* subset. Write

$$K(F) = \{(s_i) \in \mathbb{C}^n \,|\, |P(s)| \leqslant \sup_{z \in K} |P(z)| \, \forall \, P \in F\}.$$

These $K(F)$ are called *polynomial polyhedra* and it is trivial to see that

$$K \subset K(F), \qquad \forall \, F \subset P(\mathbb{C}^n)$$

and that $K(F)$ is compact.

Then

$$\tilde{K} = \bigcap K(P) = \bigcap K(F)$$

where the intersections are over all $P \in P(\mathbb{C}^n)$, and all finite subsets F of $P(\mathbb{C}^n)$, respectively. By compactness if U is an open neighbourhood of $\tilde{\mathrm{sp}}(a_i)$ there exist a finite number of polynomials $(P_k)_{k=1}^N$ and numbers $r_1, \ldots, r_n, R_1, \ldots, R_N$ such that $r_i > \| a_i \|$, $R_k > \| P_k(a) \|$ and $\Delta \subset U$, where Δ is as in Theorem 1.

1.5. So now we have a map from $\mathcal{O}(U)$ into $\mathcal{O}(\Delta)$ (it is just a restriction map) and then a map from $\mathcal{O}(\Delta)$ into A. We have just seen that a (compact) polynomially convex set has a fundamental neighbourhood system consisting of polynomial polyhedra. So I could take one of these, U, say, and notice that the polynomials are dense in $\mathcal{O}(U)$ by Runge's Theorem. So there cannot be more than one continuous homomorphism of $\mathcal{O}(U)$ into A which takes $z \leftrightarrow 1$ into 1, $z \leftrightarrow z_i$ onto a_i. But we have constructed one and so that is the unique homomorphism of $\mathcal{O}(U)$ into A such that $z \leftrightarrow 1$ goes to 1, $z \leftrightarrow z_i$ goes to a_i.

Why is the uniqueness important? Well, it is useful for showing that this mapping is independent of the choice of Δ. Now consider the following diagram: $\tilde{\mathrm{sp}}(a_i) \subset V \subset U$

$$\begin{array}{ccccc} \mathcal{O}(U) \longrightarrow \mathcal{O}(V) & & \longrightarrow & & A \\ & \searrow & & \nearrow & \\ & & \mathcal{O}(\Delta) & & \end{array} \qquad \text{(some } \Delta\text{)}$$

This diagram is commutative and so there is a continuous homomorphism of $\varinjlim \mathcal{O}(U) \to A$; $\varinjlim \mathcal{O}(U)$ is the direct limit of $\mathcal{O}(U)$, U polynomially convex neighbourhood of $\tilde{\mathrm{sp}}(a_i)$, with the direct limit topology. We will denote this direct limit by $\mathcal{O}(\tilde{\mathrm{sp}}(a_i))$.

1.6. However, we are not at the end of the road yet, because $\tilde{\mathrm{sp}}(a_i)$ is not the joint spectrum, but is its polynomially convex hull. By the way, we have not shown that, but I trust that you believe me.

Now if we assume that, it is clear that $\tilde{\mathrm{sp}}(a_i)$ is not empty, since $\mathrm{sp}(a_i)$ is not empty. But it is not clear if we just take the original definition of $\tilde{\mathrm{sp}}(a_i)$ that this set should be non-empty. Well, assume it is empty. Then if $f(z) \equiv 1$ we get by the holomorphic functional calculus $1 = f[a]$ and if we take $g(z) \equiv 0$ we get $0 \equiv g[a]$. But $f = g$ on an empty set and so $1 = 0$. This is of course a contradiction, if we assume $1 \neq 0$. This is true in most Banach algebras, except one, and that one is not very exciting. It is the Banach algebra which has just one element, namely 0, and there of course $0 \cdot 0 = 0$ so 0 is also the identity. But this case has been excluded from Barry Johnson's lectures when he says that in Banach algebras we will always require $1 \neq 0$. So $\mathrm{sp}(a)$ is never empty (in a Banach algebra).

1.7. We still have to cut off some part of the set $\tilde{\mathrm{sp}}(a_i)$, and this is an ingenious device in Banach algebra theory called the *Arens-Calderón trick*.

We consider $a_i \in A$, $i = 1, \ldots, n$ and we take some further elements $b_j \in A$, $j = 1, \ldots, k$. Consider $\tilde{\mathrm{sp}}(a_1, \ldots, a_n, b_1, \ldots, b_k) \subset \mathbb{C}^{n+k}$, and the projection

$$P: \mathbb{C}^{n+k} \to \mathbb{C}^n, (z_1, z_2, \ldots, z_{n+k}) \mapsto (z_1, \ldots, z_n).$$

It turns out that $P(\tilde{\mathrm{sp}}(a, b)) \subset \tilde{\mathrm{sp}}(a)$ and that $\tilde{\mathrm{sp}}(a)$ is exactly the polynomially convex hull of $P(\tilde{\mathrm{sp}}(a, b))$.

To prove this result I shall have to introduce the *spectral radius* of $a \in A$, which I denote by $\rho(a)$. From elementary Banach algebra theory it is easy to prove that the following definitions are equivalent:

$$\rho(a) = \sup\{|s| : s \in \mathrm{sp}(a)\} = \sup\{|\hat{a}(m)| : m \in \mathfrak{M}_A =$$

$$= \limsup_n \| a^n \|^{1/n} = \lim_{n\to\infty} \| a^n \|^{1/n}$$

$$= \inf \{ M \in \mathbb{R}^+ \text{ and } a^k/M^k \text{ is a bounded sequence}\}.$$

First we show $P(\tilde{\mathrm{sp}}(a, b)) \subset \tilde{\mathrm{sp}}(a)$, i.e. if $(s, t) \in \tilde{\mathrm{sp}}(a, b)$ then $s \in \tilde{\mathrm{sp}}(a)$, i.e. $|Q(s)| < \| Q(a) \|$ $\forall$ polynomials Q. But $(s, t) \leftrightarrow Q \circ P(s, t) = Q(s)$ is a polynomial in (s, t) and so

$$|Q \circ P(s, t)| = |Q(s)| \leqslant \| Q \circ P(a, b) \| = \| Q(a) \|$$

because $\tilde{\mathrm{sp}}(a, b)$ is polynomially convex).

Now we show that $[P(\tilde{\mathrm{sp}}(a, b))]^{\sim}$ is just $\tilde{\mathrm{sp}}(a)$: First we show that if Q is a polynomial in $\mathbb{C}^n$, then

$$\rho(Q(a)) = \max_{s \in \tilde{\mathrm{sp}}(a)} |Q(s)|$$

Let

$$M > \max_{s \in \tilde{\mathrm{sp}}(a)} |Q(s)|; \quad \text{then} \quad F_k(z) = Q(z)^k/M^k$$

is a bounded sequence in $\mathcal{O}(\tilde{\mathrm{sp}}(a))$. The holomorphic functional calculus maps this sequence onto a *bounded* sequence in A (since this map is a continuous homomorphism), i.e. $Q(a)^k/M^k$ a bounded sequence in A implies that $\rho(Q(a)) \leqslant M$, which in turn implies that

$$\rho(Q(a)) \leqslant \max_{s \in \tilde{\mathrm{sp}}(a)} |Q(s)|.$$

Conversely, let $s \in \mathbb{C}^n$ be such that $|Q(s)| > \rho(Q(a))$. We must show $s \notin \tilde{\mathrm{sp}}(a)$. Take M such that

$$|Q(s)| > M > \rho(Q(a)).$$

Then $\rho(Q(a))/|Q(s)| < M < 1$, so $Q(a)^k/Q(s)^k \to 0$ as $k \to \infty$. So, choosing k large enough, we can arrange $\| Q(a)^k \| < |Q(s)^k|$ (*strict* inequality). If we put $Q_1 = Q^k$ as functions then the above inequality takes the form

$$\| Q_1(a) \| < |Q_1(s)|$$

and

$$Q_1 \text{ is a polynomial} \Rightarrow [s \notin \tilde{\mathrm{sp}}(a) \Rightarrow \rho(Q(a)) \geqslant \max_{s \in \tilde{\mathrm{sp}}(a)} |Q(s)|]$$

and so combined with the previous estimate we have

$$\rho(Q(a)) = \max_{s \in \tilde{\mathrm{sp}}(a)} |Q(s)|.$$

Now, $\tilde{\mathrm{sp}}(a)$ will be the polynomial hull of $P(\tilde{\mathrm{sp}}(a, b))$ if

$$\max_{s \in \tilde{\mathrm{sp}}(a)} |Q(s)| = \max_{(s, t \in \tilde{\mathrm{sp}}(a, b)} |Q \circ P(s, t)|$$

i.e. if $\rho(Q(a)) = \rho(Q \circ P(a, b))\ \forall$ relevant polynomials Q, and this is the case since $Q \circ P(a, b) = Q(a)$.

1.8. Let us now consider the family of compact sets

$$\{\tilde{\mathrm{sp}}(a, b) \,|\, b \in A \times \ldots \times A \text{ (k factors, k varying with b)}\}$$

i.e. b ranges over all finite k-tuples of elements of A; and associate with each $\tilde{\mathrm{sp}}(a, b)$ a projection

$$P_b : \mathbb{C}^{n+k} \to \mathbb{C}^n \ (z_1, \ldots, z_n, z_{n+1}, \ldots, z_{n+k}) \mapsto (z_1, \ldots, z_n)$$

if b is a k-tuple.

Then for all b,

$$P_b\, \tilde{\mathrm{sp}}(a, b) \subset \tilde{\mathrm{sp}}(a)$$

and the polynomial hull of $P_b\, \tilde{\mathrm{sp}}(a, b)$ is $\tilde{\mathrm{sp}}(a)$.

So, as $\tilde{\mathrm{sp}}(a)$ is not empty, $P_b\, \tilde{\mathrm{sp}}(a, b)$ is non-empty $\forall\, b$, and we can consider

$$X = \bigcap_b P_b(\tilde{\mathrm{sp}}(a, b)). \tag{3}$$

X, being the intersection of compact sets, is compact and is non-empty as the family $\{P_b(\tilde{\mathrm{sp}}(a, b)) \,|\, b \text{ some } k\text{-tuple}\}$ has the finite intersection property. This is clear from the observation that

$$P_{b,c}(\tilde{\mathrm{sp}}(a, b, c)) \subset P_b(\tilde{\mathrm{sp}}(a, b) \cap P_c\, \tilde{\mathrm{sp}}(a, c)$$

and the left hand side is a non-empty compact set.

So X is a non-empty compact subset of $\mathbb{C}^n$, where n is the number of elements in (a). Later we shall identify X with $\mathrm{sp}(a_i)$ as defined previously, and for the matter in hand we will assume that $X = \mathrm{sp}(a_i)$ have been shown.

From (3) it is an easy application of compactness to show that if U is a neighbourhood of $\mathrm{sp}(a_i)$ then $\exists\, (b_1, \ldots, b_k)$ so that $\mathrm{sp}(a_i) \subset P_b(\tilde{\mathrm{sp}}(a, b)) \subset U$. The proof is standard and will be omitted.

We can now proceed as follows:

We are given $f \in \mathcal{O}(\mathrm{sp}(a_i))$; take $f \in \mathcal{O}(U)$, where U is a neighbourhood of $\mathrm{sp}(a_i)$ and f is the function which extends the original f. Now take $(b_1, \ldots b_k)$ such that $P_b(\tilde{\mathrm{sp}}(a, b)) \subset U$. Then $f \circ P_b \in \mathcal{O}(\tilde{\mathrm{sp}}(a, b))$ and so we can define

$$f \circ P_b[a, b].$$

We notice that the above expression involves b. We would like to define $f[a]$ to be $f \circ P_b[a, b]$, but to do this we must show that

$$f \circ P_b[a, b] = f \circ P_c[a, c]$$

$\forall\, b, c$ such that $P_b\, \tilde{\mathrm{sp}}(a, b) \subset U$, $P_c\, \tilde{\mathrm{sp}}(a, c) \subset U$.

First we show that

$$f \circ P_{b,c}[a, b, c] = f \circ P_b[a, b].$$

Notice that $f \circ P_b \in \mathcal{O}(\tilde{\mathrm{sp}}(a, b))$, i.e. the ring of holomorphic functions on a polynomially convex set. We have already shown that in this case there is a unique holomorphic functional calculus which sends $z \mapsto z_i$ onto a_i, $i = 1, \ldots, n$, $z \mapsto z_{n+j}$ onto b_j, $j = 1, \ldots, k$ and $z \mapsto 1$ onto 1, i.e. any map that sends $g \in \mathcal{O}(\tilde{\mathrm{sp}}(a, b))$ onto $g[a, b] \in A$ is of the above type.

Now consider $g \circ P_c[a, b, c]$. But this is of the same type as above and we know there is just one and so it must be equal to $g[a, b]$. So

$$g \circ P_c[a, b, c] = g[a, b] \; \forall \, c.$$

So

$$f \circ P_b[a, b] = f \circ P_{b,c}[a, b, c] = f \circ P_c[a, c].$$

Hence we have constructed a mapping from $\mathcal{O}(U)$ into A such that $z \mapsto z_i$ goes to a_i, $z \mapsto 1$ goes to 1.

Things are not as satisfactory as they were before, where the question of uniqueness is concerned. Basically, things can go wrong if $\mathrm{sp}(a_i)$ does not have a fundamental system of neighbourhoods $\{U\}$, which are domains of holomorphy. In this case one can have more than one homomorphism from $\mathcal{O}(U)$ into A and such that all the conditions of holomorphic functional calculus are satisfied. So if we want a uniqueness theorem, there is just one way of getting it, and that is to consider all of the homomorphisms together.

We have associated with each $(a_1, \ldots, a_n)$ a compact set $\mathrm{sp}(a_i)$ and we have the condition that $P_b(\mathrm{sp}(a, b)) = \mathrm{sp}(a)$. Now if

$$f \in \mathcal{O}\partial(\mathrm{sp}(a)), \text{ if } f \circ P_b \in \mathcal{O}(\mathrm{sp}(a, b)) \text{ and if } f[a] = f \circ P_b[a, b] \; \forall \, b$$

(these are the compatibility conditions to be satisfied by all the continuous homomorphism into A), then the system of the continuous homomorphisms we have constructed is *the* holomorphic functional calculus. So there is just one such system, but at each step there may be more than one homomorphism.

1.9. There now follows the proof that $\mathrm{sp}(a_i) = X$. Let now U be a neighbourhood of $\mathrm{sp}(a)$. By compactness, it is possible to choose $b = (b_1, \ldots, b_k)$ in such a way that $U \supset P_b \, \tilde{\mathrm{sp}}(a, b)$. If $f \in \mathcal{O}(U)$, $f \circ P_b \in \mathcal{O}(\tilde{\mathrm{sp}}(a, b))$, and we can define $f \circ P_b[a, b]$.

The system $(b_1, \ldots, b_k)$ is rather indeterminate. We could have chosen other elements $(c_1, \ldots, c_l)$. and considered $f \circ P_c \in \mathcal{O}(\tilde{\mathrm{sp}}(a, c))$ and $f \circ P_c[a, c]$.

Now, if f has been defined on a neighbourhood of $\tilde{\mathrm{sp}}(a)$, we would have had $f[a] = f \circ P_b[a, b]$, the mapping $f \to f \circ P_b$ being in these circumstances a continuous mapping of $\mathcal{O}(\tilde{\mathrm{sp}}(a))$ into A mapping z_i on a_i and unit on unit. We may apply this remark to $f \circ P_b$ and to $f \circ P_c$. It follows that

$$f \circ P_b[a, b] = f \circ P_{b,c}[a, b, c] = f \circ P_c[a, c].$$

Weaving our results together, we get the

PROPOSITION: *It is possible to associate to each system* $(a_1, \ldots, a_n)$ *of elements of* A *a compact set* $\mathrm{sp}(a_1, \ldots, a_n) \subset \mathbb{C}^n$, *and a homomorphism* $\mathcal{O}(\mathrm{sp}(a)) \to A$ *mapping* z_i *on* a_i *and unit on unit.*

1.10. I have chosen to present Banach algebra theory as I have because I do not believe that I have used the axiom of choice, or any such exotic axiom, along the way. I have not investigated in detail the theorems of complex analysis that I use, but these theorems look fairly constructive. Now, what I mean by constructivity should be clearly understood. I do not maintain that a computer would be able to carry out any moderately complicated "construction". The computer firms have not yet marketed the instruments dreamt up by constructivists.

Now, how do we get the maximal ideals? We consider for each a the compact space $\mathrm{sp}(a)$. The definition of $\mathrm{sp}(a)$ shows that P_b maps $\mathrm{sp}(a, b)$ on $\mathrm{sp}(a)$. It is an easy application of Tychonov's theorem to show that functions $\chi(a)$ exist with, for all $(a_1, \ldots, a_n)$:

$$(\chi(a_1), \ldots, \chi(a_n)) \in \mathrm{sp}(a_1, \ldots, a_n),$$

and better, if $\mathfrak{M}$ is the set of these functions

$$\mathrm{sp}(a_1, \ldots, a_n) = \{(\chi(a_1), \ldots, \chi(a_n)) \mid \chi \in \mathfrak{M}\}.$$

This set $\mathfrak{M}$ is of course compact for the projective limit topology, i.e. for the topology of simple convergence.

PART II

ALGEBRAS WITH BOUNDED STRUCTURES

2. SPACES AND ALGEBRAS WITH BOUNDED STRUCTURES

2.1. These talks are about non-Banach algebras to which the holomorphic functional calculus can be extended. We shall see as we go along that the bounded subsets of a topological algebra turn out to be more important than the topology—at least for the constructions which I wish to carry out. It is reasonable to investigate what properties a boundedness must have if we want to prove a few decent theorems.

It is usual, when E is a topological vector space, and B is a subset of E, to say that B is *bounded* if $sx \to 0$ uniformly for $x \in B$, when $s \to 0$. This means that we can associate to every neighbourhood U of 0 in E an $\varepsilon > 0$ in such a way that $sB \subset U$ when $|s| < \varepsilon$. The bounded structure we obtain in this way is the *standard bounded structure* of E, sometimes called the "Mackey boundedness". Such standard bounded structures are interesting examples, but we shall consider them only as examples. It is also possible to find interesting bounded structures that are not standard.

A topological vector space is *locally bounded* if the origin has a bounded neighbourhood. It is not hard to see that the only locally convex and locally bounded topological vector spaces are the normed spaces (or seminormed, if we do not assume the Hausdorff condition). There is also a more difficult theorem: if E is a locally bounded topological vector space then there is a $p > 0$ such that E is p-seminormable (p-normable if E is Hausdorff). This was first established by Tosio Aoki and later rediscovered independently by S. Rolewicz: see G. Köthe [**41**] pp. 165–166 for a proof.

Relations which can exist between a bounded structure and a topology can be important. If A is a topological algebra, if $a \in A$ or if $a \in A^n$, it may be that we can define $f[a]$ for some holomorphic f. There are cases where $f[a]$ is defined for a ranging over an open set, where $f[a]$ depends continuously on a. These properties are important when they are true.

2.2. If E is a set, a *bounded structure* or *boundedness* on E is a set $\mathscr{B}$ of subsets of E such that $B_1 \cup B_2 \in \mathscr{B}$ if $B_1 \in \mathscr{B}$, $B_2 \in \mathscr{B}$, such that $B' \in \mathscr{B}$ if $B' \subset B$ and $B \in \mathscr{B}$, and such that $\bigcup_{B \in \mathscr{B}} B = E$. The sets $B \in \mathscr{B}$ are the *bounded sets* of the bounded structure. If $(E, \mathscr{B})$ and $(E', \mathscr{B}')$ are two sets with boundednesses, a mapping $f: E \to E'$ is a *morphism* when $fB \in \mathscr{B}'$ when $B \in \mathscr{B}$.

Let now E be a real or a complex vector space. A boundedness $\mathscr{B}$ on E is a *vector space boundedness* if $B_1 + B_2$ is bounded when B_1 and B_2 are, and when every bounded B is contained in a balanced bounded B'. (B' is balanced when $sb \in B'$ as soon as $b \in B'$, $s \in \mathbb{C}$, $|s| \leqslant 1$). The boundedness $\mathscr{B}$ is *separated* if the only bounded subspace of E is the null space. It is *convex* if the convex hull of a bounded set is bounded. I shall not investigate here why it is important that the topologies and bounded structures we meet be convex, though I can assure you that it is. The subject is discussed in Section 7. But you are probably ready to believe that some sort of completeness axiom is important. Completeness says that a sequence converges, that a limit exists, if it is a Cauchy sequence, if it has some readily verifiable property.

We need a little bit more than this. If we consider a set of Cauchy sequences which all come from a given bounded set, if these sequences are in a way uniformly Cauchy, then the set of limits should be bounded. A sequence u_n is a *Cauchy sequence* if some bounded set B exists and a sequence $\varepsilon_n \to 0$ such that $u_n - u_m \in \varepsilon_n B$ when $m \geqslant n$. It tends to u if $u_n - u \in \varepsilon_n B$, again with B bounded, $\varepsilon_n \to 0$.

We come in this way to the following axiom. Assume that a sequence of mappings $x \mapsto u_n(x)$ is given, of a set X into E. Assume that

$$\{u_n(x) \mid n \in \mathbb{N}, x \in X\}$$

is bounded, and that a sequence $\varepsilon_n \to 0$ and a bounded set B can be found such that $u_n(x) - u_m(x) \in \varepsilon_n B$ for $x \in X$, $m \geqslant n$. A bounded mapping (i.e. a mapping with a bounded range) $x \to u(x)$ should then exist, and a bounded set B' such that $u_n(x) - u(x) \in \varepsilon_n B'$ for all $n \in \mathbb{N}$ and $x \in X$. (Strictly speaking, we should require the existence of B', and a new sequence $\delta_n \to 0$ such that $u_n(x) - u_n(x) \in \delta_n B$, but it is not difficult to show that the original sequence ε_n will do as soon as some sequence δ_n has the satisfactory properties.)

2.3. The above description of the completeness axiom must be considered as heuristic. It is in fact a complete description but it turns out that, in the convex case, this axiom is equivalent to the following one: *every bounded set is contained in a completant bounded set.*

Since this new formulation is easier to handle, I shall adopt it. Since time is short, and the main point of this series of talks is different, I shall not prove the equivalence of the two definitions. But I should say what a completant set is.

An absolutely convex set B has a *gauge*, a Minkowski functional

$$\nu_B(x) = \inf\{\lambda > 0 \mid x \in \lambda B\}.$$

This is a semi-norm on the vector space generated by B. It is a norm if B does not contain a non-zero vector subspace, in particular if B is bounded in a space with a separated bounded structure.

Definition: B is *completant* if its gauge is a Banach space norm.

The vector spaces and algebras that we shall consider will be fitted out with a completant bounded structure, i.e. with a bounded structure in which each bounded set is contained in a completant bounded set. Completant sets are convex, and do not contain any non-zero subspace. Completant bounded structures are therefore separated and convex.

2.4. I have already said that I would consider algebras with a completant bounded structure. The boundedness must be an *algebra boundedness;* i.e., $B_1 \cdot B_2$ must be a bounded set when B_1 and B_2 are. These completant bounded structures on vector spaces and algebras will be objects that we shall often have reason to consider. Therefore, I shall need an abbreviation. I shall speak of *b-spaces* and *b-algebras.*

2.5. We will meet in this second part of the course *b*-algebras with a more special property. These are the b-algebras with an *idempotent boundedness* (Allan's terminology), [2], or multiplicatively convex b-algebras (according to Hogbe-Nlend). It turns out that these can be realized as direct limits of Banach algebras. A set B in an algebra is idempotent if $B^2 \subset B$. If $B' = MB$, we see that $B'^2 \subset MB'$. This remark shows that b-algebras in which all bounded sets are contained in idempotent bounded sets tend to be uninteresting: the property can only be verified when $A \cdot A = \{0\}$.

What is assumed is that it is possible to associate to every bounded set B an $M > 0$ and a bounded B' such that $B \subset B'$ and $B'^2 \subset MB'$, i.e. that each bounded set B is contained in a set B' which is idempotent up to a scalar factor.

Arens and Michael have considered a different class of topological algebras, the so called *locally multiplicatively convex algebras*. They can be described as inverse limits of Banach algebras.

As is well known, if we have Banach algebras A_1, A_2 and a morphism $\phi: A_1 \to A_2$ then this induces a continuous map $\hat{\phi}$ from $\mathfrak{M}_{A_2}$ into $\mathfrak{M}_{A_1}$ (the maximal ideal spaces of A_2, A_1 respectively). These remarks lead one to associate with the inverse limits of Banach algebras (Michael) the direct limits of their corresponding maximal ideal spaces—these are K-spaces of some sort, usually quite nasty and with the direct limit of Banach algebras one associates the inverse limit of the corresponding maximal ideal spaces (and we know the inverse limits of compact spaces are fairly well behaved).

The locally multiplicatively convex algebras will not be considered further in these lectures.

3. EXAMPLES

3.1. When a new type of structure is introduced, it is normal to investigate its stability properties. The category of b-spaces and b-algebras has many trivial stability properties. Direct limits of b-spaces for injective mappings and inverse limits of b-spaces are in a natural way b-spaces. As an example, let A be a directed set, and consider the monic direct limit of $\{E_\alpha\}_{\alpha \in A}$. That is $E_\alpha \subset E_\beta$ if $\alpha < \beta$, each E_α has a b-structure $\mathscr{B}_\alpha$; and $\mathscr{B}_\alpha \subset \mathscr{B}_\beta$ if $\alpha < \beta$, i.e. the injections $u_{\beta\alpha}: E_\alpha \to E_\beta$, $\alpha < \beta$ are bounded mappings. Then we can define on the direct limit

$$E = \bigcup_\alpha E_\alpha = \lim_{\rightarrow} E_\alpha$$

the natural boundedness $\mathscr{B}_E = \bigcup_\alpha \mathscr{B}_\alpha$, and it is easy to check that we get a b-space structure if each $\mathscr{B}_\alpha$ is a b-space boundedness.

It is more difficult to get a b-structure on a direct limit of vector spaces when the structural mappings are not injective. To be precise, the boundedness which one introduces naturally on the direct limit is not necessarily separated.

It has a separated quotient, and the quotient is the categorical direct limit; but it can very well be zero.

Similarly, if we take a space with a convex, non complete bounded structure, and try to complete it, we may run into trouble. The completion can always be defined as the solution of a universal problem. It is trivial that the universal problem has a solution, but a non-zero space may very well have a vanishing completion.

Consider as an example $C_0[x]$, the space of all polynomials in one real variable x, say, vanishing at 0. Then we can define on $C_0[x]$ a natural boundedness $\mathscr{B}$ by

$$B \in \mathscr{B} \text{ if } \exists\, \varepsilon > 0\, \exists\, M > 0, M < \infty \text{ such that } \forall\, P \in \mathscr{B} \subset C_0[x]$$
$$\text{we have } \sup\{|P(x)| : x \in [+\varepsilon, \varepsilon]\} < M.$$

This is obviously a separated boundedness on $C_0[x]$. However the completion can be computed in two steps: first complete $C_0[x]$ on $[-\varepsilon, \varepsilon]$ and then take a direct limit. In the first step we get the continuous functions on $[-\varepsilon, \varepsilon]$ vanishing at 0. The limit is the space of germs of continuous functions at the origin and these do not have a nice bounded structure. The completion of our space $C_0[x]$ is just $\{0\}$. For further information on this example see [**29**].

Let E be a vector space. Let E_1 and E_2 be vector subspaces, each with a boundedness, say $\mathscr{B}_1$, $\mathscr{B}_2$. We may consider $E_1 + E_2$, and on $E_1 + E_2$ the boundedness $\mathscr{B}_1 + \mathscr{B}_2$ generated by the sets $B_1 + B_2$, $B_1 \in \mathscr{B}_1$, $B_2 \in \mathscr{B}_2$. This is in general not a separated bounded structure. It is however completant if it is separated, and it is separated if and only if $E_1 \cap E_2$ is equipped with a completant bounded structure by the boundedness $\mathscr{B}_1 \cap \mathscr{B}_2$, where $\mathscr{B}_1 \cap \mathscr{B}_2$ is the set of all B such that $B \in \mathscr{B}_1$ and $B \in \mathscr{B}_2$, or equivalently the set of all $B_1 \cap B_2$, $B_1 \in \mathscr{B}_1$, $B_2 \in \mathscr{B}_2$.

Many spaces of E-valued functions can be defined, when E is a b-space. These spaces can usually be equipped with a bounded structure in such a natural way that one even forgets to say what the bounded sets are. But—and this is always the case with natural things—it is completely obvious that the structure obtained is that of a b-space.

3.2. Let E be a Hausdorff locally convex space, and $\mathscr{B}$ its bounded sets, in the natural boundedness. Then $\mathscr{B}$ is a separated, convex boundedness on E. For $(E, \mathscr{B})$ to be a b-space, it is necessary and sufficient that a sequence x_n converge as soon as a bounded set B exists, and a sequence $\varepsilon_n \subseteq 0$ such that $x_n - x_m \in \varepsilon_n B$ when $m \geqslant n$. This condition means in fact that the closed, absolutely convex, bounded sets are completant. This is then the first concrete example of a b-space.

We could start with a topological vector space (E, τ) that is not locally convex, and let $\mathscr{B}$ be the set of $B \subset E$ whose convex hull is bounded. Then $\mathscr{B}$ is a convex bounded structure. For $(E, \mathscr{B})$ to be a b-space, it is again necessary and sufficient that the bounded, closed, absolutely convex sets be completant. And this happens exactly when x_n converges as soon as some $B \in \mathscr{B}$ exists, and $\varepsilon_n \to 0$ such that $x_n - x_m \in \varepsilon_n B$ when $m \geqslant n$.

Worse, we may start out with any vector space topology, and consider the sets B which are contained in some bounded completant set. This is a completant bounded structure. Unfortunately, the closure of a completant set is in general not completant. The bounded structures we meet are getting further away from the topology, the more we weaken the requirements.

3.3. Equicontinuous bounded structures are also important. Let E be a topological vector space and F a locally convex space. The set $\mathscr{B}$ of equicontinuous subsets of $\mathscr{L}(E, F)$ is a bounded structure. With this bounded structure, $\mathscr{L}(E, F)$ is a b-space when the natural boundedness of F is completant (closed, absolutely convex, bounded sets are completant).

We could obtain completant bounded structures on $\mathscr{L}(E, F)$, when F is not locally convex, by considering sets $B \subset \mathscr{L}(E, F)$ with an equicontinuous convex hull, when F has completant closed absolutely convex bounded subsets. Or, dropping all conditions about F, we could simply consider the $B \subset \mathscr{L}(E, F)$ such that some $B' \subset \mathscr{L}(E, F)$ exists which is equicontinuous, completant, and contains B.

However, these generalizations tend to vanish into thin air when it comes to building a theory. The essential property of the equicontinuous boundedness is its multiplicative property. If $B_1 \subset \mathscr{L}(E, F)$ and $B_2 \subset \mathscr{L}(F, G)$ are equicontinuous, then $B_2 \circ B_1 \subset \mathscr{L}(E, G)$ is equicontinuous. However, the fact that B_1 and B_2 are both contained in a completant bounded set does not imply that $B_2 \circ B_1$ is, unless of course $\mathscr{L}(E, G)$ has a completant equicontinuous bounded structure.

So we will note the fact that if E is a locally convex vector space with a completant natural bounded structure, then $\mathscr{L}(E, E)$ with its equicontinuous boundedness is a b-algebra.

3.4. Let A be a locally convex algebra, with separately continuous multiplication. Assume also that its natural bounded structure is completant. Then A is a b-algebra with its natural bounded structure.

This follows from uniform boundedness. We must show that $B_1 \cdot B_2$ is bounded when B_1 and B_2 are. We may assume that B_1 and B_2 are completant, A_{B_1} and A_{B_2} are Banach spaces.

Multiplication is a separately continuous mapping: $A_{B_1} \times A_{B_2} \to A$.

The Banach–Steinhaus theorem shows that this mapping is continuous, hence that $B_1 \cdot B_2$ is bounded.

We are stuck if we drop the condition that the natural bounded structure of A is completant. We may start out with completant bounded sets B_1 and B_2. Our proof shows that $B_1 \cdot B_2$ is bounded. But there is no reason why $B_1 \cdot B_2$ should be contained in a completant bounded set.

3.5. The considerations of Sections 3.3 and 3.4 may be applied to a class of spaces and algebras which are not locally convex, but a little more general. These are the locally pseudo-convex algebras. And instead of considering the standard bounded structure of these spaces and algebras, we must introduce the rapidly decreasing bounded structure.

Let $0 < p \leqslant 1$. A *p-semi-norm* on a real or a complex vector space E is a mapping $v: E \to \mathbb{R}^+$ such that

$$v(x + y) \leqslant v(x) + v(y)$$

$$v(\lambda x) = |\lambda|^p v(x)$$

($x, y \in E$, λ scalar). A *p-norm* is a p-semi-norm which does not vanish off the origin.

A vector space topology is *locally p-convex* if it can be defined by a family of p-semi-norms (equivalently if it can be defined by a family of p'-semi-norms, $p' \geqslant p$). It is *locally pseudo-convex* if it can be defined by a family of p-semi-norms, p depending on the semi-norm.

A sequence of elements e_n of a topological vector space E is said to be *rapidly decreasing* if $n^k e_n$ is a bounded sequence for all choices of k. Similarly a sequence of linear mappings $u_n: E \to F$ is rapidly decreasing if $n^k u_n$ is an equicontinuous sequence for all choices of k. The absolutely convex hull of a rapidly decreasing sequence of elements of E is bounded, if E is locally pseudo-convex. Similarly, the absolutely convex hull of a rapidly decreasing sequence of linear mappings $u_n: E \to F$ is equicontinuous if F is locally pseudo-convex.

The first statement can be considered to be a special case of the second, so I shall only prove the second one. We assume that F is locally pseudo-convex, and that u_n is rapidly decreasing. We consider a p-semi-norm v, continuous on F and look for a neighbourhood V of zero in E such that $v \circ u(x) \leqslant 1$ for every u in the convex hull of the sequence u_n and every $x \in V$. We choose an integer $k > 1/p$; since $n^k u_n$ is equicontinuous, we can find a neighbourhood V of zero in E such that $v(n^k u_n) = n^{pk} v(u_n) \leqslant 1$, i.e. such that $v(u_n) \leqslant n^{-pk}$. Also, u is in the convex hull of the sequence u_n; $u = \sum \lambda_n u_n$ with real numbers λ_n which are certainly all less than one. So $v(\sum \lambda_n u_n) \leqslant \sum_n n^{-pk} = A_{pk}$ and A_{pk} is finite. The fact that we wind up with a finite constant where I had promised to furnish the constant 1 is of course immaterial.

Definition; A bounded subset of a topological vector space is *rapidly decreasing* if it is contained in the closed absolutely convex hull of a rapidly decreasing sequence. An equicontinuous set of linear mappings of one topological vector space into another is rapidly decreasing if it is contained in the pointwise closed, absolutely convex hull of a rapidly decreasing sequence of linear mappings.

We have in fact proved the

PROPOSITION: *If F is a locally pseudo-convex space, the rapidly decreasing boundedness of F, and of $\mathscr{L}(E, F)$, where E is a topological vector space, are convex bounded structures.*

It is clear that with these bounded structures, F and $\mathscr{L}(E, F)$ are b-spaces if the closed bounded absolutely convex subsets of F are completant.

PROPOSITION: *Let A be a locally pseudo-convex algebra with separately continuous multiplication. Assume that the bounded, closed, absolutely convex subsets of A are completant. The rapidly decreasing boundedness defines then a b-algebra structure on A.*

The proof given in Section 3.4 shows that $B_1 \cdot B_2$ is a bounded set when B_1 and B_2 are bounded sets whose convex hulls are bounded. Let now x_n and y_n be rapidly decreasing sequences. The sequences $n^k x_n$ and $n^k y_n$ have a bounded convex hull. Let $z_j = x_m y_n$ when $j = \frac{1}{2}(m + n)(m + n + 1) + n$. The set of $m^k n^k z_j$, thus of $j^{k/2} z_j$ is bounded for all k. This means that the countable set $x_m y_n$ can be ordered in such a way that it becomes rapidly decreasing.

PROPOSITION: *Let E be a locally pseudo-convex space whose closed, absolutely convex, bounded subsets are completant. The rapidly decreasing boundedness of $\mathscr{L}(E)$ is that of a b-algebra.*

The proof is practically that given above. The set $\{x_m \circ y_n\}$ is equicontinuous if x_m and y_n are equicontinuous sequences. We must give an ordering of $x_m \circ y_n$ such that this becomes rapidly decreasing when x_m and y_n are rapidly decreasing and the ordering given above is satisfactory.

3.6. Let E be a b-space. $\mathscr{L}(E, E)$, or $\mathscr{L}(E)$ is the set of linear mappings $E \to E$ which map bounded sets onto bounded sets. A subset $B_1 \subset \mathscr{L}(E, E)$ is bounded, or equibounded when $B_1(B) = \{u(x) | u \in B_1, x \in B\}$ is bounded in E for all bounded subsets B of E. It is clear that the boundedness defined in this way on $\mathscr{L}(E, E)$ is that of a b-algebra.

3.7. We have described above several b-spaces and b-algebras. None of the algebras was equipped with an idempotent boundedness.

The first, obvious class of algebras with an idempotent bounded structure is that of Banach algebras. Direct limits of Banach algebras also have idempotent bounded structures. In fact, we can prove the following:

THEOREM: *Every algebra with an idempotent bounded structure is expressible as a direct limit of Banach algebras.*

Let A be an algebra. Assume that $A = \bigcup_{i \in I} A_i$ where I is a directed set, where each A_i is a subalgebra of A, and $A_i \subset A_j$ when $i \leqslant j$. Assume also that each A_i has a Banach algebra norm, and that the embedding $A_i \to A_j$ $(i \leqslant j)$ is a bounded linear mapping. A subset $B \subset A$ is bounded when it is possible to find $i \in I$ such that $B \subset A_i$ and is bounded in A_i. Then A has clearly an idempotent bounded structure.

Conversely, let A be a b-algebra with an idempotent bounded structure. We shall see that A is the union of a directed family of Banach algebras. Start out with the set of idempotent completant bounded subsets, say $\mathscr{B}_0$ of A. We can think of those as unit balls in some Banach algebra. For B_1, $B_2 \in \mathscr{B}_0$, define $B_2 < B_1$ (B_2 stronger than B_1) by $\exists M : B_2 \subset MB_1$, and $B_2 \sim B_1$ (B_2 is equivalent to B_1) by $B_2 < B_1$ and $B_1 < B_2$. Choose an element B of each equivalence class and let $\mathscr{B}_1$ be the set of representatives thus chosen. (We have temporarily abandoned here our principle that the axiom of choice should be avoided where possible.) For $B \in \mathscr{B}_1$, let $A_B = \bigcup_{M>0} MB$, normed by the Minkowski functional of B. This is a Banach algebra. If $B_1 < B_2$, $A_{B_1} \subset A_{B_2}$, the inclusion is bounded. If $B_1 \in \mathscr{B}_1$, $B_2 \in \mathscr{B}_1$, then $B_1 \cup B_2$ is bounded, and is contained in MB_3 where B_3 is completant, bounded, and idempotent. We can even assume that $B_3 \in \mathscr{B}_1$; anyway $B_3 \sim B_3' \in \mathscr{B}_1$ and $B_1 \cup B_2 \subset M'B_3'$. This shows that $\mathscr{B}_1$ is a directed set, and that A has the direct limit structure of the A_B, $B \in \mathscr{B}_1$.

If A is commutative, the situation is easier because the class of idempotent bounded sets is directed by inclusion. If B_1 and B_2 are two such sets, $B_1 \cup B_2 \cup B_1 B_2$ is bounded and idempotent. We consider now simply the set $\mathscr{B}_0$ of bounded, idempotent, and completant sets. This is directed by inclusion. A is the direct limit of a family of Banach algebras A_i, and the inclusion maps $A_i \to A_j$ are norm-decreasing.

In the noncommutative case we can have $B_1 \cup B_2 < B_3$ with B_3 an idempotent and yet $B_1 \cup B_2 \not\subset B$ for any idempotent B.

3.8. The problem is not so much to have a general structure theorem for our algebras, as we have done in Section 3.7, but to have good examples of such algebras. Most interesting cases are discussed in [3].

If S is a compact subset of $\mathbb{C}^n$, and if A is a Banach algebra, we let $\mathscr{O}(S, A)$

be the direct limit of the algebras of A-valued holomorphic functions on neighbourhoods of S. Of course, $\mathcal{O}(S, A)$ is also the direct limit of the algebras $\mathcal{O}_\infty(U, A)$ where $\mathcal{O}_\infty(U, A)$ denotes the bounded holomorphic functions on U, a neighbourhood of S. The direct limit boundedness of $\mathcal{O}(S, A)$ is that of a b-algebra with idempotent boundedness.

This can be generalized. We can let A be a b-algebra with an idempotent boundedness; then $A = \bigcup_{i \in I} A_i$ where I is a directed set and each A_i is a Banach algebra. With any of the obvious definitions for holomorphic A-valued functions, $\mathcal{O}(S, A) = \bigcup \mathcal{O}_\infty(U, A_i)$ and this defines an idempotent bounded structure on $\mathcal{O}(S, A)$.

3.9. Let A be a commutative b-algebra. An element $a \in A$ is called *regular* if some $M > 0$ exists such that a^n/M^n is a bounded sequence. Let A_r be the set of regular elements of A.

A set $B \subset A_r$ is bounded for the Allan boundedness of A_r if $B \subset MB_1$ for some $M \in \mathbb{R}^+$ and some idempotent B_1. With the Allan boundedness, A_r is a b-algebra with multiplicatively convex boundedness.

If B_1 is idempotent, and B_1' is the convex hull of B_1, then B_1' is clearly idempotent. We assume that B_1 is bounded. The series $\sum \lambda_n b_n$ converges in A as soon as $\sum |\lambda_n| < \infty$, $\forall n: b_n \in B_1$. Let

$$B_1'' = \{\sum \lambda_n b_n \,|\, \sum |\lambda_n| \geqslant 1, \forall n: b_n \in B_1\}.$$

Then B_1'' is completant and idempotent.

We know that $B_1 \cup B_2$ is contained in an idempotent set,

$$B_1 \cup B_2 \cup B_1 \cdot B_2,$$

when B_1 and B_2 are idempotent. The set of subsets which are absorbed by some bounded idempotent set is thus an idempotent b-algebra boundedness.

3.10. Let A be a complete p-normed algebra. It is possible to describe in more than one way an idempotent b-algebra boundedness on A, such that the bounded subsets for the b-algebra structure are topologically bounded. Allan, Dales, and McClure [3] describe such a boundedness. Its elements are the closed absolutely convex idempotent hulls of finite sets of elements whose norms are strictly less than unity, and multiples of such hulls. An easy computation involving p-norms shows that the convex hulls of the set of elements $a_1^{k_1} \dots a_n^{k_n}$ $(k_1, \dots, k_n \in \mathbb{N})$ is bounded when, for all i, $\|a_i\| < 1$.

3.11. The following b-algebra with idempotent bounded structure is interesting when we deal with functions which can be approximated locally by elements of a sup-norm algebra.

We let X be a compact space, and A be a uniform algebra on X. A function

$f \in C(X)$ is *A-holomorphic* if each $x \in X$ has a neighbourhood U such that the restriction of f to U is a uniform limit of restrictions of elements of A to U.

Let $\mathscr{U}$ be a finite open covering of X, and let A be the set of $f \in C(X)$ whose restriction to U is a uniform limit of elements of A, for each $U \in \mathscr{U}$. The algebra of A-holomorphic functions is of course the union of the algebras $A_{\mathscr{U}}$, $\mathscr{U}$ an open covering of X. We have clearly on the union the structure of a b-algebra with an idempotent bounded structure.

3.12. Let E be a locally convex space. Let $\beta(E)$ be the set of "bornifying" linear mappings, i.e. of linear mappings which map some neighbourhood of the origin of E into some bounded set in E. It is natural to say that a set $B \subset \beta E$ is *equibornifying* if some neighbourhood U of the origin in E exists such that $\bigcup_{f \in B} f(U)$ is bounded in E. The equibornifying boundedness is clearly idempotent on $\beta(E)$.

The existence of this bounded structure is important in P. Uss's theory of bornifying operators [**26**], though Uss does not speak of bounded structures and does not introduce the equibornifying sets. Hogbe–Nlend [**10**] credits Akkar [**1**] with the description of this bounded structure.

4. COMMUTATIVE *b*-ALGEBRAS WITH IDEMPOTENT BOUNDEDNESS

4.1. The considerations that were developed in Section 1 can be repeated when A is a commutative b-algebra with an idempotent boundedness. Of course we do not have a single norm defining the structure of A, but we observe that the condition (essential in the definition of $\tilde{\text{sp}}(a)$): for every polynomial P, $|P(s)| \leqslant \|P(a)\|$ implies, for every k, that $|P(s)^k| \leqslant \|P(a)^k\|$ and thus that $|P(s)| \leqslant \rho(P(a))$, where ρ is the spectral radius. Since $\rho(P(a)) \leqslant \|P(a)\|$ anyway, we see that the following definition is equivalent, in the Banach algebra case, with the one we gave:

Definition:

$$\tilde{\text{sp}}(a_1, \ldots, a_n) = \{(s_1, \ldots, s_n) \in \mathbb{C}^n \mid \forall P \text{ a polynomial } |P(s)| \leqslant \rho(P(a))\}$$

As in the Banach algebra case, we define the spectral radius of an element a as the infimum of the real numbers M such that a^k/M^k is a bounded sequence.

If U is a neighbourhood of $\tilde{\text{sp}}(a_1, \ldots, a_n)$, it is possible to find a finite number of polynomials $P_1, \ldots, P_N$, and real numbers $r_i > \rho(a_i)$, $R_k > \rho(P_k(a))$, such that $U \supset \Delta$ if

$$\Delta = \{(s_1, \ldots, s_n) \mid \forall i \colon |si| \leqslant r_i, \forall j \colon |P_j(s)| \leqslant R_j\}.$$

We next consider the sequences a_i^k/r_i^k, $P_j(a)^k/R_j^k$; These are bounded sequences.

We let B be a bounded, completant, idempotent set, containing all the elements a_i^k/r_i^k and $P_j(a)^k/R_j^k$. And we consider the Banach algebra A_B; let its norm be ν_B. We see that

$$\tilde{\text{sp}}_{A_B}(a_1, \ldots, a_n) \subset \Delta$$

and that the results proved in Section 2 allow us to construct a homomorphism $\mathcal{O}(\Delta) \to A_B$, *a fortiori* $\mathcal{O}(\Delta) \to A$, mapping z_i on a_i and unit on unit.

Repeating the argument of Sections 1.7, 1.8, 1.9., i.e. the Arens–Calderón trick, we see that

$$\text{sp}\,(a_1, \ldots, a_n) = \bigcap_B \text{sp}_{A_B}(a_1, \ldots, a_n)$$

when we define

$$\text{sp}(a_1, \ldots, a_n) = \{(s_1, \ldots, s_n) \in \mathbb{C}^n \,|\, 1 \notin \text{idl}\,(a_1 - s_1, \ldots, a_n - s_n)\};$$

hence every neighbourhood U of $\text{sp}(a_1, \ldots, a_n)$ contains $\text{sp}_{A_B}(a_1, \ldots, a_n)$ for some bounded, comptetant, idempotent B.

The generalization of the holomorphic functional calculus to commutative b-algebras with idempotent bounded structures is completely straightforward.

4.2. Something must be said about the spectrum of the whole algebra. We shall identify this first with the set of multiplicative linear functionals. We know how to identify this (cf. Section 1.10) with the projective limit of the spectra of finite sets of elements.

For each bounded, completant, and idempotent B we have a Banach algebra A_B, and a structure space for A_B; call this $\mathfrak{M}_B$. If $B \subset B'$, we have an imbedding $A_B \to A_{B'}$ and a restriction mapping $\mathfrak{M}_{B'} \to \mathfrak{M}_B$. The structure space of A is the projective limit of the spaces $\mathfrak{M}_B$. This is clear. Consider a multiplicative linear functional χ on A. This has a restriction χ_B to each A_B, and if $B_1 \supset B_2$, the restriction of χ_{B_1} to A_{B_2} is χ_{B_2}, i.e. the system of χ_B belongs to the projective limit.

Conversely, assume that we are given a multiplicative linear functional χ_B on each A_B and that the restriction of χ_{B_1} to A_{B_2} is χ_{B_2} when $B_1 \supset B_2$. We can then consider the union of the multiplicative linear forms χ_B. This union is a multiplicative linear form χ on A.

4.3. To show that we have generalized completely the Gelfand theory of Banach algebras, we must still prove that each maximal ideal is the kernel of a multiplicative linear form.

We must first generalize the theorem that maximal ideals are closed. Let m be maximal, and let

$$m' = \bigcup_B \bigcap_{\varepsilon > 0} (m + \varepsilon B)$$

where we let B range over all bounded (or equivalently bounded idempotent) sets. The operation associating m' to m is not in general a closure operation in b-space theory (see e.g. [**28**], p. 34. Proposition 10). But m has good reason to be called closed if $m = m'$, as is the case here. m' is in any case an ideal. It contains m, and it does not contain the unit, otherwise $1 \in m + \varepsilon B$ for some completant, idempotent m, but $1 - \varepsilon B \cap m = \varnothing$ since all elements of $1 - \varepsilon B$ are invertible ($\varepsilon < 1$) and no element of m is.

This proves that $m = m'$, hence $m \cap A_B$ is closed for the norm of A_B for all B. On A/m, we can consider the quotient bounded structure, in which the bounded sets are the quotient images of those of A. This is a separated, idempotent and completant boundedness.

Of course the quotient of a commutative algebra with unit by a maximal ideal is a field. The quotient here must be the complex field, for if it were not we would choose $\lambda \in A/m$, $\lambda \notin \mathbb{C}$, the spectrum of λ would be empty, and this is not the case.

4.4. Application of the above results, hence of most of Banach algebra theory to the algebras described in Sections 3.8, 3.9, 3.10, and 3.11 is a non-negligible generalization of Banach algebra theory.

4.5. We make one last remark. Let $(A, \mathscr{B})$ be a b-algebra. Assume that it is possible to associate to each $a \in A$ a compact set $S_a \subset \mathbb{C}$ and a bounded homomorphism $\mathcal{O}(S_a) \to A$ mapping unit on unit and z on a. Then all elements of A are regular: if $M_a > \max_{z \in S_a} |z|$, we see that a^k/M_a^k is a bounded sequence.

We can reduce the boundedness $\mathscr{B}$, and consider only the Allan boundedness $\mathscr{B}_1$ of A. Then $(A, \mathscr{B}_1)$ is a b-algebra with an idempotent boundedness, and considerations of this section apply.

5. REGULAR ELEMENTS AND EQUIREGULAR SETS

5.1. We assume that A is a complex b-algebra with unit. An element $a \in A$ has been called regular if a^k/M^k is bounded for some large enough M. Let us show that a is regular if and only if a neighbourhood of infinity exists, such that $(a - s)^{-1}$ is defined and bounded for s in that neighbourhood.

Start out with the following form of the resolvent identity:

$$\frac{(a - s)^{-1} - (a - t)^{-1}}{s - t} = (a - s)^{-1}(a - t)^{-1}.$$

A first application of this identity shows that $(a - s)^{-1}$ is continuous; it is even Lipschitz on any set where it is bounded. And once we know that the function is continuous, we let $t \to s$, and the right-hand side tends to $(a - s)^{-2}$

hence the left-hand side tends to the same limit. But the left-hand side is a differential quotient.

Hence $(a - s)^{-1}$ is holomorphic on any open set where it is locally bounded. Of course, some readers might know what a holomorphic E-valued function on U is when E is a Banach space and U a complex manifold, but might not know what such a function would be when E is a space with a bounded structure. Here E is a b-space; it contains many Banach spaces. An E-valued function f on U is holomorphic if we can associate to every $z \in U$ a neighbourhood V of z and a bounded completant subset B of E in such a way that $fV \subset E_B$, $f: V \to E_B$ being holomorphic.

Next, we observe that $(a - s)^{-1}$ is analytic near infinity. and bounded on a neighbourhood of infinity. The removable singularities theorem shows that $(a - s)^{-1}$ is clearly holomorphic at infinity. The limited geometric expansion

$$(a - s)^{-1} = -s^{-1} - s^{-2}a - \dots - s^{-r-1}a^r + s^{-r-1}a^{r+1}(a - s)^{-1}$$

is an asymptotic expansion for the resolvent near infinity, which is formally similar to a Laurent expansion. But $(a - s)^{-1}$ cannot have two different Laurent expansions, i.e.

$$(a - s)^{-1} = -\sum_0^\infty s^{-r-1}a^r,$$

and the series converges.

Looking at what has been proved, and at results which follow trivially from considerations similar to those of Banach algebra theory, we see on the one hand that a^k/M^k is a bounded sequence if $(a - s)^{-1}$ is defined and bounded for $|s| \geqslant M$; and on the other hand that $(a - s)^{-1}$ is defined and bounded for $|s| \geqslant M + \varepsilon$ if a^k/M^k is a bounded sequence.

If a is regular, we define the spectral radius $\rho(a)$ of a as the infimum of the M such that a^k/M^k is bounded. We see that

$$\rho(a) = \inf\{M \,|\, (a - s)^{-1} \text{ is defined and bounded for } |s| \geqslant M\}.$$

5.2. I have not said enough about the regular elements in Section 3.9. It is not difficult to show directly that the set of regular elements is a subalgebra of A. Clearly λa is regular, $\rho(\lambda a) = |\lambda|\rho(a)$ if a is regular and λ is scalar. Also, if a_1, a_2 are regular, $M_1 > \rho(a_1)$, $M_2 > \rho(a_2)$ then a_1^k/M^k and a_2^k/M_2^k are bounded sequences, hence $(a_1a_2)^k/(M_1M_2)^k$ also, and $M_1M_2 \geqslant \rho(a_1a_2)$. Similarly

$$\frac{(a_1 + a_2)^k}{(M_1 + M_2)^k} = \sum_l \binom{k}{l} \frac{M_1^l M_2^{k-l}}{(M_1 + M_2)^k} \frac{a_1^l}{M_1^l} \frac{a_2^{k-l}}{M_2^{k-l}}$$

is a sequence of elements of the convex hull of a bounded set, so it is a bounded sequence and $M_1 + M_2 > \rho(a_1 + a_2)$.

These considerations show that $a_1 + a_2$ and $a_1 . a_2$ are regular elements, also that ρ is a submultiplicative semi-norm on the algebra of regular elements.

5.3. It is quite natural to say that a set B of regular elements is *equiregular* if some M exists such that $\{a^k/M^k | a \in B, k \in \mathbb{N}\}$ is bounded. It is just as natural to say that B is equiregular if some M exists such that $a - s$ is invertible as soon as $a \in B$, $|s| \geqslant M$, $\{(a - s)^{-1} | a \in B, |s| \geqslant M\}$ being a bounded set.

And an analysis of the proof of the equivalence of our two definitions of a regular element shows that the two definitions of an equiregular set are equivalent. An analysis of the proof of the fact that the set of regular elements is an algebra shows that the equiregular boundedness is an algebra boundedness.

It is also clear that an equiregular set is bounded, since

$$B \subset M\{a^k/M^k | k \in \mathbb{N}, a \in B\}.$$

5.4. The equiregular boundedness is as closed as it can be. Let B be equiregular, and let B_1 be bounded and absolutely convex. The set

$$\bigcap_{\varepsilon > 0} B + \varepsilon B_1$$

is then equiregular, i.e. (when $B \subset B_1$) the A_{B_1} closure of B is equiregular. Or again, let u_n be a sequence of mappings of some set T into a given equiregular set B, and assume that u_n is a Cauchy sequence of bounded mappings of T into A (for the obvious boundedness of the space of mappings of T into a b-space). Then u, the limit of the sequence u_n, has a equiregular range.

We consider $a_0 \in \bigcap (B + \varepsilon B_1)$ and choose a sequence of $a_n \to a_0$, $a_n \in B$ such that $a_n - a_0 \in 2^{-n} B_1$. Then $a_n - a_{n+1} \in 2^{-n+1} B_1$. We choose M in such a way that $(a - s)^{-1}$ exists when $a \in B$, $|s| \geqslant M$, the set of inverses $\{(a - s)^{-1} | \, |s| \geqslant M\}$ being bounded. Let this set of inverses be B_2. We observe that

$$\begin{aligned} (a_n - s)^{-1} - (a_{n+1} - s)^{-1} &= (a_n - a_{n+1})(a_n - s)^{-1}(a_{n+1} - s)^{-1} \\ &\in 2^{-n+1} B_1 . B_2^2 \end{aligned}$$

The set $B_1 . B_2^2$ is bounded, and is independent of the element $a_0 \in B$ we started from. If B_3 is a bounded completant set containing $B_2 + B_1 . B_2^2$, we see that $(a - s)^{-1} \in B_3$ when $a_0 \in \bigcap B + \varepsilon B_1$ and $|s| > M$.

Assume for a moment that a locally convex topology τ is given on A along with its bounded structure $\mathscr{B}$. Is the closure of an equiregular set equiregular? I don't believe that this is the case in general. In any case, the proof does not work.

It is possible to get the proof to work, but we must start out with a bounded structure that is topologically complete (there is a fundamental system of complete bounded sets). And we must make a hypocontinuity assumption on the product. If B is bounded (i.e. $B \in \mathscr{B}$), the set of mappings $y \to xy$ $(x \in B)$ is equicontinuous.

With these hypotheses, $(a_n - a_m)(a_n - s)^{-1}(a_m - s)^{-1} \to 0$, hence $(a_n - s)^{-1}$ is a Cauchy (generalized) sequence; the limit exists, and is an inverse of $(a - s)$, and the closure of B is equiregular.

We notice, the notations I have used notwithstanding, that $a \in \bar{B}$ is not the limit of a sequence (in the usual sense) of elements of B but of a generalized sequence. It is not sufficient to assume sequential completeness of A to have a limit for $(a_n - s)^{-1}$.

5.5. *A set B has an equiregular convex hull if and only if it is contained in MB_1 for some bounded idempotent B_1.*

In one direction this is trivial. Bounded idempotent sets, and their multiples, are equiregular.

Consider conversely B, which we may take absolutely convex, and assume that B is equiregular, say $\{a^k | a \in B, k \geqslant 1\} \subset B_1$ with B_1 bounded. We apply a relatively well known polarization formula

$$(-1)^n n! a_1 \dots a_n = \textstyle\sum_0^n (-1)^k w_k$$

where

$$w_k = \textstyle\sum_{I \in \mathscr{I}_k} (\sum_{i \in I} a_i)^n,$$

if $\mathscr{I}_k$ designates the set of subsets of $\{1, \dots, n\}$ which have k elements. This expresses $a_1 \dots a_n$ as a linear combination of nth powers of linear combinations of the factors.

A straightforward evaluation shows that if B is absolutely convex, if B_1 is absolutely convex, and if $a^k \in B_1$ when $a \in B$, $k \in \mathbb{N}$, then

$$a_1 \dots a_n \in \frac{(2n)^n}{n!} B_1 \subset (2e)^n B_1,$$

by Stirling's formula, with $e = 2{\cdot}71828\dots$. And $B/2e$ is contained in a bounded idempotent set.

5.6. So the only difference between the equiregular boundedness of the algebra of regular elements of a b-algebra, and the Allan boundedness is the fact that the equiregular boundedness may be non-convex. But can it really be non-convex? It can.

It turns out that the straightforward, universal way of constructing a b-algebra with an equiregular bounded set is also the good one if we wish the equiregular sets not to be convex.

Our algebra will be generated commutatively by elements $a_0, \ldots, a_n, \ldots$. These elements will contitute an equiregular sequence. We wish therefore the set $\{a_i^k | i \in \mathbb{N}, k \in \mathbb{N}\}$ to be bounded. We also wish the sets

$$\{a_{i_1}^{k_1} \ldots a_{i_n}^{k_n} | i_1, \ldots, i_n, k_1, \ldots, k_n \in \mathbb{N}\}$$

to be bounded (when $n \in \mathbb{N}$) because the boundedness of A must be multiplicative.

The construction of an algebra having these properties is standard. The elements of this algebra will be not necessarily finite formal linear combinations

$$\sum_{n, i_1, \ldots, i_n, k_1, \ldots, k_n} s_{n, i_1, \ldots, i_n, k_1, \ldots k_n} a_{i_1}^{k_1} \ldots a_{i_n}^{k_n}$$

where

$$\sum |s_{n, i_1, \ldots i_n, k_1 \ldots k_n}| < \infty$$

and where some n_0 can be found for each element such that $s_{n, i, k} = 0$ when $n > n_0$.

A subset of B this algebra is bounded if $\sum_{n, i, k} |s_{n, i, k}| \leqslant M$ for some finite M and all $(s_{n, i, k}) \in B$, and if furthermore some n_0 can be found with $s_{n, i, k} = 0$ when $n > n_0$ for all $s \in B$. It is clear that we have constructed a b-algebra in this way (multiplication derives from the standard Cauchy product).

The algebra is really constructed in such a way that the sequence $\{a_1, \ldots, a_n, \ldots\}$ would be equiregular. But is its convex hull equiregular? It is not, otherwise the sequence $a_1, (\frac{1}{2}(a_1 + a_2))^2, (\frac{1}{3}(a_1 + a_2 + a_3))^2, \ldots$ would be bounded, which is not the case.

The equiregular boundedness of A is important when we investigate the continuity properties of the holomorphic functional calculus in b-algebras.

Let B be an equiregular set, and f an entire function of one variable. We can define $f[a]$ for every $a \in B$. To every bounded completant set B_1 we may associate a bounded completant B_2 in such a way that the mapping $f \leftrightarrow f[a]$ is a continuous mapping of B, topologized by the gauge of B_1, into A_{B_2}, topologized by the gauge of B_2.

This result can be refined by considering the equiregular spectrum of an equiregular set B. This is the intersection of the closed sets $F \subset \mathbb{C}$ such that $a - s$ has an inverse when $a \in B$, $s \notin F$, $\{(a - s)^{-1} | a \in B, s \in \mathbb{C} \backslash F\}$ being a bounded set. The equiregular spectrum of B is compact: it is bounded, contained in the disc of radius R if $\{(a - s)^{-1} | a \in B, |s| \to R\}$ is bounded. It is closed since it is an intersection of closed sets.

Let now S be the equiregular spectrum of B. Let f be holomorphic on a neighbourhood of S. Then $f[a]$ is defined for every $a \in B$. As before, it is possible to associate with every completant bounded set B_1 a completant bounded set B_2 in such a way that the mapping $f \leftrightarrow f[a]$ maps B into A_{B_2}, continuously when B is topologized by the gauge of B_1 and A_{B_1} by that of B_2.

We observe that

$$f[a] = \frac{1}{2\pi i} \int_j f(\zeta)(\zeta - a)^{-1} \, d\zeta$$

where j is the boundary of V, a neighbourhood of sp (a) which has a smooth boundary and is relatively compact in U. We also observe that

$$f[a] - f[a'] = \frac{(a' - a)}{2\pi i} \int_j f(\zeta)(\zeta - a)^{-1}(\zeta - a')^{-1} \, d\zeta.$$

If B'_1 is such that $(\zeta - a)^{-1} \in B'_1$ for $\zeta \in j$, $a \in B$, we see that

$$f[a] - f[a'] \in \frac{L(j)}{2\pi} v(a' - a) \,.\, B_1 \,.\, B_1'^2$$

where $L(j)$ is the length of the contour j and $v(a' - a)$ the value of the gauge of B_1 at $a' - a$. We choose B_3 completant and such that $B_3 \supset B_1 B_1'^2$.

5.7. The mapping $a \leftrightarrow f[a]$ is not only continuous, it is also of class C_∞, again when we restrict it to the equiregular sets. Better, it is of A-class C_∞, i.e. the differentials are A-linear and A-homogeneous forms. If B is equiregular, $a' \to a$, $a \in B$, $a' \in B$, we again have

$$f[a'] = f[a] + (a' - a) f'[a] + \ldots + \frac{(a' - a)^r}{r!} f^{(r)}[a] + [o(a' - a)]^r$$

when f is holomorphic on a neighbourhood of the equiregular spectrum of a.

The expression "$[o(a' - a)]^r$" must be interpreted: for every bounded set B_1 we can find a bounded set B_2 such that $\nu_{B_2}(\phi(a'))$ tends to zero faster than $[\nu_{B_1}(a' - a)]^r$ when $\nu_{B_1}(a' - a) \to 0$, here, of course, $\phi(a')$ is the Taylor remainder we are investigating. This result follows from the limited geometric expansion formula

$$(\zeta - a')^{-1} = (\zeta - a)^{-1} + (a' - a)(\zeta - a)^{-2} + \ldots + (a' - a)^r(\zeta - a)^{-r-1} + (a' - a)^{r+1}(\zeta - a)^{-r-1}(\zeta - a')^{-1}$$

and the fact that a continuous homomorphism of $\mathcal{O}(S)$ into A mapping z on a maps

$$\frac{r!}{2\pi i} \int_j f(\zeta)(\zeta - z)^{-r-1} \, d\zeta = f^{(r)}(z)$$

on

$$\frac{r!}{2\pi i} \int_j f(\zeta)(\zeta - a)^{-r-1} \, d\zeta,$$

hence this expression is $f^{(r)}[a]$.

Morally speaking, we do not yet have the right to say that f is of class C_∞. H. Whitney [36], [37] showed that a differentiable function on a closed set was one which has a limited Taylor expansion with remainder, and whose derivatives have similar expansions. Here, we must have

$$f^{(s)}[a'] = f^{(s)}[a] + (a' - a)f^{(s+1)}[a] + \dots + \frac{(a' - a)^{r-s}}{(r-s)!} f^{(r)}[a] + [o(a' - a)]^{r-s}$$

but this is a special case of the expression already obtained, just put $f^{(s)}$ instead of f in the former relation.

5.8. Conditions have been described at the end of Section 5.4, which are sufficient to prove that the closure of an equiregular set is equiregular, when we consider an algebra where a topology and a boundedness coexist. The condition was that the set of mappings $x \to ax$ ($a \in B$) be equicontinuous when B is bounded.

Assume the condition satisfied, let $a' \to a$, and let ζ range over a contour j which does not meet the equiregular spectrum of an equiregular set B that we consider (and to which a and a' belong); we have

$$(\zeta - a')^{-1} - (\zeta - a)^{-1} = (a' - a)(\zeta - a')^{-1}(\zeta - a)^{-1}.$$

We see that the right hand side tends to zero, and uniformly for $\zeta \in j$. This proves that $f[a]$ depends continuously on a when a ranges over B, if f is holomorphic on a neighbourhood of the equiregular spectrum of B.

Similarly, we could prove A-differentiability of class C_∞ for $f[a]$.

6. CONTINUOUS INVERSE ALGEBRAS

6.1. I shall say that A is a *continuous inverse algebra* if A is a topological algebra with unit with jointly continuous multiplication in which the set of invertible elements is open, and the map $a \leftrightarrow a^{-1}$ is continuous on its domain. Philippe Turpin showed that it was possible to weaken the hypotheses, at least in the commutative case.

(i) Let A be an algebra, and τ a vector space topology on A. Assume that the set of invertible elements of A is a neighbourhood of the unit and that $a^{-1} \to 1$ when $a \to 1$. The set of invertible elements of A is then open, a^{-1} is continuous on its domain, and the Jordan multiplication of A, $ab + ba$ is jointly continuous.

If A is commutative, multiplication and Jordan multiplication do not differ essentially. To prove that ab is continuous, it is sufficient to assume that a^{-1} is defined near the unit and continuous there. Turpin also proved that one could strengthen the requirements, in the locally convex case.

(ii) A locally convex continuous inverse algebra is locally multiplicatively convex.

These two results of Turpin can be found in [25], or in [28], pp. 87–88 and 122–123. It is not my purpose here to sketch their proofs.

I will mention in passing that one cannot weaken the hypothesis to locally pseudo-convex algebras. This also has been shown by Turpin in his paper. The proof uses ideas of Mitjagin, Rolewicz and Żelazko [38], who show that M is locally multiplicatively convex if M is locally convex, Fréchet, and $\sum \lambda_n a^n$ converges whenever $\sum \lambda_n z^n$ is an entire function.

6.2. We shall thus assume that A is a complete, locally convex, continuous inverse algebra, commutative and with unit.

It is easy to see that in such A every element $a \in A$ is regular. All one has to do is notice that $(a - s)^{-1} = -s^{-1}(1 - s^{-1}a)^{-1}$ and as $s \to \infty$, $1 - s^{-1}a \to 1$; and since $a \leftrightarrow a^{-1}$ is continuous on some neighbourhood of 1 the assertion follows, using some results of Section 5.

So $A = A_r =$ set of all regular elements, and it makes good sense to talk about spectral radii and joint spectrum of elements of A.

If $U \subset \mathbb{C}^n$ is open, then

$$\{(a_1, \ldots, a_n) \in A^n \,|\, \mathrm{sp}\,(a_1, \ldots, a_n) \subset U\} = D_U$$

is open. This follows from the fact that the spectral radius is a continuous semi-norm, and the spectral radius is a continuous semi-norm because the set $\rho(a) < 1$ is the largest balanced open set X such that $(1 + x)^{-1}$ exists for all $x \in X$, and this is a neighbourhood of zero.

If $a \in D_U$, i.e. $\mathrm{sp}(a) \subset U$, if ε is small enough so that $s + t \in U$ when $t = (t_1, \ldots, t_n)$ is such that $|t_i| < \varepsilon$ for all i, and $s \in \mathrm{sp}(a)$, if $(h_1, \ldots, h_n) \in A^n$ are such that $\rho(h_i) < \varepsilon$ then $\mathrm{sp}(a + h) \subset U$. As a matter of fact, $\mathrm{sp}(a + h)$ is the image of $\mathrm{sp}(a, h)$ by the mapping $(s, t) \leftrightarrow s + t$. Now $\mathrm{sp}(h)$ is contained in the polydisc of polyradius $(\varepsilon, \ldots, \varepsilon)$ and $\mathrm{sp}(a, h) \subseteq \mathrm{sp}(a) \times \mathrm{sp}(h)$.

6.3. The mapping $a \leftrightarrow f[a]$, $D_U \to A$ is continuous. This is nearly clear; the construction of the holomorphic functional calculus given in Section 1 shows that it is sufficient to consider the case where U is a polydisc. As a matter of fact, some function F on a polydisc of large dimension has been described (that is the Arens–Calderón trick—to be used when $\widetilde{\mathrm{sp}}(a) \subset U$) such that

$$f(z) = F(z, y, P(z, y))$$

(the right hand side is independent of y) and $f[a]$ was defined by

$$f[a] = F[a, b, P(a, b)].$$

If a' is near enough to a, we can define $f[a']$ by

$$f[a'] = F[a', b, P(a', b)].$$

We define $F[a]$ by the Cauchy integral formula

$$F[a] = \frac{1}{(2\pi i)^n} \int F(z_1, \ldots, z_n)(z_1 - a_1)^{-1} \ldots (z_n - a_n)^{-1} dz_1 \ldots dz_n.$$

So all that remains to show is that $F[a]$ is a continuous function of a when F is a holomorphic function on a polydisc. Of course

$$(\zeta_i - a'_i)^{-1} = (\zeta_i - a_i)^{-1} + (a'_1 - a_i)(\zeta_i - a'_1)^{-1}(\zeta_i - a_i)^{-1}.$$

When we substitute this expression in the Cauchy integral formula, we see that $F[a'] - F[a]$ tends to zero when a' tends to a.

6.4. We can do better. We consider $\mathcal{O}(\mathrm{sp}(a_1, \ldots, a_n))$, the holomorphic functions on neighbourhoods of a compact set with the direct limit topology. We consider also a small closed polydisc T of polyradius ε, and $\mathcal{O}(\mathrm{sp}(a) \times T)$. The positive number ε is chosen small enough so that $\mathrm{sp}(a) + T \subset U$.

We observe that $f(z + y) \in (\mathrm{sp}(a) \times T)$ can be written

$$f(z + y) = \sum \frac{f^{(k)}(z)}{k!} y^k$$

with the usual multi-index notation,

$$f^{(k)} = f^{(k_1, \ldots, k_n)}, \; k! = k_1! \ldots k_n! \quad \text{and} \quad y^k = y_1^{k_1} \ldots y_n^{k_n}.$$

Now, the mapping $F(z, y) \leftrightarrow F[a, b]$ will map $f(z + y)$ on $f[a + b]$ hence

$$f[a + b] = \sum \frac{f^{(k)}[a]}{k!} b^k;$$

this when $\mathrm{sp}(b) \subset T$, i.e. $\rho b_1 \leqslant \varepsilon, \ldots, \rho b_n \leqslant \varepsilon$.

The function $a \leftrightarrow f[a]$ can thus be expanded locally into a power series. It is "Lorch analytic". (see [**11**]).

7. LOCALLY PSEUDO-CONVEX SPACES AND ALGEBRAS

7.1. There seem to be reasons to show that the theory sketched above really does not generalize to the non-locally convex case. I shall rather give examples where several of the lemmas we use do not generalize, and also show how one can generalize the theory anyway to the locally pseudo-convex topological algebras. On the other hand things go wrong when we try to study algebras with a pseudo-convex bounded structure.

We shall consider cases where (E, τ) is a topological vector space (algebra) with the topology τ given by a family J of p-seminorms. If $\inf_{p \in J} p \geqslant \varepsilon > 0$ then (E, τ) is locally ε-seminormed.

If I is the unit interval, if $p < 1$, if $L_p(I)$ is the p-normed space of p-summable functions, we put $u_s(t) = 1$ for $t < s$, $u_s(t) = 0$ for $t \geqslant s$ and observe that $u_{s+h} = u_s + |h|^{1/p}$. This shows that the mapping $s \leftrightarrow u_s$ is differentiable, and has derivative zero. A differentiable $U: I \to L_p(I)$ is not determined any more by its derivative and its initial value.

Letting $v_{s+it} = u_s$, we obtain a function of a complex variable, say $s + it$, which is complex differentiable but does not look in the least like a holomorphic function.

7.2. Mazur and Orlicz [**12**] (see also [**30**], §7) give a positive result, showing that not all continuous functions can be integrated when E is metrizable and non-locally convex.

More precisely, let E be a metrizable topological vector space. Assume that a continuous linear mapping $C(I, E) \to E$ can be found, which maps $u(x) \cdot e$ onto $\int u(x)\, dx.\, e$. Then E is locally convex.

The proof of this result will be found in the references suggested. The fact that I is a diffuse measure space is important.

7.3. D. O. Etter Jr. [**5**] has described what is the most general class of E-valued functions on a compact space X which can be integrated with respect to a bounded measure m, if we want the mapping $m \leftrightarrow \int u\,.\, dm$ to be continuous. These functions are continuous mappings $u: X \to E$ where $u\, X \subset B$, with B compact, absolutely convex, and such that the topology induced by E on B is locally convex. In other words, the linear forms on E, (or on E_B) whose restriction to B is continuous should separate B. Once we make this assumption, we do not have any problem defining an integral, since the approximands do not get out of B. This integral is discussed by D. O. Etter Jr. (loc. cit.) and by the author ([**27**], §7).

7.4. Gramsch [**8**], and Przeworska-Rolewicz and Rolewicz [**15**] have investigated the integrals of functions taking their values in a complete p-normed space E and belonging to $C(X) \hat{\otimes}_p E$, for a reasonable definition of the p-projective tensor product norm. The difficult part of their proof is the proof of the fact that the natural mapping $C(X) \hat{\otimes}_p E \to C(X, E)$ is an injection. This uses the fact that $C(X)$ has the approximation property. We can identify $C(X) \hat{\otimes}_p E$ with a subspace of $C(X, E)$.

Integration maps $C(X)$ into the scalars, and, functorially, $C(X) \otimes_p E$ into $C \otimes_p E = E$. The integrable functions defined in this way are Etter-integrable. What Gramsch, Przeworska-Rolewicz and Rolewicz do is give a useful criterion for Etter integrability.

7.5. Differentiability is another criterion for Etter integrability in complete locally pseudo-convex spaces. This was investigated by Turpin and the author [**22, 23, 24, 31**].

If U is a differentiable manifold (presumably of class C_∞), if E is a topological vector space, a mapping $f: U \to E$ will be called of *class C_r on U* if it has a limited Taylor expansion with remainder to order r, while its derivatives of order s ($|s| \leqslant r$) have similar expansions, to order $r - |s|$. We thus require, once a local coordinate system has been chosen, that functions $f^{(s)}$, $s = (s_1, \ldots, s_n) \in \mathbb{N}^n$, $|s| = s_1 + \ldots + s_n \leqslant r$ exist, and that for all s ($|s| \geqslant r$) we have

$$f^{(s)}(x + h) = \sum_{|l| \leqslant r - |s|} \frac{f^{(s+l)}(x)}{l!} h^l + w_s(x, h)$$

where, of course, $f^{(0)} = f$ and $|h|^{|s|-r} w_s(x, h)$ is a continuous function of (x, h) on its domain, which vanishes when $h = 0$ (we require convergence to zero when $h \to 0$ for all x, and uniformity when x ranges over a compact set).

A function is of class C_∞ if it is of class C_r for all r. We topologize $C_r(U, E)$ by uniform convergence of the $|h|^{|s|-r} w_s(x, h)$ on compact subsets of the relevant domains.

We notice that it is not sufficient to require that a function be of class C_1, and have derivatives of class C_1 iteratively to check that it is of class C_∞. The function u defined in Section 7.1 is of class C_1; it has derivatives of class C_∞, but it is not of class C_r for any $r \geqslant 1/p$.

Turpin and I show that a function of class C_r, for r large enough, and taking its values in a complete p-normed space, p constant, can be integrated. It follows that a function of class C_∞ taking its values in a complete locally pseudo-convex space can be integrated.

Part of the completeness requirement can be dispensed with. And the statement that the integral exists can be strengthened. It turns out that it is sufficient that the absolutely convex, closed bounded sets be completant. And the function is of class C_∞ when we replace the topology of E by its rapidly decreasing boundedness (this is proved in [**31**]).

7.6. All these remarks are relevant for the following reasons. The resolvent is a function of class C_∞ wherever it is bounded. The limited geometric expansion

$$(a - s - h)^{-1} = (a - s)^{-1} + h(a - s)^{-2} + \ldots + h^r(a - s)^{-r-1} + h^{r+1}(a-s)^{-r-1}(a - s - h)^{-1}$$

has the form of a Taylor expansion of arbitrarily high order. To show that the resolvent is of class C_∞, we must also see that its derivatives, i.e. the functions $s!(a-s)^{-s-1}$, have the required Taylor expansions. And this is proved by raising the previous geometric expansions to the power $s+1$:

$$(a-s-h)^{-s-1} = (a-s)^{-s-1} + (s+1)h(a-s)^{-s-2} + \dots + \frac{(s+r+1)!}{r!} h^r(a-s)^{-s-r} + \phi,$$

where ϕ tends to zero faster than h^r when $h \to 0$.

This shows that the resolvent is of class C_∞ on each neighbourhood of an $s \in \mathbb{C}$ on which it is bounded. We also wish the resolvent to be of class C_∞ at infinity if it is bounded there. This leads us to the consideration of $(a-s^{-1})^{-1} = -s(1-sa)^{-1}$ for s near zero. Again

$$(1-sa-ha)^{-1} = (1-sa)^{-1} + ha(1-sa)^{-2} + \dots + h^r a^r (1-sa)^{-r-1} + h^{r+1}a^{r+1}(1-sa)^{-r-1}(1-sa-ha)$$

is a limited Taylor expansion of the function, and the $(s+1)$st power of this expression is a limited expansion of the derivative of order s (modulo the trivial factor $s!\,a^s$).

7.7. These considerations show that an element a of a locally pseudo-convex algebra whose convex boundedness is completant happens to be regular for the rapidly decreasing boundedness of the algebra if $(a-s)^{-1}$ is defined and bounded for s near to infinity. The application of the theorems developed in Sections 4 and 5 is now straightforward.

7.8. To prevent any misunderstanding, I must now give a counterexample. I have been insisting on the fact that bounded structures were important in functional analysis. In the proof of results about topological algebras, I have introduced bounded structures as tools. This was especially remarkable in the above treatment of locally pseudo-convex algebras.

The reader would conjecture that a reasonable theory of algebras with a pseudo-convex bounded structures could be developed along the lines sketched here, and that is not the case.

A set B is *absolutely p-convex* if its p-homogeneous Minkowski functional (gauge) is a p-semi-norm, more precisely if $sx + ty \in B$ when $x \in B$, $y \in B$, $|s|^p + |t|^p \leqslant 1$. A boundedness is *pseudo-convex* if a fundamental system of pseudo-convex bounded sets exists. Such a boundedness is completant if the corresponding p-normed spaces can be made complete.

Let Γ be the unit square in the complex plane. It is clear that $\bigcup_{p>0} L_p(\Gamma)$ is a very good algebra, as complete algebras with pseudo-convex bounded structures go. Yet the resolvent of any rational function on Γ is defined and bounded everywhere on ℂ. So the spectrum of a rational function should be empty. And empty spectra are objects we do not accept.

An analysis of where the proof does not work shows that $1/(z - s)$ is effectively a function of class C_∞, and A-valued (if A is our algebra). But the integration theorems of Turpin and myself allow us to integrate functions of class C_r with values in a p-convex vector space when r is large enough as a function of p.

Here, our function will have class C_r with values in some L_p, but p is consistently just not quite large enough for the theorems to hold (p depends on r, decreases when r increases).

PART III

HOLOMORPHIC FUNCTIONAL CALCULUS IN ALGEBRAS WHERE SPECTRA ARE NOT COMPACT

8. HEURISTIC CONSIDERATIONS

8.1. In the preceding sections, we have studied non-Banach algebras, but the algebras were "very-nearly-Banach" algebras, and we twisted the theory around so that the spectra would be compact sets. This twisting around led us to the consideration of direct limits of Banach algebras with unit. The rules of the game could be described as follows: find a suitable Banach algebra with unit, carry out whatever construction you are interested in in that Banach algebra, and embed the intermediate Banach algebra in the algebra which was given *a priori*. The compactness of whatever spectra we defined followed from the application of this construction.

Now it is possible to define a holomorphic functional calculus in cases where the operators have an essentially non-compact spectrum. Perhaps the best known example is the Heaviside operational calculus. Let D be the operator d/dt. We consider D to be an element of the ring, say $S^*(\mathbb{R}^+)$ of tempered distributions with support in $\mathbb{R}^+$ (multiplication is convolution), or alternatively in $D_b^*(\mathbb{R}^+)$, the distributions of bounded type on $\mathbb{R}^+$.

In the Heaviside calculus, the Laplace transform allows us to define $f(D)$ when f is holomorphic in a half-plane $\operatorname{Re} z > -\varepsilon$ and has polynomial growth at infinity. This restriction is crucial, if we allowed holomorphic functions with greater growth at infinity we might not get an element of $S^*(\mathbb{R}^+)$ for the value of $f(a)$, for some $a \in S^*(\mathbb{R}^+)$, but instead a hyperfunction.

The next example is less of a joke than it seems to be. Let $a_1, \ldots, a_n$ be commuting elements of a topological algebra. Let $f(z_1, \ldots, z_n)$ be an entire function of its arguments, and assume that f has polynomial growth at infinity. We can then define $f[a_1, \ldots, a_n]$. This is clear since f is a polynomial.

In both these examples, we have a domain in $\mathbb{C}^n$, or a filter with open basis. We also have growth conditions on f at infinity. And when f is a good function, we can define $f[a]$. A little thought shows that the point at infinity is not *a priori* different from any other point on the boundary of the domain. In the general formulation, growth conditions should be given not only at infinity, but also at the boundary of the domain. We may expect algebra of functions that are holomorphic on an open set and have suitably limited growth at the boundary of the set.

In the case we have studied up to now, the spectrum was a compact subset of $\mathbb{C}^n$. A family of domains was associated to this compact set. These were the neighbourhoods of the spectrum. Holomorphic functions on these neighbourhoods were interesting. Since every neighbourhood of a compact set in $\mathbb{C}^n$ contains a relatively compact neighbourhood, the growth conditions did not have to be considered. Henceforth, in general, the spectrum will be a means of describing a family of domains, and growth conditions on the boundary of these domains, in such a way that the holomorphic functions satisfying the relevant growth conditions can be used to define the operational calculus.

I should also mention the spectrum of the algebra. In the Banach algebra case we have projective limits and intersections of compact sets, which behave rather well. But if we do not have compact sets all kinds of peculiar things can happen. The inverse limit of the joint spectra of finite sets of elements in the general case is a very complicated object indeed. I shall refrain from trying to describe it here.

8.2. This is unfortunately not the only heuristic argument I have to go through. In my original paper [**32**], I considered the quotient of a b-algebra by a not necessarily closed b-ideal. Banach ideals of Banach algebras are well known and well studied; for example in $\mathscr{L}(H)$ or in connection with sequence spaces. See Schatten [**40**], more recently Pietsch [**39**]. Trace class operators are a Banach ideal of the algebra of linear operators on a Hilbert space. They are an ideal, they have their own norm $\| A \|_{\mathrm{tr}}$, and if $X \in \mathscr{L}(H)$ and A is trace class,

$$\| TA \|_{\mathrm{tr}} \leqslant \| T \|_{\mathrm{tr}} \| A \|,$$

similarly

$$\| AT \|_{\mathrm{tr}} \leqslant \| A \| \, \| T \|_{\mathrm{tr}}.$$

Also, $\| T \|_{\mathrm{tr}} \geqslant \| T \|$ when T is of trace class. In general, a Banach ideal is an ideal α in a Banach algebra A with a norm that is stronger than the Banach algebra norm, such that multiplication is a continuous bilinear map of $A \times \alpha \to \alpha$ (for α a right ideal: similarly for left, 2-sided ideals).

If A is a b-algebra, and if α is an ideal of A, we shall say that α is a *b-ideal of A* if a b-structure has been defined in some way on α, such that the inclusion mapping $\alpha \to A$ is a bounded mapping, and if the relevant products of A-bounded and α-bounded sets are α-bounded (e.g. $B_1 \cdot B_2$ is α-bounded if B_1 is A-bounded and B_1 α-bounded when α is a left ideal).

We notice in passing that the left ideal generated by a finite number of elements has an obvious bounded structure. If $a_1, \ldots, a_n$ are the generators of the ideal, the bounded sets of the ideal are contained in sets $B_1 a_1 + \ldots + B_n a_n$ where $B_1, \ldots, B_n$ are bounded in A. With this bounded structure the ideal would become a b-ideal. The right ideal is just as good as the left ideal, the two-sided b-ideal is not quite as good, because $B_1 a_1 B'_1 + \ldots + B_n a_n B'_n$ is not necessarily contained in any bounded completant set of the ideal. The idea should be to define a two-sided b-ideal generated by the a_i in such a way that it is a little bit larger than the two-sided ideal.

As far as applications of the theory go, we shall usually consider commutative algebras. At worst, we shall consider ideals which are generated in the centre. In any case, the generated left and right ideals will coincide.

Now look how good it is, if we can study A/α, and if we can show that some element of A/α is zero, we show that some element of A belongs to α, to the ideal generated by $a_1, \ldots, a_n$. Standard techniques not involving the quotient of A by a not necessarily closed ideal would only show that the element considered belongs to the closure of α. That is not as good.

I know even one proof where it is not excessively difficult to show that α is a closed ideal if we can at an intermediate step consider A/α without knowing that α is closed. This proof will not be given in these lectures.

8.3. It may be worth while to mention here a category, which I usually call q, whose elements are formal quotients of b-spaces, see e.g. Noël [**14**]. If E is a b-space, and F is a vector subspace of E, with a completant bounded structure stronger than that induced by E, then we say that F is a *b-subspace* of E. The couple (E, F), which we shall henceforth write as $E|F$, will be an object of our category, and we shall call it a *q-space*, although of course it is not a space in the usual sense.

If $E|F$ and $E'|F'$ are two such objects, a bounded linear mapping $E \to E'$ whose restriction to F is a bounded linear mapping $F \to F'$ induces a morphism $E|F \to E'|F'$. If u happens to be a bounded morphism $E \to F'$, the induced morphism $E|F \to E'|F'$ must be the zero morphism. Defining $\tilde{q}(E|F, E'|F')$ in this way, we already have a category, that we shall call $\tilde{q}$.

The category q is larger than $\tilde{q}$. It has the same objects. The elements of $\tilde{q}(E|F, E'|F')$ are morphisms of q. But q must contain further morphisms.

If $E|F$ is an object of q, let X be a closed subspace of E which is contained in F (this may be elliptic; we say that X is a closed subspace of E if X is a b-space, contained in E, whose bounded subsets are exactly the bounded subsets of E contained in X; we also ask that the bounded subsets of X are bounded in F, so that X is closed subspace of F). When that is the case, we want $E|F$ to be isomorphic with $E/X|F/X$. The isomorphism does not usually hold in $\tilde{q}$, e.g. $E|X$ should be isomorphic with $E/X|0$, but $\tilde{q}$ only contains a morphism inverting the natural mapping $E|X \to E/X|0$ when X is a complemented subspace. So we must add to $\tilde{q}$ the inverses of such natural isomorphisms.

And q is well determined if we say that q contains $\tilde{q}$, has the same objects as $\tilde{q}$, and a functor $q \to K$ (another category) extends to $\tilde{q}$ if and only if the morphisms $E|F \to (E/X)|(F/X)$ described above are mapped by the functor on isomorphisms of K. We know thus what the functors from q are.

If $\mathbf{\Phi}_1 : \tilde{q} \to K$ and $\mathbf{\Psi}_1 : \tilde{q} \to K$ are two functors which extend to q, then any functor homomorphism $\Phi_1 \to \Psi_1$ will be a functor homomorphism of the extensions. The category q turns out to be a category. And it is exact, exact sequences with good properties can be defined in it.

8.4. But this is enough of linear algebra in the category q. And it is sufficient to say that multilinear algebra can also be constructed in q, hence q-algebras can be defined. If you wish, a q-algebra could be an object $E|F$ of $\tilde{q}$, with a multiplication on E for which F is a two-sided b-ideal, multiplication being furthermore associative modulo F, the trilinear map $(x, y, z) \leftrightarrow (x \cdot y). z - x \cdot (y - z)$ is a bounded trilinear map of $E \times E \times E$ into F, and also commutative modulo F if the q-algebra we consider is commutative, i.e. $(x, y) \leftrightarrow x \cdot y - y \cdot x$ is a bounded bilinear map of $E \times E$ into F.

The results I shall prove could be proved when $a_1, \ldots, a_n$ are elements of the centre of the q-algebra, i.e. elements of E such that the mapping $x \leftrightarrow a_i x - x a_i$ is a bounded linear mapping of E into F, where $A = E|F$ is an associative q-algebra. The consideration of elements of the centre of an associative q-algebra does more to complicate the results and the theory than to increase the number of applications. And it does involve the introduction of non-obvious techniques. I shall not go into this.

So we are left with the consideration of a quotient E/F, where E is a non-associative algebra, where F is a two-sided ideal, and where E is commutative and associative modulo F. At each step in the proof, we shall be obtaining relations between elements of E, or differentiable E-valued functions or forms; and at the end of relation i, we would put $+ \phi_i$, where ϕ_i belongs to F, or is a function or a form of the right kind and F-valued, and ϕ_i is a conglomerate of the ϕ_k $(k < i)$, and commutators and associators obtained when deriving relation i.

We shall consider therefore A/α, where A is a commutative b-algebra, where α is a b-ideal, and we shall leave it to the reader to supply the remainder ϕ_i, when A is not commutative, nor associative, but is commutative and associative modulo α.

9. IDENTITIES INVOLVING DIFFERENTIAL FORMS

9.1. In this section, we shall sketch the path that will be followed in the construction of the holomorphic functional calculus. There are however technical details that I wish to leave aside in this first draft. I shall therefore assume that A is a Banach algebra. It is only later that I shall show what happens when A is non-Banach.

If A is a Banach algebra, if α is an ideal of A, if n_1 and n_2 are two Banach space norms on α, both stronger than the norm induced by A, then n_1 and n_2 are equivalent. This follows from the closed graph theorem. The uniform boundedness theorem shows that such a Banach space norm is an ideal norm. And for A to be associative, or commutative modulo α, it is sufficient that A/α be associative, or commutative, again by uniform boundedness.

We assume that A is associative and commutative, but we assume that the reader can supply the missing details below when A is only commutative and associative modulo α.

9.2. Let thus A be a commutative Banach algebra with unit and let α be a Banach ideal of A. Let $a_1, \ldots, a_n$ be elements of A. The *joint spectrum* of $(a_1, \ldots, a_n)$ modulo α will be the set

$$\mathrm{sp}_\alpha(a_1, \ldots, a_n) = \{(s_1, \ldots, s_n) \in \mathbb{C}^n \,|\, 1 \notin \alpha + \mathrm{idl}\,(a_1 - s_1, \ldots, a_n - s_n)\}.$$

For $(s_1, \ldots, s_n) \notin \mathrm{sp}_\alpha(a_1, \ldots, a_n)$ it is thus necessary and sufficient that elements $u_{i,s}$, v_s in A, α respectively exist such that

$$1 = \sum (a_i - s_i)\, u_{is} + v_s.$$

If s' is near to s, $1 + \sum(s_i - s_i')u_{i,s}$ has an inverse, we define

$$u_{i,s}(s') = u_{i,s}[1 + \sum(s_i - s_i')u_{i,s}]^{-1}, v_s(s') = v_s[1 + \sum(s_i - s_i')u_{i,s}]^{-1},$$

functions which satisfy locally the relation

$$1 = \sum(a_i - s_i')u_{i,s}(s') + v_s(s')$$

These functions are holomorphic and respectively A-valued and α-valued.

It is now obvious—for those who had not noticed it—that $\mathrm{sp}_\alpha(a)$ is compact. Consideration of maximal ideals proves the set is not empty, though we shall give a proof later independent of any obnoxious axioms.

With a partition of unity, we can define functions $u_i(s)$, $v(s)$ which are respectively A-valued and α-valued, of class C_∞, and such that

$$1 = \sum (a_i - s_i)u_i(s) + v(s)$$

on the complement of the spectrum.

It will be easier to introduce a function $y(s)$, with compact support in a neighbourhood U of the spectrum, and such that

$$1 = \sum (a_i - s_i)u_i(s) + v(s) + y(s). \tag{1}$$

It would be sufficient to start with the functions u_i, v above, let y be a function equal to unity on a neighbourhood of $\mathrm{sp}_\alpha(a)$ and with compact support in an arbitrarily small neighbourhood of that set. We let $u_i' = u_i(1 - y)$ and $v' = v(1 - y)$.

9.3. We shall be using the functions u_i as though they were inverses of $(a_i - s_i)$. But they are not inverses; in particular, they are not unique. We must have relations which replace the theorems stating that the inverse is unique when it exists. Let thus u_i', v', y' be new solutions of our relations

$$\sum (a_i - s_i)u_i'(s) + v'(s) + y'(s) = 1.$$

We then have

$$\begin{aligned} u_i' - u_i &= u_i'[\sum (a_j - s_j)u_j + v + y] - [\sum (a_j - s_j)u_j' + v' + y']u_i \\ &= \sum (u_i'u_j - u_j'u_i)(a_j - s_j) + (u_i'v - v'u_i) + (u_i'y - y'u_i) \\ &= \sum \phi_{ij}(a_j - s_j) + \psi_i + \eta_i \\ y' - y &= y'[\sum (a_j - s_j)u_j + v + y] - [\sum (a_j - s_j)u_j' + v' + y']y \\ &= -\sum \eta_j(a_j - s_j) + \psi' \end{aligned}$$

where ϕ_{ij}, η_i are A-valued functions of class C_∞, η_i having compact support within the neighbourhood U of the spectrum, and $\phi_{ij} = -\phi_{ji}$, while the ψ_i, ψ' are α-valued functions of class C_∞.

Now if we start from u, v, y satisfying relation (1), and if we put

$$\begin{aligned} u_i' &= u_i + \sum \phi_{ij}(a_j - s_j) + \psi_i + \eta_i \\ y' &= y - \sum \eta_j(a_j - s_j) + \psi' \\ v' &= 1 - \sum (a_j - s_j)u_j' - y' \end{aligned}$$

where $\phi_{ij} = -\phi_{ji}$ and η_i are A-valued of class C_∞, where ψ_i, ψ' are α-valued of class C_∞, while the η_i and ψ' have compact support in U, then u', v', y' have all the nice properties of u, v, y.

This shows that it is possible to go from any admissible system u, v, y to another, say U, V, Y, in a finite number of steps, each one being of one of the following forms:

A_{ij}: Replace u_i by u_i', v by v', y by y' in the following way

$$u_i' = u_i + \phi(a_j - s_j), \quad u_j' = u_j - \phi(a_i - s_i)$$
$$u_k' = u_k \, (k \neq i, j), \quad y' = y, \quad v' = v$$

where ϕ is an A-valued function of class C_∞.

B_i: Replace (u, v, y) by (u', v', y') in the following way

$$u_i' = u_i + \eta, \quad y' = y - \eta(a_i - s_i)$$
$$u_j' = u_j \, (j \neq i), \quad v' = v$$

where η is an A-valued function of class C_∞ with compact support in U.

C: Replace (u, v, y) by (u', v', y') where

$$u_i' = u_i + \psi_i, y' = y + \psi', \quad v' = 1 - \sum (a_i - s_i) u_i' - y'$$

where ψ_i, ψ' are α-valued functions of class C_∞ and ψ' has compact support in U.

9.4. This is now the time to write down the differential form

$$\frac{(n + k)!}{k!} y^k \bar{\partial} u_1 \wedge \ldots \wedge \bar{\partial} u_n = \omega_k(u, v, y). \tag{2}$$

It has degree n in $d\bar{z}_1, \ldots, d\bar{z}_n$, and is of class C_∞, with compact support in U. This form depends on the choice of (u, v, y), but does not depend too much on this; if (u', v', y') is another solution of (1), if k' is another strictly positive integer

$$\omega_k(u, v, y) - \omega_{k'}(u', v', y') = \psi + \bar{\partial}\phi \tag{3}$$

where ψ is an α-valued form of class C_∞ with compact support within U and ϕ is an A-valued form of class C_∞, also with compact support within U.

(a). Consider first the case where $u = u', v = v', y = y'$. We must compare $\omega_k(u, v, y) \, (k > 0)$ with $\omega_{k+1}(u, v, y)$. Our proof will involve the form of degree $n - 1$

$$\varpi(u, v, y) = \sum (-1)^{i-1} u_i \bar{\partial} u_1 \wedge \ldots \wedge \bar{\partial} \hat{u}_i \wedge \ldots \wedge \bar{\partial} u_n \tag{4}$$

where the symbol $\bar{\partial}\hat{u}_i$ means, as it often does, that the factor $\bar{\partial} u_i$ is omitted. It is well known that

$$\bar{\partial}\varpi = n \bar{\partial} u_1 \wedge \ldots \wedge \bar{\partial} u_n. \tag{5}$$

We may rewrite relation (1):

$$1 = \sum (a_i - s_i)u_i + v + y$$

and differentiate it:

$$0 = \sum (a_i - s_i)\bar{\partial} u_i + \bar{\partial} v + \bar{\partial} y. \tag{6}$$

Of course

$$\sum (a_i - s_i)\bar{\partial} u_i \wedge \varpi(u, v, y) = \sum (a_i - s_i)u_i \cdot \bar{\partial} u_1 \wedge \ldots \wedge \bar{\partial} u_n. \tag{7}$$

We may also replace in relation (7), $\sum (a_i - s_i)\bar{\partial} u_i$ and $\sum (a_i - s_i)u_i$ by $-\bar{\partial} y - \bar{\partial} v$ and $1 - y - v$ respectively, hence

$$\bar{\partial} y \wedge \varpi(u, v, y) = (y - 1)\bar{\partial} u_1 \wedge \ldots \wedge \bar{\partial} u_n + \psi_1 \tag{8}$$

where ψ_1 is a α-valued form of class C_∞.

It follows that

$$\begin{aligned}\bar{\partial}[y^k \varpi(u)] &= k y^{k-1} \bar{\partial} y \wedge \varpi(u) + n y^k \bar{\partial} u_1 \wedge \ldots \wedge \bar{\partial} u_n \\ &= (n + k) y^k \bar{\partial} u_1 \wedge \ldots \wedge \bar{\partial} u_n - k y^{k-1} \bar{\partial} u_1 \wedge \ldots \wedge \bar{\partial} u_n + \psi_2 \end{aligned} \tag{9}$$

where ψ_2 is a form of class C_∞ with compact support in U. Multiplying this relation by $(n + k - 1)!/k!$ we see that

$$\omega_k(u, v, y) - \omega_{k-1}(u, v, y) = \bar{\partial}\phi + \psi$$

as required.

(b). We must next go on to consider the dependence of $\omega_k(u, v, y)$ on the specific u, v, y we consider. It will be sufficient to consider

$$\omega_k(u, v, y) - \omega_k(u', v', y')$$

when u', v', y' are deduced from u, v, y by one of the specific transformations described at the end of Section 9.3. Transformations of type C are rather trivial. We clearly have $\omega_k(u', v', y') - \omega_k(u, v, y) = \psi$ where ψ is α-valued of class C_∞, and with compact support in U.

Transformations of type A are not much more difficult. It will be sufficient to consider A_{12}, i.e.

$$u'_1 = u_1 + \phi(a_2 - s_2), u'_2 = u_2 - \phi(a_1 - s_1), u'_3 = u_3, \ldots, u'_n = u_n, y' = y.$$

This being written, we see that

$$\begin{aligned} &y'^k \bar{\partial} u'_1 \wedge \ldots \wedge \bar{\partial} u'_n - y^k \bar{\partial} u_1 \wedge \ldots \wedge \bar{\partial} u_n \\ &\quad = y^k \cdot (a_2 - s_2)\bar{\partial}\phi \wedge \bar{\partial} u_2 \wedge \ldots \wedge \bar{\partial} u_n \\ &\qquad\qquad - y^k \bar{\partial} u_1 \wedge (a_1 - s_1)\bar{\partial}\phi \wedge \bar{\partial} u_3 \wedge \ldots \wedge \bar{\partial} u_n \end{aligned}$$

$$= - y^k\bar{\partial}y \wedge \bar{\partial}\phi \wedge \bar{\partial}u_3 \wedge \ldots \wedge \bar{\partial}u_n + \psi$$
$$= \bar{\partial}\phi + \psi$$

where

$$\phi = \frac{1}{k+1} y^{k+1}\bar{\partial}\phi \wedge \bar{\partial}u_3 \wedge \ldots \wedge \bar{\partial}u_n.$$

(c). We must still consider a transformation of type B, say

$$u_1' = u_1 - \eta, y' = y + \eta(a_1 - s_1), u_2' = u_2', \ldots, u_n' = u_n.$$

We observe that

$$y'^k\bar{\partial}u_1' - y^k\bar{\partial}u_1 = \sum_1^k \binom{k}{k'} (a_1 - s_1)^{k-k'}\eta^{k-k'} y^{k'}\bar{\partial}u_1$$
$$- \sum_0^k \binom{k}{k'} (a_1 - s_1)^{k-k'}\eta^{k-k'} y^{k'}\bar{\partial}\eta$$

$$= \sum_0^{k-1} \frac{k!}{k'!(k-k'+1)!} (a_1 - s_1)^{k-k'}\rho_{k'} - (a_1 - s_1)^k\eta^k\bar{\partial}\eta$$

where

$$\rho_{k'} = k'(a_1 - s_1)\eta^{k-k'+1}y^{k'-1}\bar{\partial}u_1 - (k - k' + 1)\eta^{k-k'}y^{k'}\bar{\partial}\eta.$$

We next consider $\rho_{k'} \wedge \bar{\partial}u_2 \wedge \ldots \wedge \bar{\partial}u_n$, keeping in mind that

$$(a_1 - s_1)\bar{\partial}u_1 \wedge \bar{\partial}u_2 \wedge \ldots \wedge \bar{\partial}u_n = - \bar{\partial}y \wedge \bar{\partial}u_2 \wedge \ldots \wedge \bar{\partial}u_n + \psi_1$$

where ψ is an α-valued form of class C_∞, and hence

$$\rho_{k'} \wedge \bar{\partial}u_2 \wedge \ldots \wedge \bar{\partial}u_n = - k!\eta^{k-k'+1}y^{k'-1}\bar{\partial}y \wedge \bar{\partial}u_2 \wedge \ldots \wedge \bar{\partial}u_n$$
$$- (k - k' + 1)\eta^{k-k'}y^{k'}\bar{\partial}\eta \wedge \bar{\partial}u_2 \wedge \ldots \wedge \bar{\partial}u_n + \psi_2,$$

where ψ_2 is now α-valued, of class C_∞, and has compact support in U. We see $\bar{\partial}[\eta^{k-k'+1}y^{k'}\bar{\partial}u_2 \wedge \ldots \wedge \bar{\partial}u_n]$ appearing, hence as usual

$$\omega_k(u', v', y') - \omega_k(u, v, y) = \bar{\partial}\phi + \psi.$$

9.5. We have thus associated, and in a natural way, an object which turns out to be an A/α-valued class of $\bar{\partial}$-cohomology with compact support in U, whenever U is a neighbourhood of $\mathrm{sp}_\alpha(a_1, \ldots, a_n)$. The question can be asked, is this natural object trivial? As far as we are concerned, there is only one trivial object in our space of cohomology. This is the null object. We shall show that our object is not trivial by integrating it, or more exactly integrating its product with $dz_1 \wedge \ldots \wedge dz_n$ over U, and showing that the integral (which belongs to A/α, the good old quotient algebra) is the product of the unit of A/α by $(2\pi i)^n$.

But before doing that, we shall compare the object associated to $(a_1, \ldots, a_n)$ and that associated to $(b_1, \ldots, b_m)$ with the object associated with $(a_1, \ldots, a_n, b_1, \ldots, b_m)$. A natural direct product mapping maps the product of the $\bar{\partial}$-cohomology with compact support in U and that in V into the $\bar{\partial}$-cohomology with compact support in $U \times V$. If ω is a form in one cohomology class in U and ω' one in V, then the direct product class in $U \times V$ contains the form $\omega \wedge \omega'$.

We assume that

$$\sum (a_i - s_i)u_i(s) + v(s) + y(s) = 1,$$
$$\sum (b_j - t_j)\, u'_j(t) + v'(t) + y'(t) = 1.$$

It follows that

$$\sum (a_i - s_i)U_i(s, t) + \sum (b_j - t_j)U'_j(s, t) + V(s, t) + Y(s, t) = 1$$

if

$$U_i(s, t) = u_i(s),\ U'_j(s, t) = y(s)u'_j(t),\ Y(s, t) = y(s)y'(t)$$

and

$$V(s, t) = v(s) + y(s)v'(t).$$

Our compactly supported $\bar{\partial}$-cohomology class contains therefore the form

$$\frac{(n + m + k)!}{k!}(yy')^k \bar{\partial}u_1 \wedge \ldots \wedge \bar{\partial}u_n \wedge \bar{\partial}(yu'_1) \wedge \ldots \wedge \bar{\partial}(yu'_m)$$

$$= \frac{(n + m + k)!}{k!} y^{m+k} \bar{\partial}u_1 \wedge \ldots \wedge \bar{\partial}u_n \wedge y'^k \bar{\partial}u'_1 \wedge \ldots \wedge \bar{\partial}u'_m$$

$$= \omega_{m+k}(u, v, y) \wedge \omega_k(u', v', y') \tag{10}$$

For the doubters, the first expression is equal to the second, because the difference is of degree larger than n in $d\bar{s}_1, \ldots, d\bar{s}_n$.

9.6. Let now $f(s_1, \ldots, s_n)$ be a holomorphic function on U, a neighbourhood of $\mathrm{sp}_\alpha(a_1, \ldots, a_n)$. We can consider

$$f[a] = \frac{1}{(2\pi i)^n} \int f(s_1, \ldots, s_n) \frac{(n + k)!}{k!} y^k \bar{\partial}u_1 \wedge \ldots \wedge \bar{\partial}u_n \wedge ds_1 \wedge \ldots \wedge ds_n \tag{11}$$

where of course, the a's, u's, and y are related by (1). The right-hand side of this defining equation does not depend on a, except through u and y which exist but are not determined by a. However, $f[a]$ is well defined, modulo α, by this relation because

$$\int f(s)[\bar{\partial}\phi + \psi] \wedge ds_1 \wedge \ldots \wedge ds_n$$

$$= \int d[f(s)\cdot\phi] \wedge \ldots \wedge ds_n + \int f(s)\psi \wedge ds_1 \wedge \ldots \wedge ds_n.$$

The first term is zero, the second belongs to α.

If $f(s_1,\ldots,s_n)$ and $g(t_1,\ldots,t_m)$ are holomorphic on neighbourhoods U, V of $\mathrm{sp}_\alpha(a)$, $\mathrm{sp}_\alpha(b)$, then $f \times g$ defined by $f \times g(s,t) = f(s)\, g(t)$ is holomorphic on $U \times V$, a neighbourhood of $sp_\alpha(a,b)$. We shall show that

$$f \times g[a,b] = f[a]\,.\,g[b] \tag{12}$$

and this is rather obvious because of the result proved in Section 9.5.

If $f(s_1,\ldots,s_n) = 1$, then $f[a] = 1$ is the unit of A/α. Since the constant 1 is the direct product of n functions, each constant and equal to one, it will suffice because of relation (12) to prove this when $n = 1$. We choose u in such a way that $u = (a-s)^{-1}$ when s is large, then y has compact support in the domain of our function 1, i.e. in $\mathbb{C}$. Going back to relation (9) we see that $\bar{\partial}u$ has compact support in $\mathbb{C}$ and is in the same $\bar{\partial}$-cohomology class with compact support as the form we should integrate, i.e. $(k+1)y^k\bar{\partial}u$. The integral that we are interested in is therefore

$$\frac{1}{2\pi i}\int_{\mathbb{C}} \bar{\partial}u \wedge ds = \frac{1}{2\pi i}\int_j (a-s)^{-1}ds = 1$$

and we have shown that

$$1[a_1,\ldots,a_n] = 1 \tag{13}$$

if $1(s)$ is the constant function equal to one.

We observe that we have proved, as had been announced, that

$$\int_{\mathbb{C}^n} \omega(u,v,y) \wedge ds_1 \wedge \ldots \wedge ds_n = (2\pi i)^n$$

hence, the class of cohomology ω cannot vanish, unless of course $\alpha = A$, i.e. the whole cohomology modulo α is trivial.

It is therefore important that $(a_i - s_i)\omega(u,v,y)$ is the null class of cohomology, e.g. when $i = 1$

$$(a_1 - s_1)\frac{(n+k)!}{k!}\, y^k\bar{\partial}u_1 \wedge \ldots \wedge \bar{\partial}u_n$$

$$= -\frac{(n+k)!}{k!}\, y^k\bar{\partial}y \wedge \bar{\partial}u_2 \wedge \ldots \wedge \bar{\partial}u_n + \psi_1$$

$$= \bar{\partial}\left[\frac{(n+k)!}{(k+1)!}\, y^{k+1}\bar{\partial}u_2 \wedge \ldots \wedge \bar{\partial}u_n\right] + \psi_1.$$

It follows, when $f(s) = \sum (s_i - a_i)g_i(s)$ with g_i holomorphic and A-valued, that

$$f[a] = 0.$$

In particular, if $P(s)$ is a polynomial

$$P[a] = P(a) \tag{14}$$

since $P(s) - P(a)$ belongs to the ideal generated by the functions $s_i - a_i$.

9.7. We have not yet proved that the mapping $f \leftrightarrow f[a]$, $\mathcal{O}(\mathrm{sp}_\alpha(a), A) \to A/\alpha$ is a homomorphism. This will follow from a result about linear mappings.

Let $T: \mathbb{C}^n \to \mathbb{C}^m$ be a linear, or an affine map. Then $\mathrm{sp}_\alpha(Ta) = T\mathrm{sp}_\alpha(a)$. *If* $f \in \mathcal{O}(\mathrm{sp}_\alpha(Ta))$, *then* $f \circ T \in \mathcal{O}(\mathrm{sp}_\alpha(a))$ *and* $f \circ T[a] = f[Ta]$.

An affine map $\mathbb{C}^n \to \mathbb{C}^m$ is a composition of maps, each of which is a projection $(s_1, \ldots, s_n) \leftrightarrow (s_1, \ldots, s_{n-1})$, an injection $(s_1, \ldots, s_n) \leftrightarrow (s_1, \ldots, s_n, 0)$, or an invertible affine transformation. It will be sufficient to prove the result in each of these three cases.

(a). Let us first assume that T is an invertible linear map. We assume that

$$\sum (a_i - s_i)u_i(s) + v(s) + y(s) = 1.$$

We let $b_i = (Ta)_i = \sum T_{ij}a_j$ and $S = T^{-1}$.

$$\sum (b_i - Ts_i)\,({}^tSu)_i(s) + v(s) + y(s) = 1$$

or equivalently if $t_i = (Ts)_i$,

$$\sum (b_i - t_i)\,{}^tSu(St) + v(St) + y(St) = 1.$$

We of course let $u' = {}^tS(u \circ S)$, $y' = y \circ S$. The kernel $\bar{\partial}u'_1 \wedge \ldots \wedge \bar{\partial}u'_n$ evaluated at $t = Ts$ will be the product of $\bar{\partial}u_1 \wedge \ldots \wedge \bar{\partial}u_n$ by

$$\det {}^tS \cdot \overline{\det S} = |\det S|^2,$$

but evaluated at s. A factor $\det S$ comes from the replacement of

$$\bar{\partial}u'_1 \wedge \ldots \wedge \bar{\partial}u'_n \quad \text{by} \quad \bar{\partial}u_1 \wedge \ldots \wedge \bar{\partial}u_n.$$

The factor $\overline{\det S}$ comes from the change of variables. The product, $|\det S|^2$, is the Jacobian of S seen as a linear transformation of $\mathbb{C}^n \simeq \mathbb{R}^{2n}$. We can apply the formulae for change of variables under multiple integrals.

(b). Translation operators are relatively trivial.

(c). A projection $T:(s_1, \ldots, s_n) \to (s_1, \ldots, s_{n-1})$ maps $(a_1, \ldots, a_n)$ onto $(a_1, \ldots, a_{n-1})$. We also have $T\mathrm{sp}_\alpha(a) = \mathrm{sp}_\alpha(Ta)$ by the spectral mapping theorem. Now, if $f \in \mathcal{O}(\mathrm{sp}_\alpha(Ta))$, $f \circ T$ is the direct product $f \times 1$ of f by the function, constant and equal to unit of the last variable. Hence

$$\begin{aligned} f \circ T[a_1, \ldots, a_n] &= f[a_1, \ldots, a_{n-1}] \cdot 1[a_n] \\ &= f[Ta] \end{aligned}$$

as required.

(d). We must still consider mappings $T:(s_1, \ldots, s_n) \to (s_1, \ldots, s_n, 0)$, i.e. compare the $(n+1)$-tuple $(a, 0)$ to the n-tuple a.

We notice that $\mathrm{sp}_\alpha(a_1, \ldots, a_n, 0) = \mathrm{sp}_\alpha(a_1, \ldots, a_n) \times \{0\}$. If

$$f(z_1, \ldots, z_{n+1}) \in \mathcal{O}(\mathrm{sp}_\alpha(a, 0)),$$

then

$$f(z_1, \ldots, z_{n+1}) = f(z_1, \ldots, z_n, 0) + z_{n+1} g(z_1, \ldots, z_{n+1})$$

where

$$g(z_1, \ldots, z_{n+1}) = \frac{f(z_1, \ldots, z_{n+1}) - f(z_1, \ldots, z_n, 0)}{z_{n+1}}$$

has a removable singularity for $z_{n+1} = 0$. The result of substituting $(a_1, \ldots, a_n, 0)$ in $z_{n+1}\, g(z_1, \ldots, z_{n+1})$, (or, by the way, in any product of a holomorphic function by z_{n+1}) is zero. This shows that

$$f \circ T[a] = f_0[Ta] = f[a] \cdot 1[0] = f[a]$$

where $f_0(z_1, \ldots, z_{n+1}) = f(z_1, \ldots, z_n, 0) = f \circ T \times 1$.

(e). I have used the spectral mapping theorem quite freely. This theorem follows from the axiom of choice and the fact that A is a Banach algebra. But in each special case, it is possible to justify the form of the spectral mapping theorem that we use without using the maximal ideal theory for Banach algebras.

9.8. The fact that $f \to f[a]$ is a homomorphism is now easy to prove. We consider $f \cdot g(s)$ where f and g belong to $\mathcal{O}(\mathrm{sp}_\alpha(a))$, then

$$f \times g(s, s) \in \mathcal{O}(\mathrm{sp}_\alpha(a, a)).$$

We also consider $T:(s) \to (s, s)$, $\mathbb{C}^n \to \mathbb{C}^{2n}$. We notice that

$$f \times g[s, s] = f \times g(Ts) = f \cdot g(s)$$

hence

$$f \cdot g[a] = f \times g[a, a] = f[a] g[a].$$

10. THE SPECTRUM

10.1. The next main step in this development should be a definition of the spectrum of an n-tuple $(a_1, \ldots, a_n)$ of elements of A/α where A is a b-algebra and α is a b-ideal. But before even defining the spectrum we must generalize the well-known theorem, that the spectrum is not empty. The reader must grant that the following is a good one-dimensional generalization: s is a complex variable, $\delta_0(s) = (1 + |s|^2)^{-\frac{1}{2}}$, A is a b-algebra with unit, and $\Theta(s; \delta_0; A)$ is the b-algebra of functions of s on $\mathbb{C}$, say $u(s)$, such that for some N, $\delta_0(s)^N u(s)$ is a bounded function. It holds that $(a - s)$ *does not have an inverse in* $\Theta(s; \delta_0; A)$.

This has essentially been proved. If $(a - s)$ had an inverse, the inverse must locally be a bounded function, but the resolvent is analytic on any open set where it is bounded. So the resolvent is entire. It has also polynomial growth at infinity, but the term of highest order in $(a - s)\sum_1^n c_k s^k$ is $-c_n s^{n+1}$, and that cannot be 1.

The reader will observe some feedback originating from notations that will be introduced later. What may disturb readers most is the fact that I use ordinary letters as placetakers where it is the fashion to put a dot. This is done in the calculus courses and has therefore become illegitimate. Must one write $a - .$ instead of $a - s$? How must one cope with function-valued functions, viewed sometimes as functions of one variable, sometimes as functions of the second, sometimes as functions of both? Could we introduce several types of dot?

10.2. Thinking of the definition of the spectrum of a Banach algebra element, of that of a commutative Banach algebra n-tuple, looking at the above theorem, we state the following conjecture.

Let $s = (s_1, \ldots, s_n)$ be n complex variables, let $\delta_0(s) = (1 + |s|^2)^{-\frac{1}{2}}$, let A be a commutative b-algebra with unit, and $\Theta(s; \delta_0; A)$ be the algebra of functions of s on $\mathbb{C}^n$, say $u(s)$, such that some integer N can be found with $\delta_0(s)^N u(s)$ bounded. *The ideal generated by* $(a_1 - s_1, \ldots, a_n - s_n)$ *in* $\Theta(s; \delta_0; A)$ *is proper.*

It turns out to be easier to prove a more general theorem. Let α be a b-ideal of A, and define $\Theta(s; \delta_0; \alpha)$ in an obvious way. We shall assume that

$$1 \in \operatorname{idl}(a_1 - s_1, \ldots, a_n - s_n; \Theta(s; \delta_0; A)) + \Theta(s; \delta_0; \alpha)$$

and prove that $1 \in \alpha$.

As a matter of fact, it is sufficient to prove the more general theorem when $n = 1$. We write $s = (s_1, s')$, let $A_1 = \Theta(s', \delta_0, A)$, and

$$\alpha_1 = \operatorname{idl}(a' - s'; A_1) + \Theta(s'; \delta_0; \alpha).$$

Our assumption is that

$$1 \in \text{idl}\,(a_1 - s_1 . \Theta(s_1, \delta_0, A_1)) + \Theta(s_1, \delta_0, \alpha_1).$$

Once the general theorem is proved for $n = 1$, it follows that

$$1 \in \alpha_1 = \text{idl}\,(a' - s', \Theta(s'; \delta_0; A)) + \Theta(s'; \delta_0; x)$$

and hence, by induction, $1 \in \alpha$.

In the above statements, I have used the natural bounded structures of A_1 and α_1, and I assume that the reader can reconstruct these.

10.3. The following turns out to be an interesting lemma: *Let k_1, k_2 be strictly positive real numbers, k_2 less than 1. Let $u_n \in \alpha, v_n \in A$, $u_n + v_n = 1$ where $u_n = O_\alpha(k_1^{-n})$, $v_n = O_A(k_2^n)$,* i.e. *$k_1^n u_n$ is bounded in α and $k_2^{-n} v_n$ is bounded in A. Then $1 \in \alpha$, or $\alpha = A$.*

We assert therefore that α is the improper ideal if a sequence of elements of α can be found, which does not diverge too fast in α, but tends rather fast to the unit in A.

The first remark is that we could assume that $k_2 < k_1$. Just write

$$1 = (u_n + v_n)^r = U_n + v_n^r.$$

Then $v_n^r = O_A(k_2^{nr})$, while $U_n = O_\alpha(k_1^{-n})$, write the binominal expansion of $(u_n + v_n)^r$ and remember that all the factors that occur are A-bounded ($u_n \to 0$ in $A, v_n = 1 - u_n$), and in each term (except v_n^r) we have a factor $u_n = O_\alpha(k_1^{-n})$. So we have effectively replaced k_2 by k_2^r and left k_1 unchanged.

So $k_2 < k_1$. Write now

$$\begin{aligned} u_n - u_{n+1} &= u_n(u_{n+1} + v_{n+1}) - (u_n + v_n)u_{n+1} \\ &= u_n v_{n+1} - v_n u_{n+1} = O_\alpha(k_1^{-n} k_2^n). \end{aligned}$$

This evaluation shows that u_n is an α-convergent sequence. But $u_n \to 1$ in A so $u_n \to 1$ in α.

10.4. We now assume that A is a b-algebra, that α is a b-ideal, and that functions $u(s) \in \Theta(s; \delta_0; A)$, $v(s) \in \Theta(s; \delta_0; \alpha)$ can be found such that

$$(a - s)u(s) + v(s) = 1.$$

Define $U(s) = -\bar{s}/(1 + |s|^2)$, $Y(s) = (1 + a\bar{s})/(1 + |s|)^2$, and observe that

$$(a - s)[U(s) + Y(s)u(s)] + Y(s)v(s) = 1.$$

This operation has effectively reduced the rate of growth of u and v, by one unit unless u was already tending to zero at infinity.

It is therefore no loss in generality to assume that $|s|^2[u(s) - U(s)]$ and

$|s|^2 v(s)$ are bounded functions, respectively A-valued and α-valued. If u and v were differentiable A-valued and α-valued functions, we would consider $\int_j u(s)ds = \iint_{(j)} \bar{\partial} u ds$; but $\bar{\partial} u$ can be expressed in terms of $v, \bar{\partial} v$, and the right-hand side would be in α, the left-hand side would be near $\int_j U(s)ds$, i.e. near $(2\pi i)^{-1}$. The result would follow.

Since we cannot integrate, we must introduce a makeshift integral, i.e. Simpson's formula. We know that Simpson's integral is equal to the usual integral when the function integrated is a polynomial of degree 3. Also, when f is sufficiently differentiable,

$$\int_a^b f(x)dx - S^b_{h,a} f(x)dx = O(h^3)$$

where $S^b_{h,a} f(x)dx$ is the approximate value assigned to the integral from a to b by Simpson's formula with step h (this has a meaning when $b - a$ is an integral multiple of $2h$). The remainder involves the value of $f^{(4)}$ at an intermediate point, and

$$S^b_{h,a} f = \frac{h}{3}[f(a) + 4f(a+h) + 2f(a+2h) + \ldots + 4f(b-h) + f(b)].$$

We shall integrate, or integrate approximately over a square contour Γ_N of side $2N$ and centred at the origin. The step will be N^{-1} (or iN^{-1}, $-N^{-1}$, $-iN^{-1}$ according to the side of the square we consider). It is clear that

$$S_{N^{-1},\Gamma_N} U(s)ds$$

will be very near $\int_\Gamma U(s)ds$. Since U itself is very near s^{-1} we see that $S_{N^{-1},\Gamma_N} U(s)ds$ is near to $2\pi i$. All that we need to remember is that

$$a < |S_{N^{-1},\Gamma_N} U(s)ds| < b$$

when a, b are scalars with $a < 2\pi < b$, and N is large enough.

Γ_N is a contour of length $8N$. So S_{N^{-1},Γ_N} is an integral with respect to a measure with a finite support ($8N^2$ points), but total mass $8N$. On the other hand $(u(z) - U(z)) = O_A(|z|^{-2})$; on Γ_N, we have $u - U = O_A(N^{-2})$ hence $S_{N^{-1},\Gamma_N}(u - U) = O_A(N^{-1})$.

We shall evaluate our integral otherwise, splitting up Γ_N into N^4 subsquares, each of side $2N^{-1}$. If γ_k is one of these subsquares, $S_{\gamma_k} u(s)ds$ will be the Simpson evaluation of the integral $\int_{\gamma_k} u(s)ds$ where the sides of the squares are each split into two subintervals (this is the least that one can do and apply Simpson). Because the integrals, or their Simpson evaluations on the inner sides cancel, we see that

$$S_{N^{-1},\Gamma_N} u(s)ds = \sum_k S_{\gamma_k} u(s)ds.$$

We shall use as well as we can the fact that u is the resolvent of a (modulo α) and that the resolvent is, or at least should be, holomorphic.

$$\begin{aligned} u(s') - u(s) &= [(a - s)u(s) + v(s)u(s') - u(s)(a - s')u(s') + v(s')] \\ &= (s' - s)u(s)u(s') + (v(s)u(s') - u(s)v(s')); \end{aligned}$$

this is the resolvent identity, with a term left over in α. We apply this identity recursively to obtain a limited geometric expansion

$$\begin{aligned} u(s') = u(s) + (s' - s)u(s)^2 + (s' - s)^2u(s)^3 + (s' - s)^3u(s)^4 \\ + (s' - s)^4u(s)^4u(s') + \psi(s, s') \end{aligned}$$

where ψ is α-valued, and bounded independently of (s, s') when $|s' - s| < 1$. (The expression $s' - s$ enters in the expression of ψ when we develop the expansion).

For each k, let s_k be the centre of the square γ_k, so $s' - s_k = O(N^{-1})$ when s' is on γ_k, and apply S_{γ_k}, remembering that Simpson's integral is exactly the integral if the integrand is a polynomial of the third degree, and that cyclic integrals of holomorphic functions vanish. Then

$$\begin{aligned} S_{\gamma_k}u(s')ds' &= u(s_\alpha)^4 S_{\gamma_\alpha}(s' - s_\alpha)^4u(s')ds' + S_{\gamma_\alpha}\psi(s_\alpha, s')ds' \\ &= O_A(N^{-5}) + O_\alpha(N^{-1}). \end{aligned}$$

When we sum over the N^4 squares, we see that

$$S_{N^{-1}, \Gamma_N}u(s' = O_A(N^{-1}) + O_\alpha(N^3).$$

But this same expression is $2\pi i + O_A(N^{-1})$. Hence

$$1 = O_A(N^{-1}) + O_\alpha(N^3).$$

Choose a subsequence $N = 2^n$. Then $1 = O_A((\tfrac{1}{2})^n) + O_\alpha((\tfrac{1}{8})^{-n})$ The lemma proved in Section 10.3 applies, $1 \in \alpha$.

10.5. We now know what it means to say that the spectrum is not empty. But what is the spectrum? If anyone changes some parameters in the right hand side of the relation

$$1 \notin \text{idl}\,(a_1 - s_1, \ldots, a_n - s_n; \Theta(s; \delta_0; A)) + \Theta(s; \delta_0; \alpha)$$

in such a way that $\notin$ turns in to $\in$, I am willing to grant that he has defined a spectrum. If he constructs a functional calculus with his spectrum, bravo. What I am about to describe is not a final theorem, but one of a number of good theorems. (It's the only one I know, by the way).

So the spectrum $\Delta_\alpha(a_1, \ldots, a_n; A)$ (or $\Delta(a_1, \ldots, a_n; A|\alpha)$) will be the set of functions $\delta: \mathbb{C}^n \to \mathbb{R}^+$ such that

$$1 \in \text{idl}\,(a_1 - s_1, \ldots, a_n - s_n, \delta; \Theta(s; \delta_0 A)) + \Theta(s; \delta_0; \alpha).$$

It is clear that $\Delta_\alpha(a)$ is a filter on the lattice of non-negative functions in $\mathbb{C}^n$. i.e. $\Delta_\alpha(a)$ has the property that constants are not in it and if $\delta_1, \delta_2 \in \Delta_\alpha(a)$ then $\delta_1 \wedge \delta_2(s) = \min(\delta_1(s), \delta_2(s)) \in \Delta_\alpha(a)$. Also if $\delta' \geqslant \delta$ and $\delta \in \Delta_\alpha(a)$ then $\delta' \in \Delta_\alpha(a)$. If $\delta \in \Delta_\alpha(a)$ and $\varepsilon > 0$ is a constant, $\varepsilon\delta \in \Delta_\alpha(a)$. Also $\delta_0 = (1 + |s|^2)^{-\frac{1}{2}} \in \Delta_\alpha(a)$, since δ_0 has an inverse in $\Theta(s; \delta_0; A)$. If $\delta \in \Delta_\alpha(a)$, and $N \in \mathbb{N}$, then $\delta^N \in \Delta_\alpha(a)$; just raise the relation

$$1 = \sum (a_i - s_i)u_i(s) + v(s) + \delta(s)y_0(s)$$

to the power N and collect the terms suitably, isolating $\delta(s)^N y_0^N(s)$. We proved in Section 10.4 that this was a proper filter if α was a proper ideal, i.e. $0 \in \Delta_\alpha(a) \Rightarrow 1 \in \alpha$.

The spectrum has a basis made up of Lipschitz functions. This is not too difficult. We assume that $\delta(s) \in \Delta(a; A|\alpha)$ and let

$$\delta_1(s) = \inf_{t \in \mathbb{C}^n} (\delta(t) + |t - s|);$$

δ_1 is the largest function with Lipschitz constant one that is smaller than δ. For each $s \in \mathbb{C}^n$ and each k, we choose $t_k(s) \in \mathbb{C}^n$ with

$$\delta(t_k(s)) + |t_k(s) - s| < \delta_1(s) + 2^{-k}$$

(this uses, I am afraid, the axiom of choice.) The fact that we can do this is obvious from the definition of δ_1.

We have then

$$\begin{aligned} 1 &= \sum (a_i - s_i)u_i(t_k(s)) + v(t_k(s)) + \sum (s_i - t_{ik}(s))u_i(t_k(s)) + \delta(t_k(s))y_0(t_k(s)) \\ &= \sum (a_i - s_i)u_{i\cdot k}(s) + v_k(s) + [\delta_1(s) + 2^{-k}]y_{0,k}(s) \end{aligned}$$

where $u_{i,k} = u_i \circ t_k$, $v_k = v \circ t_k$ and

$$y_{0,k}(s) = \frac{\sum(s_i - y_{ik}(s))u_i \circ t_k + (\delta \circ t_k)(y_0 \circ t_k)}{\delta_1 + 2^{-k}}.$$

The sequences $u_{i,k}, v_k, y_{0,k}$ are bounded sequences in A. We have constructed in this way a bounded sequence of elements of the ideal

$$\operatorname{idl}(a_1 - s_1, \ldots, a_n - s_n, \delta_1; \Theta(s; \delta_0; A)) + \Theta(s; \delta_0; \alpha)$$

such that $1 - \sum (a_i - s_i)u_{ik} - v_k - \delta_1(s)y_{0,k} = O_N(2^{-k})$. The result of Section 10.3 applies.

The following result is even easier than the preceding one. Assume δ_1 is Lipschitz. Then *the spectrum has a basis of functions δ_2 such that δ_2 is of class C_∞ on the open set $\delta_1 > 0$, $\varepsilon < \delta_1/\delta_2 < M$; $\varepsilon, M > 0$ and such that*

$$|D^k \delta_2| \leqslant M_k \delta_2^{1-|k|}$$

where the M_k's are constant, D^k is a differential operator and $|k|$ is its order.

This will be interesting because δ_2^r, which also belongs to the spectrum, is of class C_{r-1} on $\mathbb{C}^n$, as a matter of fact $|D^k \delta_2^r| \leqslant M_{k+r}\delta_1^{r-|k|}$.

In the proof we use the following strong form of Urysohn's theorem on $\mathbb{R}^n$: *A sequence of constants $N_k \in \mathbb{R}^+$ can be found in such a way that a function u of class C_∞, equal to one on a neighbourhood of F_1 and to zero on a neighbourhood of F_2 exists, such that $|D^k u| < N_k r^{-k}$, wherever F_1 and F_2 are closed sets at a distance at least r from each other.*

Just let G be the set of points at a distance less than $r/2$ from F_1, let g be the characteristic function of G let ϕ be a given positive function of class C_∞ with mass 1 supported in the ball of radius 1/3, let $\phi_r(x) = r^{-n}\phi(x/r)$ and let $u = g*\phi_r$ (where $*$ is convolution). It follows that if we take

$$N_{k_0} = \sup_{|k|=k_0} \| D^k \rho \|_\infty,$$

the required inequality holds, recalling that $D^k u = g*D^k\phi_r$.

We obtain the required δ_2 starting out from a Lipschitz function $\delta_1 > 0$. We consider the sets $S_N = \{x | \delta_1(x) = 2^{-N}\}$; the distance of S_{N-1} from S_N is at least 2^{-N}. By the strong form of Urysohn's theorem we can then take smooth $\delta_{2,N}$ such that $\delta_{2,N}(s) = 2^{-N}$ on S_N and interpolate as above. With the hypotheses above, the derivatives of the interpolating functions can be majorized, so that the function δ_2 obtained from pasting all these interpolants together has the required properties.

10.6. So we do not lose any generality when we consider only the $\delta \in \Delta_\alpha(a)$ which are of class C_∞ on the open set $\delta > 0$, and of class C_r on $\mathbb{C}^n$.

Can we assume that the coefficients, u_i, v, y_0 which occur in the relation

$$1 = \sum (a_i - s_i)u_i(s) + v(s) + \delta(s)y_0(s)$$

do not behave too wildly? We shall show that this is the case; in fact we can take them to be functions of class C_{r-1}.

We shall let $\theta_{r-1}(s; \delta_0; A)$ be the A-valued functions of class C_{r-1} on $\mathbb{C}^n$, whose derivatives up to and including order $r-1$ are in $\Theta(s; \delta_0; A)$. This set of functions is a b-space in a natural way. We know that†

$$\delta(z) - \delta(s) \in \text{idl}\,(z - s, \bar{z} - \bar{s}; \theta_{r-1}(s, z; \delta_0; A)).$$

We want to show that

$$1 \in \text{idl}\,(a - z, \delta(z); \theta_{r-1}(z; \delta_0; A)] + \theta_{r-1}(z; \delta_0; \alpha).$$

† More than once ([32] p. 75–79 or [33]) the author has given an erroneous proof of this result assuming that $\delta(z) - \delta(s) \in \text{idl}\,(z - s, \theta_{r-1}(z; \delta_0; A)$ which cannot be the case since δ is not holomorphic. The author excuses himself therefore and hopes that the proof given here is complete.

It will suffice to show that

$$1 \in \text{idl}\,\{z - s, \bar{z} - t, a - z, \delta(z); \Theta[s, t; \delta_0; \theta_{r-1}(z; \delta_0; A)]\} + \Theta[s, t; \delta_0; \theta_{r-1}(z; \delta_0; \alpha)].$$

We apply the fact that

$$1 \in \text{idl}\,\{z - s, \bar{z} - \bar{s}, \delta(z), a - z; \Theta[s; \delta_0; \Theta_{r-1}(z; \delta_0; A)]\} + \Theta[s; \delta_0; \theta_{r-1}(z; \delta_0; \alpha)]$$

which is true since this ideal contains $a - s$, $\delta(s)$, and $\Theta(s; \delta_0; \alpha)$, and these functions generate the improper ideal in $\Theta(s; \delta_0; A)$. We consider the function $\varepsilon(s, t) = 0$ when $s = \bar{t}$, $\varepsilon(s, t) = 1$ when $s \neq \bar{t}$ and apply the fact that

$$\Theta(s, t; \delta_0(s, t); A)/\text{idl}\;\varepsilon(s, t)$$

is canonically isomorphic with $\Theta(s; \delta_0; A)$. It follows that

$$1 \in \text{idl}\,\{z - s, \bar{z} - t, \varepsilon(s, t), a - z, \delta(z); \Theta[s, t; \delta_0; \theta_{r-1}(z; \delta_0; A)]\} + \Theta[s, t; \delta_0; \theta_{r-1}(z; \delta_0; \alpha)].$$

This relation can be rewritten

$$\varepsilon \in \Delta_{\alpha_1}(z, \bar{z}; \theta_{r-1}(z; \delta_0; A))$$

where

$$\alpha_1 = \text{idl}\,(a - z, \delta(z); \theta_{r-1}(z; \delta_0; A)) + \theta_{r-1}(z; \delta_0; \alpha).$$

But the spectrum has a basis made up of Lipschitz functions, whence

$$\varepsilon_1(s, t) \in \Delta_{\alpha_1}(z, \bar{z}; \theta_{r-1}(z; \delta_0; A)) \tag{1}$$

where

$$\varepsilon_1(s, t) = \inf(|s - \bar{t}|, \delta_0(s, t)).$$

When $s \neq \bar{t}$, we have a relation

$$\sum |z_i - s_i|^2 + \sum |\bar{z}_i - t_i|^2 \in \text{idl}\,(z - s, \bar{z} - t)$$

and

$$1/[\sum |z_i - s_i|^2 + \sum |\bar{z}_i - t_i|^2] \in \Theta(s, t; \varepsilon_1(s, t); \theta_{r-1}(z, \delta_0))$$

hence

$$1 \in \text{idl}\,[z - s, \bar{z} - t; \Theta(s, t; \varepsilon_1; \theta_{r-1}(z, \delta_0)]. \tag{2}$$

The proof is nearly complete. It will be complete if we show in general that $1 \in \alpha$ follows from

$$\delta \in \Delta_\alpha(a, A) \tag{3}$$

and

$$1 \in \text{idl}\,(a - s, \Theta(s;\delta;A)). \tag{4}$$

For a general Lipschitz function δ we define $\Theta(s;\delta;A)$ to be the A-valued functions $u(s)$ on the open set $\delta > 0$ which are such that $\delta(s)^N u(s)$ is a bounded A-valued function on its domain when N is large enough, and similarly for $\Theta(s;\delta;\alpha)$. But of course (4) implies that functions $u_i(s) \in \Theta(s;\delta;A)$, exist such that

$$1 = \sum (a_i - s_i)u_i(s)$$

and (3) implies that functions $U_i(s)$, $V(s)$, $Y_0(s)$ exist such that

$$\sum (a_i - s_i)U_i(s) + V(s) + \delta(s)^N Y_0(s) = 1.$$

We choose N large enough that $\delta(s)^N u_i(s)$ are bounded A-valued functions. We define $\delta(s)^N u_i(s) = 0$ where $\delta = 0$ (u_i is not defined there) and observe that

$$\sum (a_i - s_i)(U_i(s) + \delta(s)^N u_i(s)) + V(s) = 1$$

whence $1 \in \alpha$.

11. THE HOLOMORPHIC FUNCTIONAL CALCULUS AND NON-COMPACT SPECTRA

11.1. The idea now is to show that we have all the ingredients necessary for the following theorem, and to give a recipe to be used when baking the cake.

Let $\delta \in \Delta_\alpha(a_1, \ldots, a_n; A)$, where A is a b-algebra, α a b-ideal, and δ is Lipschitz, $\delta \leqslant \delta_0$. Let $\mathcal{O}(s;\delta)$ be the algebra of complex-valued holomorphic functions u on the open set $\delta > 0$, and such that $\delta^N u$ is bounded for N large. The b-algebra $\mathcal{O}(s;\delta)$ is the direct limit of Banach spaces ${}_N\mathcal{O}(s;\delta) = \{u \mid \delta^N u$ is bounded$\}$.

It is possible then to define a sequence of bounded linear mappings $\sigma_N : {}_N\mathcal{O}(s;\delta) \to A$, with the property that $\sigma_N - \sigma_{N+1}$ maps ${}_N\mathcal{O}(s;\delta)$ into α and is a bounded mapping, and the mapping $(f, g) \mapsto \sigma_{2N}(f \cdot g) - \sigma_N(f) \cdot \sigma_N(g)$ is a bounded bilinear mapping ${}_N\mathcal{O}(s;\delta) \times {}_N\mathcal{O}(s;\delta) \to \alpha$.

Expressed in terms of q-spaces and q-algebras (see Section 8.4) I mean that we have a morphism ${}_N\mathcal{O}(s;\delta) \to A \mid \alpha$. The reader will notice that I state a theorem for A-valued functions and prove it for scalar functions. This is not serious. Recovery of algebra-valued theorems from scalar-valued ones is quite standard thanks to the topological tensor products. We need finer tensor products than the usual ones, because we are working in the category q, but nuclear spaces, such as $\mathcal{O}(s;\delta)$ are very nice spaces, and it is true that

$\mathcal{O}(s;\delta)\otimes_q A|\alpha \simeq \mathcal{O}(s;\delta;A)|\mathcal{O}(s;\delta;\alpha)$ where here $\otimes_q$ is the natural tensor product within the category q. (See [13], [14])

11.2. We start out from $\delta\in\Delta(a;\Delta|\alpha)$, and assume that $\delta\leqslant\delta_0$, $\delta\in\theta_{r+1}(s;\delta_0)$. We can then find u, v, y_0, with $u, y_0\in\theta_r(s;\delta_0;A)$ and $v\in\theta_r(s;\delta_0;\alpha)$ such that

$$1=\sum(a_i-s_i)u_i(s)+v(s)+\delta(s)y_0(s)$$

on $\mathbb{C}^n$. We also consider $U_i=\bar{s}_i/(1+|s|^2)$, $Y=(1+\sum a_i\bar{s}_i)/(1+|s|^2)$ so that

$$1=\sum(a_i-s_i)U_i+Y.$$

Replacing u_i by $u_i'=U_i+Yu_i$, v by $v'=Yv$, and y by $y_0'=Yy_0$, we obtain u', v', y' which grow more slowly at infinity than u, v, y.

We may therefore assume that u, v, y tend faster to zero at infinity than δ_0, and that their derivatives (to the order r) have the same property. It will be reasonable to investigate again (when $y=\delta\cdot y_0$)

$$\frac{(n+k)!}{k!}y^k\bar{\partial}u_1\wedge\ldots\wedge\bar{\partial}u_n.$$

This is a form whose coefficients are of class C_{r-1} on $\mathbb{C}^n$. It vanishes on the set where δ vanishes. It tends to zero as δ^k at the boundary of the domain $\delta>0$. The derivatives of the coefficients of the form behave in a similar way.

11.3. We arrive in this way at a space, $\Lambda^p_{r,k}(s,\delta,A)$. Its elements are differential forms of class C_r and degree p in $d\bar{z}_1,\ldots,d\bar{z}_n$, whose coefficients tend to zero along with their derivatives to the order r at the border of the set $\delta>0$, the rate at which these coefficients and derivatives tend to zero being governed by δ^k. Of course

$$\bar{\partial}:\Lambda^p_{r,k}\to\Lambda^{p+1}_{r-1,k-1}$$

and insofar as $\bar{\partial}\circ\bar{\partial}$ is defined, $\bar{\partial}\circ\bar{\partial}=0$, say, if you like indices

$$\bar{\partial}^p_{r,k}\circ\bar{\partial}^{p+1}_{r-1,k-1}=0.$$

We now remember that A contains a b-ideal. We let

$$Z^p_{r,k}(s,\delta,A|\alpha)=\{\omega\in\Lambda^p_{r,k}(s,\delta,A)|\bar{\partial}\omega\in\Lambda^{p+1}_{r-1,k-1}(s,\delta,\alpha)\}.$$

We also let

$$B^p_{r,k}(s,\delta,A|\alpha)=\bar{\partial}\Lambda^{p-1}_{r+1,k+1}(s,\delta,A)+\Lambda^p_{r,k}(s,\delta,\alpha)$$

and

$$H^p_{r,k}(s,\delta,A|\alpha)=Z^p_{r,k}/B^p_{r,k}(s,\delta,A|\alpha).$$

If $r' > r$, $k' > k$, we have $Z^p_{r',k'} \subset Z^p_{r,k}$, $B^p_{r',k'} \subset B^p_{r,k}$. Inclusion induces a natural mapping $H^p_{r',k'} \to H^p_{r,k}$.

$H^p(s, \delta, A|\alpha)$ is defined as the projective limit of the spaces $H^p_{r,k}$. H is a functor in the $\tilde{q}$ category, but possibly not in q.

11.4. If we go through the proofs of Section 9, we note that we have effectively shown that all the forms

$$\frac{(n+k')!}{k'!} y^{k'}\bar{\partial}u_1 \wedge \ldots \wedge \bar{\partial}u_n \qquad (k' \geqslant k)$$

that we must consider belong to a single element of $H^n_{r-1,k-1}(s, \delta, A|\alpha)$, i.e. that the difference between two such forms is in $B^n_{r-,k-1}(s, \delta, A|\alpha)$. And when we take the projective limit, we find a single element of $H^n(s, \delta, A|\alpha)$ which is associated naturally to $a \in A$. This element will be called $\tau(a)$.

We next consider $\mathcal{O}(s, \delta, A)$. This the direct limit of spaces ${}_N\mathcal{O}(s, \delta, A)$. The elements of $\mathcal{O}(s, \delta, A)$ are $\bar{\partial}$-cocycles with limited growth, of degree zero. There are no $\bar{\partial}$-coboundaries, except zero, in dimension zero. So we have genuine classes of cohomology.

Now a class of cohomology with limited growth can be multiplied with a class of cohomology with rapidly decreasing coefficients to yield a class of cohomology with rapidly decreasing coefficients. In our instance, the product of a holomorphic function $f \in {}_N\mathcal{O}(s, \delta, A)$ and a class of cohomology $\omega \in H^p_{r,k}(s, \delta, A|\alpha)$ will be a well determined $f\omega \in H^n_{r,k-N}(s, \delta, A|\alpha)$. Since f is given, N is constant. Letting $r, k \to \infty$, we find, e.g. $f \cdot \tau(a) \in H^n(s, \delta, A|\alpha)$.

We can integrate, as we did in Section 9

$$\frac{1}{(2\pi i)^n} \int f \cdot \tau(a) ds_1 \ldots ds_n = f[a].$$

The proof that $\tau(a, b) = \tau(a) \times \tau(b)$, the meaning even of this statement is very similar to the proof given in Section 9. So

$$f \times g[a, b] = f[a]g[b].$$

To prove that $1[a] = 1$, that $f[a] = 1$ when $f = 1$ is constant, we observe that $1 \in \mathcal{O}(s, \delta_0)$, that it will be sufficient to find U, Y_0 such that

$$(a - s)U + \delta_0(s)Y_0 = 1,$$

and that

$$U(s) = -\bar{s}/(1 + |s|^2),\ Y_0(s) = (1 + a\bar{s})/(1 + |s|^2)^{\frac{1}{2}},$$

i.e.

$$Y = \delta_0 Y_0 = (1 + a\bar{s})/(1 + |s|^2)$$

will do. We can even reduce the problem to the computation of

$$\frac{1}{2\pi i}\int_{C} dU \wedge ds = 1.$$

The multiplicativity of the mapping $f \mapsto f[a]$ follows from a proof similar to that found in Section 9. The only point where we must be a little bit careful is in the investigation of an injection mapping $(s_1, \ldots, s_n) \mapsto (s_1, \ldots, s_n, 0)$, i.e. of $(a_1, \ldots, a_n, 0)$ when $(a_1, \ldots, a_n)$ is known. It is easy to show that $\Delta_\alpha(a_1, \ldots, a_n, 0)$ contains all Lipschitz functions $\delta_1(s_1, \ldots, s_n, t)$ whose restriction $\delta(s_1, \ldots, s_n) = \delta_1(s_1, \ldots, s_n, 0)$ belongs to $\Delta_\alpha(a_1, \ldots, a_n)$. In particular, δ being given, this set contains

$$\delta_1(s_1, \ldots, s_n, t) = \max\,[\delta(s_1, \ldots, s_n) - |t|, 0].$$

Also, for this choice of δ_1, if $f(s_1, \ldots, s_n, t) \in \mathcal{O}(s, t; \delta_1)$ and $f(s, 0) = 0$ then $f(s, t)/t$ is not only holomorphic for $t = 0$, also, straightforwardly, it belongs to $\mathcal{O}(s, t; \delta_1)$ hence $f(s, t) = f(s, 0) + tg(s, t)$, and $f[a, 0] = f_0[a]$ if $f_0(s) = f(s, 0)$.

So, our result can be proved.

12. APPLICATIONS

12.1. It is now time to speak of applications of the theory developed. But before mentioning these, there is one point on which I would like to straighten the historical record.

If A is a commutative locally convex algebra, which is complete enough, Allan [**2**] considered the bounded structure generated on the algebra of its regular element by the bounded idempotent sets. He already showed then how Gelfand theory and the holomorphic functional calculus could be applied to the algebra of regular elements through what I call the Allan boundedness of the algebra.

He later formalized the theory somewhat, and in a joint paper with H. G. Dales and J. P. McClure, he considered the b-algebra with an idempotent bounded structure [**3**]. The authors called these algebras *pseudo-Banach algebras*. This paper is not interesting so much because it contains the theory, but because it contains many examples. Also, for the first time, it was shown how the holomorphic functional calculus for complete p-normed algebras followed from the standard theory for Banach algebras. After all, the theory was between the lines in Allan's original paper; it can be found, attributed to Allan, in my lecture notes [**28**] which were "in production" at Springer Verlag's when I received Allan, Dales and McClure's preprint.

These things go without saying, except that there is a preprint in circulation by Hogbe-Nlend [**10**]. This preprint is a good survey of the theory. I judge the presentation better than that of Allan, Dales, and McClure if we accept the Bourbaki canon. The only problem is that the reader might think that this, largely expository paper, is an original one. References to work done previously on the subject are, in my opinion, insufficient.

12.2. This series of talks has been so far largely divided into two parts. In one we considered algebras with an idempotent bounded structure. In the other we considered general b-algebras and non-compact spectra. All affirmative results I obtained concerned commutative algebras with unit. It is to be expected that the applications will be divided in a similar way.

There is one class of applications about which something must be said. The results about Banach algebras that Taylor used in his lectures at this meeting, [**17**], i.e. Arens' theorem about $G_n(A)$, are true when A has an idempotent, completant boundedness rather than a complete norm.

This may require a little explanation. $G_n(A)$ is defined as

$$GL(n, A)/GL_0(n, A);$$

$GL(n, A)$ is the group of invertible elements of $M_n(A)$—the algebra of $n \times n$ matrices with entries from A, $GL_0(n, A)$ is the connected component of the identity in $GL(n, A)$, i.e. $GL_0(n, A) = \exp(M_n(A))$, exp is the exponential map

$$a \mapsto \sum_{i=0}^{\infty} \frac{a^i}{i!}.$$

Arens' theorem then says, loosely speaking, that $G_n(A)$ depends only on the maximal ideal space of A.

In the case of b-algebras A we have no topology on $GL(n, A)$, so we have to say the two elements of $GL(n, A)$ are in the same component if they are in the same component in $GL(n, A_B)$, for some idempotent completant set in A, so that then A_B is just a Banach algebra and $GL(n, A_B)$ is then given the natural topology. It is also true that the components of $GL(n, A)$ are clopen (closed and open) with respect to the spectral radius seminorm on A_B. These results are to be found in [**28**].

This is important since the Arens' theorem is the vital link between algebraic K-theory and K-theory for Banach algebras. It seems likely that most of what J. L. Taylor did in his course for Banach algebras would still be true for b-algebras.

This then gives a wide class of algebras which are non-Banach, and whose K-theory can be computed. This may not be remarkable. What would be interesting, (I do not have a concrete example), would be to find an algebra whose maximal ideal space has an easily computable cohomology, so that we would have some information about the associated K-theory; or, conversely, whose K-theory would be directly computable, yielding non-trivial properties of the maximal ideal space.

While we are at it, I must mention the fact that I am convinced that a large part of what Johnson has been doing extends to non-Banach algebras, especially to algebras with an idempotent bounded structure. The problem is, to find the exact statements of the theorems, and even more, find algebras where the relevant H' are computable and not unwieldy.

12.3. Complex analysis is an obvious field of application of the theory of algebras with an idempotent boundedness. If S is a compact subset of $\mathbb{C}^n$, $\mathcal{O}(S)$ is a direct limit of Banach algebras. What is the spectrum of the variables, of $(z_1, \ldots, z_n)$ in $\mathcal{O}(S)$? And what is the maximal ideal space of $\mathcal{O}(S)$? It is well known that this maximal ideal space is the inverse limit of the envelopes of holomorphy $\tilde{U}_k$ of the fundamental neighbourhood system U_k of S. I say inverse limit, not intersection, because the natural mappings $\tilde{U}_{k+1} \to \tilde{U}_k$ are not one-to-one.

There is a case where $S \sim \tilde{S}$, this is when S has a fundamental system of neighbourhoods which are domains of holomorphy. There are cases nearly as good: S can have a fundamental system of neighbourhoods U_n with a schlicht envelope of holomorphy $\tilde{U}_n$, and we can define $\tilde{S} = \bigcap_n \tilde{U}_n$. This $\tilde{S}$ is now a compact subset of $\mathbb{C}^n$, with a fundamental system of neighbourhoods of holomorphy, and $\mathcal{O}(S)$ is isomorphic with $\mathcal{O}(\tilde{S})$.

But what good is this? Assume that

$$S = \mathrm{sp}(z_1, \ldots, z_n; \mathcal{O}(S)).$$

We can define $f[z]$ all right, but $f[z] = f(z)$, so what does the holomorphic functional calculus tell us?

Let $u: A \to B$ be a unital b-algebra morphism, and assume $\mathrm{sp}(a) \subset S$. Of course $\mathrm{sp}(b) \subset S$ if $b = ua$, since $\mathrm{sp}(b) \subseteq \mathrm{sp}(a)$. Since the construction of the holomorphic functional calculus is a natural one, we see that $f[b] = uf[a]$ for $f \in \mathcal{O}(\mathrm{sp}(a))$.

Let now $S = \mathrm{sp}(z, \mathcal{O}(S))$. Let $u: \mathcal{O}(S) \to A$ be a morphism of unital b-algebras, and let $a = u(z)$. *Then* $\mathrm{sp}(a) \subset S$ *and* $uf = f[a]$. In other words, a homomorphism $\mathcal{O}(S) \to A$ mapping unit on unit is determined by the image of the z_i.

This is not altogether obvious. It is obvious if S is polynomially convex (Runge's theorem) or rationally convex. But there are domains of holomorphy which are not polynomially or rationally convex. The rational functions are not necessarily dense in $\mathcal{O}(S)$.

It should be observed that the above homomorphism is unique if and only if S has a fundamental system of neighbourhoods whose envelopes of holomorphy are schlicht.

Observe that we gave a statement about homomorphisms, but that the same result holds essentially for derivations. Let M be an $\mathcal{O}(S)$-module. A derivation $D:\mathcal{O}(S) \to M$ determines a homomorphism $f \mapsto (f, Df)$, $\mathcal{O}(S) \to \mathcal{O}(S) \oplus M$, where

$$(f_1, m_1) \cdot (f_2, m_2) = (f_1 f_2, f_1 m_2 + f_2 m_1).$$

Hence, again when $S = \mathrm{sp}(z, \mathcal{O}(S))$, we see that

$$Df = \sum \frac{\partial f}{\partial z_i} \cdot Dz_i$$

since there can be at most one derivation $D:\mathcal{O}(S) \to M$ mapping z_i on Dz_i, and the expression written is such a one.

Somebody may have noticed that I have not assumed that the bounded structure of A was idempotent. Yet I apply a theory that was developed when the bounded structure has that property. This is not a slip of the pen.

It is rather an application of the following obvious remark. Assume that B is a b-algebra with an idempotent boundedness. Let A be any b-algebra and $u:B \to A$ be a b-algebra morphism. Then $uB \subset A_r$, and $u:B \to A_r$ is a b-algebra morphism if A_r is the algebra of regular elements of A, with the Allan boundedness. Of course the boundedness of A_r is idempotent.

12.4. It is time now to speak of the applications of the holomorphic functional calculus with non-compact spectra. The first application one can think of is the theory of partial differential equations. It was indeed considerations about partial differential equations that led me to the study of the holomorphic functional calculus.

I did manage to show that the partial differential operators had the right sorts of properties, so that $f(\partial/\partial x_1, \ldots, \partial/\partial x_n)$ could be defined when f is a holomorphic function on a tube with polynomial growth at infinity. And $f(\partial/\partial x)$ is a distribution with support in a cone. (cf. [**32**], pp. 133–137) But the mapping $f \mapsto f(\partial/\partial x)$ is nothing else than the inverse Laplace transform, defined on functions holomorphic on a tube with polynomial growth at infinity and into the convolution algebra of distributions with support in a cone.

The situation becomes worse when we try to generalize these considerations to more general partial differential equations. The problem is that multiplication and differentiation do not commute, so it may be reasonable to study single operators. There are many operators, elliptic ones, or hyperbolic, for which something is known of the resolvent. If the resolvent is bounded on a set X, the characteristic function of the complement U of X

is spectral. So is the largest function on U which is Lipschitz and tends to zero at infinity like δ_0. This gives a holomorphic functional calculus.

Sebastião e Silva has developed such applications, involving a single operator or several commuting partial differential operators [**16**]. Unfortunately, it does not seem that such results have entered the mainstream of partial differential equation theory, nor are they due to enter it. The main difficulty seems to be obtaining estimates on the rate of growth of the resolvent (in the general case).

If somebody has qualms because partial differential operators are not defined everywhere on a Banach space, but are closed and have a dense domain, they can be put at ease when the operator has a resolvent which is defined somewhere. The Banach space E then contains a dense Fréchet subspace E_∞ and is contained in a b-space $E_{-\infty}$; the given operator operates on $E_{-\infty}$, and leaves E_∞ invariant. The domain of the operator in E is the set of $x \in E$ where the image is in E. In a way, the elements of E_∞ are C_∞-functions, those of $E_{-\infty}$ are distributions.

12.5. Semi-group theory and the Hille–Yosida theorem are standard applications of the Banach algebra holomorphic functional calculus. It is natural to ask oneself what happens in the non-Banach case.

It is easier to consider only differentiable semi-groups. Their resolvent is a genuine operator. (An operator belongs to an algebra of operators.) It turns out that $a \in A$ is the generator of a differentiable semi-group if and only if $\inf(\delta_0(s), e^{-\mathrm{Re}\, s}) \in \Delta(a)$. The differentiable semi-group arises from exponentiating a. (See [**34**]).

I have obtained, myself and with Tits, further result on differentiable semi-groups [**35**] and differentiable group representations [**21**]. These are not applications of the holomorphic functional calculus. But they are applications of non-Banach algebra theory.

12.6. The main application of the holomorphic functional calculus with non-compact spectra that I presented here is to the theory of holomorphic functions with bounds, i.e. to the study of the algebras $\mathcal{O}(s, \delta)$ themselves. A very important theorem is due to Cnop [**4**]. This theorem states that

$$\delta \in \Delta(z, \mathcal{O}(z, \delta))$$

when δ is Lipschitz, $-\log \delta$ is plurisubharmonic. The Lipschitz condition crops up everywhere in the theory. The plurisubharmonicity is essential in complex analysis. Up to equivalence, Cnop's result is the best we can hope for.

This result has been used by J.-P. Ferrier and other authors to obtain results involving these holomorphic functions. The best survey of these results is the book of Ferrier, [7]. We can just quote the following application to convince the reader that one does have non-trivial results.

THEOREM: *Assume that δ, δ' are Lipschitz and that the open set $\delta' > 0$ is pseudo-convex (a domain of holomorphy), and also that $\delta' > \delta$. The following conditions are equivalent in the sense that either implies that $\mathcal{O}(\delta')$ is dense in $\mathcal{O}(\delta)$.*

(i) *up to equivalence, $1/\delta$ is the supremum on $\{\delta > 0\}$ of a family of $|f_\alpha|$ where each $f_\alpha \in \mathcal{O}(\delta')$.*

(ii) *up to equivalence, $1/\delta$ is the supremum on $\{\delta > 0\}$ of a family of π_α, where $\pi_\alpha \in \mathcal{O}(\delta')$, $\log \pi_\alpha$ plurisubharmonic on $\{\delta' > 0\}$.*

This theorem, due to Ferrier, is a Runge theorem.

Remarks: $\mathcal{O}(\delta)$ is equipped with the topology of inverse limit $\lim_N \mathcal{O}(\delta)$. ${}_N\mathcal{O}(\delta)$ as in Section 11.

We say δ_1 is equivalent to δ_2 if $\exists\, \varepsilon, N$ such that

$$\varepsilon\delta_1^N \leqslant \delta_2 \quad \text{and} \quad \varepsilon\delta_2^N \leqslant \delta_1.$$

12.7. Some investigations of Taylor [**18**], [**19**] must be at least mentioned. He considers n commuting operators on a Banach space, say $a_1, \dots, a_n$ on E. A point is out of the spectrum if it is possible to construct an exact sequence of the right kind, the "Koszul complex". It is in the spectrum if this is not exact.

The spectrum is compact, not empty, smaller (and can be strictly smaller) than the set of $(s_1, \dots, s_n)$ for which no $u_1, \dots, u_n$ commuting with $a_1, \dots, a_n$ (but not necessarily commuting with each other) can be found such that $1 = \sum (a_i - s_i)u_i(s)$.

Introducing a double complex, I do not know which, and chasing through a diagram, this is a favourite occupation of homological algebraists, and can be great fun; Joe Taylor finds an object in all respects analogous to

$$\frac{(n+k)!}{k!} y^k \bar{\partial} u_1 \wedge \dots \wedge \bar{\partial} u_n.$$

He integrates this and obtains a holomorphic functional calculus.

REFERENCES

1. M. Akkar, Sur la théorie spectrale des algèbras d'opérateurs. *Séminaire d'Analyse Fonctionnelle*. Bordeaux 1972.
2. G. R. Allan, A spectral theory for locally convex algebras. *Proc. London Math. Soc.* (3) **15** (1965), 399–421.

3. G. R. Allan, H. G. Dales and J. P. McClure, Pseudo-Banach algebras. *Studia Math.* **15** (1971), 55–69.
4. I. Cnop, Spectral study of holomorphic functions with bounded growth. *Ann. Inst. Fourier (Grenoble).* **22** (1972), 293–309.
5. D. O. Etter, Jr, Vector-valued analytic functions. *Trans. Amer. Math. Soc.* **119** (1965), 352–366.
6. J.-P. Ferrier, *Séminaire sur les algèbres complètes.* Lecture Notes in Mathematics, vol. 164. Springer-Verlag, 1970.
7. J.-P. Ferrier, *Spectral theory and complex analysis.* North-Holland Mathematics Studies, 4. North-Holland, Amsterdam, 1973.
8. B. Gramsch, Integration und holomorphe Funktionen in lokal-beschränkten Raümen. *Math. Ann.* **162** (1965/66), 190–210.
9. R. C. Gunning and H. Rossi, *Analytic functions of several complex variables.* Prentice Hall, 1965.
10. H. Hogbe-Nlend, *Les fondements de la théorie spectrale des algèbres bornologiques.* Université de Bordeaux, Départment de Mathématique, 1972.
11. E. R. Lorch, The theory of analytic functions in normed abelian vector rings. *Trans. Amer. Math. Soc.* **53** (1942), 238–248.
12. S. Mazur and W. Orlicz, Sur les espaces métriques linéaires I. *Studia Math.* **10** (1948), 184–208.
13. G. Noël, Une immersion de la catégorie des espaces bornologiques convexes séparés dans une catégorie abélienne. *C. R. Acad. Sci. Paris* **269** (1969), 238–240.
14. G. Noël, Produit tensoriel et platitude des Q-espaces. *Bull. Soc. Math. Belg.* **22** (1970), 119–142.
15. D. Przeworska-Rolewicz and S. Rolewicz, On integrals of functions with values in a complete linear metric space. *Studia Math.* **26** (1966), 121–131.
16. J. Sebastião e Silva, Sur le calcul symbolique d'opérateurs permutables à spectre vide ou non borné. *Ann. Mat. Pura Appl.* **58** (1962), 219–276.
17. J. L. Taylor, Banach algebras and topology. These *Proceedings,* p. 118.
18. J. L. Taylor, A joint spectrum for several commuting operators. *J. Functional Analysis* **6** (1970), 172–191.
19. J. L. Taylor, The analytic functional calculus for several commuting operators. *Acta. Math.* **125** (1970), 1–38.
20. J. L. Taylor, Several variable spectral theory. *Functional Analysis (Proc. Sympos., Monterey, Calif., 1969).* 1–10, Academic Press, 1970.
21. J. Tits and L. Waelbroeck, The integration of a Lie algebra representation. *Pacific J. Math.* **26** (1968), 595–600.
22. P. Turpin and L. Waelbroeck, Sur l'approximations des fonctions différentiables à valeurs dans les espaces vectoriels topologiques. *C.R. Acad Sci. Paris* **267** (1968), 94–97.
23. P. Turpin and L. Waelbroeck, Intégration et fonctions holomorphes dans les espaces localement pseudo-convexes. *C.R. Acad. Sci. Paris* **267** (1968), 160–162.
24. P. Turpin and L. Waelbroeck, Algèbres localement pseudo-convexes à inverse continu. *C.R. Acad. Sci. Paris* **267** (1968), 194–195.
25. P. Turpin, Une remarque sur les algèbres à inverse continu. *C.R. Acad. Sci. Paris* **270** (1970), 1686–1689.
26. P. Uss, Sur les opérateurs bornés dans les espaces localement convexes. *Mémoire* 31–32 *de la Société Mathématique de France.* 1972. (*Colloque d'Analyse Fonctionnelle de Bordeaux*, 1971.) 405–406.

27. L. Waelbroeck, Some theorems on bounded structure. *J. Functional Analysis* **1** (1967).
28. L. Waelbroeck, *Topological vector spaces and algebras*. Lecture Notes in Mathematics. vol. 230. Springer-Verlag, 1971.
29. L. Waelbroeck, Le complété et le dual d'un espace "à bornés". *C.R. Acad. Sci. Paris* **253** (1961), 2827–2828.
30. L. Waelbroeck, Topological vector spaces. *Summer School on Topological Vector Spaces*, 1–40. Lecture Notes in Mathematics, vol. 331. Springer-Verlag, 1973.
31. L. Waelbroeck, Fonctions différentiables et petite bornologie. *C.R. Acad. Sci. Paris* **267** (1968), 220–222.
32. L. Waelbroeck, Etude spectrale des algèbres complètes. *Acad. Roy. Belg. Cl. Sci. Mém. Coll. in* 8° (2) **31** (1960) no. 7.
33. L. Waelbroeck, About a spectral theorem. *Function Algebras*. (*ed. F. Birtel*), 310–321. Scott, Foresman, and Co. 1966.
34. L. Waelbroeck, Les semi-groupes différentiables. *Deuxième Colloque sur l'Analyse Fonctionnelle*, 97–103. C.B.R.M., Liège, 1964.
35. L. Waelbroeck, Differentiability of Hölder-continuous semi-groups. *Proc. Amer. Math. Soc.* **21** (1965), 451–454.
36. H. Whitney, Analytic extensions of differentiable functions defined on closed sets. *Trans. Amer. Math. Soc.* **36** (1934), 63–89.
37. H. Whitney, On the extensions of differentiable functions. *Bull. Amer. Math. Soc.* **50** (1944), 76—81.
38. B. Mitjagin, S. Rolewicz and W. Zelazko, Entire functions in B_0-algebras. *Studia Math.* **21** (1961/62), 291–306.
39. A. Pietsch, Ideals of operators on Banach spaces and nuclear locally convex spaces. *Proc.* III *Sympos. General Topology*. Prague 1971.
40. R. Schatten, *Norm ideals of completely continuous operators*. Ergebnisse der Math. und Ihrer Grenzgebiete, Heft 27. Springer-Verlag 1960.
41. G. Köthe, *Topologische lineare Räume*. I. Grundlehren der Math. Wiss. Bd. 107. Springer-Verlag, 1960.

PART II

ABSTRACTS AND SEMINAR TALKS

A Stone-Weierstrass Type Theorem for Annihilator Banach Algebras†

G. F. Bachelis

Department of Mathematics, Wayne State University, Detroit, Michigan, U.S.A.

Abstract

Let B be a semi-simple annihilator Banach algebra and let A be a closed semi-simple subalgebra which separates the maximal modular left ideals of B. Barnes [**1**] has proved that $A = B$ under the additional hypothesis $(\varepsilon_1): I \oplus \mathscr{L}(I) = B$ for all closed two-sided ideals I (where $\mathscr{L}(I) = \{x : xI = (0)\}$). A proof is given that $A = B$ when (ε_1) is replaced by (ε_2): B is strongly semi-simple. In particular, this latter theorem applies to $\mathscr{L}^p(G)$, where G is a compact group and $1 \leqslant p < \infty$. When G is abelian, this is a special case of a theorem of Katznelson and Rudin [**2**].

References

1. Barnes, B. A. Density theorems for algebras of operators and annihilator Banach algebras, *Michigan Math. J.* **19** (1972), 149–155.
2. Katznelson Y., and Rudin, W. The Stone–Weierstrass property in Banach algebras, *Pacific J. Math.* **11** (1961), 253–265.

†A paper entitled "Density theorems for group algebras of a compact group and other annihilator Banach algebras", on the above, has been submitted for publication to *J. London Math. Soc.*

Adjoining Inverses to Commutative Banach Algebras

B. BOLLOBÀS

Department of Pure Mathematics and Mathematical Statistics, University of Cambridge, Cambridge

Abstract

Let A be a commutative unital Banach algebra and let $a \in A$. It was proved by Shilov [**4**] that there is an extension B of A such that $a^{-1} \in B$ if and only if a is not a topological zero divisor. In fact, one can have $\| a^{-1} \| \leqslant 1$ if and only if $\| ax \| \geqslant \| x \|$ for every $x \in A$.

The following natural question was raised by Arens [**1**] about the possibility of adjoining the inverses of a set of elements. If $S \subset A$ and no element of S is a topological zero divisor, is there a commutative Banach algebra $B \supset A$ in which every element of S has an inverse?

I proved in [**2**] that the answer to this question is in the *affirmative* if S is *countable*. Furthermore, there is an algebra A_0 and a set $S_0 \subset A_0$ such that $|S_0| = \aleph_1$ and no element of S_0 is a topological zero divisor, but there is no extension B of A_0 in which every element of S_0 is invertible. In [**3**] I also determined the best possible bounds one can give for the norms of the inverses of a set of elements and so obtained the results of [**2**] in a considerably sharper form.

In the example mentioned above the set S_0 consists of independent elements, so one might reasonably expect that if, e.g., S is in a finite dimensional subspace then one can always find a suitable extension B of A. In particular, I think it would be very interesting to answer the following question.

Let $a \in A$ and let

$$T = \{\lambda \in \mathbb{C} : a - \lambda \text{ is a topological zero divisor}\}.$$

Is there an extension B of A such that $(a - \lambda)^{-1} \in B$ whenever $\lambda \notin T$? In other words, is there a smallest set among the sets $\{\mathrm{sp}_B a : B \text{ is an extension of } A\}$?

References

1. Arens, R. Inverse-producing extensions of normed algebras, *Trans. Amer. Math. Soc.* **88** (1958), 536–548.
2. Bollobàs, B. Adjoining inverses to commutative Banach algebras, *Trans. Amer. Math. Soc.* **181** (1973) 165–174.
3. Bollobàs, B. Best possible bounds of the norms of inverses adjoined to normed algebras, *Studia Math.* **51** (1974) 87–96.
4. Shilov, G. E. On normed rings with one generator, *Mat. Sb.* **21** (1947), 25–46.

Restrictions of Algebras of Holomorphic Functions with Growth

IVAN CNOP†

Department of Mathematics, Free University of Brussels, Belgium

Abstract

The holomorphic functional calculus for elements in non-Banach algebras by Waelbroeck [3] has applications to certain algebras of holomorphic functions with a growth condition near the boundary (see Ferrier [2] and the references cited there). As one more application, we investigate the restriction of these algebras to certain graphs.

If δ is a bounded non-negative continuous function on $\mathbb{C}^n$, consider the domain Ω where δ is (strictly) positive and the algebra $\mathcal{O}(\delta)$ of all functions f such that for some positive number N, $f \cdot \delta^N$ is bounded on Ω. For this algebra we have:

THEOREM: *Let δ be a bounded non-negative Lipschitz function on δ^n with $\delta(s).\ |s|^{N_0}$ bounded at infinity for some N_0, and $-\log \delta$ plurisubharmonic on the domain Ω. Let M be a bounded subset of Ω, δ' be the restriction of δ to M, and Ω' be a pseudoconvex (bounded) domain in $\mathbb{C}^{n-k}$ $(1 \leqslant k < n)$ such that*

(i) *Ω is contained in $\Omega' \times \mathbb{C}^{n-k}$*

(ii) *M is the graph of a k-tuple $h = (h_1, \ldots, h_k)$ of (bounded) holomorphic functions on Ω'. Then, for each function f' in $\mathcal{O}(\delta')$, we can find a function f in $\mathcal{O}(\delta)$ such that $f|_M = f'$, and for all N' there exists N such that $|f'|.\ \delta'^{N'} \leqslant 1$ implies $|f|\delta^N \leqslant 1$.*

Moreover, the algebra $\mathcal{O}(\delta')$ is isomorphic to the quotient $\mathcal{O}(\delta)/\alpha$, where α is the ideal generated by

$$z_{n-k+1} - h_1, \ldots, z_n - h_k$$

in $\mathcal{O}(\delta)$.

† "Aangesteld Navorser" of the Belgian N.F.W.O.

In the case where $\Omega' = \mathbb{C}^{n-k}$, and the $h_1, \ldots, h_k$ are polynomials in $n-k$ variables, we have the same result for unbounded M.

The proofs can be found in [**1**].

References

1. Cnop, I. Extending holomorphic functions with bounded growth from certain graphs. *Proceedings of the Semester in Value Distribution Theory at Tulane University*. Edited by A. Vitter III and R. O. Kujala. Publ. M. Dekker, 1974, 109–117.
2. Ferrier, J.-P. *Spectral Theory and Complex Analysis*. North-Holland Mathematics Studies 4, 1973.
3. Waelbroeck, L. The holomorphic functional calculus and non-Banach algebras. These *Proceedings*, p. 187.

The Gleason Parts of $A(D^A)$†

I. G. CRAW

Department of Mathematics, University of Aberdeen

Abstract

In 1956 Gleason generalized the classical Schwarz lemma in the disc algebra by showing that for any uniform algebra A, one can define an equivalence relation on Φ_A by

$$\phi \sim \psi \Leftrightarrow \| \phi - \psi \| < 2 \qquad (\phi, \psi \in \Phi_A).$$

His proof, although short, is rather artificial. We give an alternative proof which obtains the result more directly from the disc algebra, effectively by showing that the equivalence relation is preserved when taking (infinite weak) tensor products and quotients.

Spectral synthesis for M(G)

COLIN GRAHAM

Department of Mathematics, Northwestern University, Evanston, Illinois, U.S.A.

† The complete version appears in these *Proceedings*, p. 265.

A Silov Boundary for Systems?†

R. E. HARTE

Department of Mathematics, University College, Cork, Ireland

Abstract

For an arbitrary system of elements $\boldsymbol{a} = (a_1, a_2, \ldots, a_n)$ in a complex Banach algebra A with identity 1 recall the *left spectrum*

$$\sigma_A^{\text{left}}(\boldsymbol{a}) = \left\{ s \in \mathbb{C}^n : 1 \notin \sum_{j=1}^{n} A(a_j - s_j) \right\}$$

and the *left approximate point spectrum*

$$\tau_A^{\text{left}}(\boldsymbol{a}) = \left\{ s \in \mathbb{C}^n : \inf_{\|b\| \geqslant 1} \sum_{j=1}^{n} \|(a_j - s_j)b\| = 0 \right\}.$$

As a generalization of the topological boundary of the *right* spectrum of an element we propose

$$\partial_A^{\text{left}}(\boldsymbol{a}) = \left\{ s \in \mathbb{C}^n : \inf_{|b| \geqslant 1} \sum_{j=1}^{n} |(a_j - s_j)b| = 0 \right\},$$

replacing the norm $\|c\|$ of an element $c \in A$ by its spectral radius $|c|$.

† A fuller version of this lecture appears in these *Proceedings*, p. 268.

Some Questions About Hilbert Space Operators

ALAN MCINTOSH

Department of Mathematics, Macquarie University, N.S.W., Australia.

Abstract

Consider the following two theorems about bounded linear operators in a Hilbert space:

(i) If A and B are self-adjoint and $AB - BA = 0$ then $|A|B - B|A| = 0$, where $|A| = (A^2)^{\frac{1}{2}}$;

(ii) If the numerical range $W(Z)$ of Z is $[-1, 1]$, then $\{-1, 1\} \subset \sigma(Z)$, the spectrum of Z.

Do the conclusions remain almost true if the hypotheses are slightly changed? That is,

(i) If A and B are self-adjoint and $\varepsilon > 0$, does there exist $\delta(\varepsilon) > 0$ such that $\|AB - BA\| < \delta$ implies $\||A|B - B|A|\| < \varepsilon$?

(ii) If $|\mathrm{Im}\, W(Z)| < \varepsilon \ll 1$, and $[-1, 1] \subset W(Z)$, what can be said about $\sigma(Z)$? (Can $\sigma(Z)$ be purely imaginary? Can $\mathrm{Re}\, \sigma(Z) \geqslant 1 - \varepsilon$?).

These and other questions will be answered.

References

1. McIntosh, A. Counterexample to a question on commutators, *Proc. Amer. Math. Soc.*, **29** (1971), 337–340.
2. McIntosh, A. On the comparability of $A^{1/2}$ and $A^{*1/2}$, *Proc. Amer. Math. Soc.*, **32** (1972), 430–434.
3. Kato, T. Continuity of the map $S \to |S|$ for linear operators, *Proc. Japan Acad.*, **49** (1973), 157–160.

Orthogonal Measures on the State Space of a C^*-algebra†

C. F. SKAU

Department of Mathematics, University of Oslo, Norway

Abstract

Let $\mathfrak{A}$ be a separable C^*-algebra with a unit and let $\mathscr{S}$ be the state space of $\mathfrak{A}$. With $p \in \mathscr{S}$ let π_p be the representation of $\mathfrak{A}$ engendered by p by the GNS-construction. Let μ be a normalized positive Borel measure on $\mathscr{S}$ representing p, i.e. $\int_{\mathscr{S}} s(a)d\mu(s) = p(a)$ for each $a \in \mathfrak{A}$. Then μ gives rise to a direct integral representation π_μ of $\mathfrak{A}$ in a natural way. We show that π_μ is unitarily equivalent to π_p if and only if μ belongs to a very restrictive class of measures representing p, the so-called orthogonal measures. We give several equivalent characterizations of these. Furthermore, there is a one-to-one order-preserving correspondence between the orthogonal measures representing p and abelian von Neumann sub-algebras of the commutant of $\pi_p(\mathfrak{A})$. (Here the measures are (partially) ordered by the usual ordering studied in Choquet theory and the subalgebras are (partially) ordered by inclusion). We give a detailed analysis of this correspondence and show finally how some of these concepts can be generalized to compact convex sets in a locally convex Hausdorff space.

† A full version appears in these *Proceedings*, p. 272.

Commutative Twisted Group Algebras†

HARVEY SMITH

Department of Mathematics, Oakland University, Rochester, Michigan, U.S.A.

Abstract

The L^1 group algebra of a second countable locally compact group extension can be written as an algebra of Bochner integrable functions on the quotient taking values in L^1 of the kernel. The quotient group acts on L^1 of the kernel by inner-automorphism T and a cocycle α is introduced into multiplication on the extension L^1 by choosing a measurable cross-section over the quotient. A twisted group algebra $L^1(A, G; T, \alpha)$ generalizes this construction, replacing L^1 of the kernel by an arbitrary separable Banach *-algebra A. We show that a twisted group algebra is commutative iff A and G are, T is trivial, and α is symmetric: $\alpha(\gamma, g) = \alpha(g, \gamma)$. We show that the maximal ideal space of a commutative twisted group algebra $L^1(A, G; \alpha)$ is a principal $G\hat{}$ bundle over $A\hat{}$. We define a class of principal $G\hat{}$ bundles over second countable locally compact M which is in $1 - 1$ correspondence with the $C_\infty(M)$-valued commutative twisted group algebras on G. If G is finite, only locally trivial bundles can be such duals, but in general, the duals need not be locally trivial.

† A paper of the same title will appear in *Trans. American Math. Soc.*

The Gleason Parts of $A(D^{\Lambda})$

I. G. CRAW

Department of Mathematics, University of Aberdeen, Aberdeen

Abstract

In this note we determine the Gleason parts of $A(D^{\Lambda})$. As a consequence we obtain a proof that Gleason's relation *is* an equivalence relation for an arbitrary uniform algebra. Our proof differs from the usual proofs of this fact in using directly the result for the disc algebra; we observe that the relation remains an equivalence relation when passing to tensor products and quotients, and that this gives the general result.

Let A be a uniform algebra with maximal ideal space Spec A. In addition to its usual topology, Spec A inherits a norm as a subset of A^*. For $\phi, \psi \in$ Spec A, define $\phi \sim \psi$ if and only if $\| \phi - \psi \| < 2$. It is clear that this relation is reflexive and symmetric, while for $A(D)$, the disc algebra, with maximal ideal space $\bar{D}$, the closed unit disc in $\mathbb{C}$, it is a consequence of Schwarz's lemma that $\sim$ is also transitive. In this case the equivalence classes, or *Gleason parts* consist of D, the open unit disc, together with each point of the unit circle. Since there is no direct analogue of Schwarz's lemma available in a general uniform algebra, it is something of a surprise that $\sim$ is always transitive. The result is due to Gleason [**3**], and a proof, together with the easier proof for the disc algebra can be found in Browder [**1**, 2.6]. This also contains a discussion of the utility of Gleason parts.

We need some results on the geometry of the disc. For $t \in \bar{D}$, $s \in \mathbb{C} \backslash \{\bar{t}^{-1}\}$ define $[s, t] = (s - t)(1 - s\bar{t})^{-1}$ and conventionally write $[t, t] = 0$ if $|t| = 1$. Then if $|t| < 1$, the map $s \rightsquigarrow [s, t]$ is in $A(D)$ and has norm 1. Also, since the map takes $\bar{D}$ onto $\bar{D}$, by the maximum principle, if $s \in \mathbb{C} \backslash \{\bar{t}^{-1}\}$ and $|[s, t]| < 1$ then $s \in D$.

LEMMA (Lewittes [**4**]). *Given* $0 < k < k + \theta < 1$, *put* $\varepsilon = \theta/k(1 + k + \theta)$ *and define* $I_{\varepsilon} = \{z \in \mathbb{C} : d(z, [0, 1]) < \varepsilon\}$. *Let* $s, t \in D$ *with* $|[s, t]| \leqslant k$; *then* $s + z(t - s) \in D$ *whenever* $z \in I_{\varepsilon}$.

Proof. Pick $z \in I_\varepsilon$ and choose $\gamma \in [0, 1]$ such that $|z - \gamma| < \varepsilon$. Then

$$|[s + z(t - s), t]| = \left| \frac{1 - z}{[s, t]^{-1} - z\bar{t}} \right|$$

$$\leqslant \frac{1 + \varepsilon - \gamma}{k^{-1} - \varepsilon - \gamma} \leqslant \frac{1 + \varepsilon}{k^{-1} - \varepsilon} = k + \theta < 1.$$

Thus by the remark above, $s + z(t - s) \in D$. ■

Let Λ be an index set and write $A(D^\Lambda)$ for the closure in $C(\bar{D}^\Lambda)$ of the algebra of polynomials in a finite number of the co-ordinate functions. Then just as for the disc algebra, $A(D^\Lambda)$ has maximal ideal space $\bar{D}^\Lambda$. We write a typical point $\mathbf{s} \in \operatorname{Spec} A(D^\Lambda)$ as $\mathbf{s} = \{s_\lambda\}$ with $s_\lambda \in \bar{D}$ for each $\lambda \in \Lambda$. We can now describe the Gleason parts of Spec $A(D^\Lambda)$.

THEOREM. *Let* $\mathbf{s}, \mathbf{t} \in \operatorname{Spec} A(D^\Lambda)$. *Then* $\|\mathbf{s} - \mathbf{t}\| < 2$ *if and only if there is some* $k < 1$ *such that* $|[s_\lambda, t_\lambda]| \leqslant k$ *for each* $\lambda \in \Lambda$.

Proof. Assume first there is some $c < 1$ such that $\|\mathbf{s} - \mathbf{t}\| \leqslant 2c$. Then for each $\lambda \in \Lambda$, $\|s_\lambda - t_\lambda\| \leqslant 2c$. Suppose there is no $k < 1$ as stated in the theorem. Then we can choose sequences $\{s_n\}$ and $\{t_n\}$ from $\mathbf{s}$ and $\mathbf{t}$ such that $|[s_n, t_n]| \to 1$. If $|s_n| = 1$ or $|t_n| = 1$ then since $\|s_n - t_n\| < 2$, we have $s_n = t_n$ and so $[s_n, t_n] = 0$ by convention. Thus we may suppose that $\{s_n\}$ and $\{t_n\}$ are sequences in D. Let $g_n(z) = -[[z, t_n], c]$; then for each n, $g_n \in A(D)$ and $\|g_n\| = 1$. Thus for each n,

$$|g_n(s_n) - g_n(t_n)| \leqslant \|s_n - t_n\| \leqslant 2c.$$

But $g(t_n) = c$, $g(s_n) \to -1$ $(n \to \infty)$ so

$$|g(s_n) - g(t_n)| \to 1 + c > 2c \qquad (n \to \infty).$$

This contradiction establishes the existence of a suitable $k < 1$.

Conversely, suppose there is some $k < 1$ such that $|[s_\lambda, t_\lambda]| \leqslant k$ $(\lambda \in \Lambda)$. We assume also that $s_\lambda \neq t_\lambda$ $(\lambda \in \Lambda)$; it will be clear from the proof that this involves no loss. Choose θ such that $k + \theta < 1$ and then $\varepsilon > 0$ as in the lemma. Let f be a conformal map of $\{z : |z| < 1 + \delta\}$ onto I_ε and choose δ small enough so that $\alpha = f^{-1}(0) \in D$, $\beta = f^{-1}(1) \in D$. Define

$$\Phi(z) = \mathbf{s} + f(z)(\mathbf{t} - \mathbf{s}),$$

so by the lemma $\Phi : \bar{D} \to D^\Lambda$ and Φ is analytic. Finally, choose $g \in A(D^\Lambda)$ with $\|g\| = 1$. Then

$$|g(\mathbf{s}) - g(\mathbf{t})| = |g \circ \Phi(\alpha) - g \circ \Phi(\beta)|$$

$$\leqslant \|\alpha - \beta\|, \text{ since } g \circ \Phi \in A(D) \text{ and } \|g \circ \Phi\| \leqslant 1$$

$$= 2c < 2 \text{ since } \alpha, \beta \in D.$$

This holds, with the same constant, for each such g, so $\|\mathbf{s} - \mathbf{t}\| \leqslant 2c < 2$. ■

Remarks. 1. An easy calculation shows that $|[s, u]| = |[[s, t], [u, t]]|$; thus it is clear from the theorem that we do have an equivalence relation on Spec $A(D^{\Lambda})$.

2. Let R be any quotient of $A(D^{\Lambda})$ with quotient map $q: A(D^{\Lambda}) \to R$. Then if $\phi, \psi \in \text{Spec } R$

$$\|\phi - \psi\|_{\text{Spec } R} = \|\phi \circ q - \psi \circ q\|_{\text{Spec } A(D^{\Lambda})}$$

Thus, by 1, defining $\phi \sim \psi$ if and only if $\|\phi - \psi\| < 2$ gives an equivalence relation on the maximal ideal space of every such quotient algebra R. Gleason's result then follows by noting that every uniform algebra can be written as such a quotient; indeed, more generally, every quotient of a uniform algebra by a closed ideal can be written in this way [**2**, Lemma 3.1].

References

1. Browder, A. *Introduction to Function Algebras,* W. A. Benjamin, New York, 1969.
2. Davie, A. M. Quotient algebras of uniform algebras, *J. London Math. Soc.* **7** (1973) 31–40.
3. Gleason, A. M. Function algebras, *Seminar on Analytic Functions,* vol. 2, Institute for Advanced Study, Princeton, N.J. 1957.
4. Lewittes, J. A note on parts and hyperbolic geometry, *Proc. Amer. Math. Soc.* **17** (1966) 1087–1090.

A Silov Boundary for Systems?

R. E. HARTE

Department of Mathematics, University College, Cork, Ireland

1. Suppose A is a complex Banach algebra, with identity 1: then for an n-tuple $\mathbf{a} = (a_1, a_2, \ldots, a_n)$ of its elements we have defined [2, 3] the *joint spectrum*

$$\sigma_A^{\text{joint}}(\mathbf{a}) = \sigma_A^{\text{left}}(\mathbf{a}) \cup \sigma_A^{\text{right}}(\mathbf{a}), \tag{1.1}$$

where

$$\sigma_A^{\text{left}}(\mathbf{a}) = \left\{ \mathbf{s} \in \mathbb{C}^n : 1 \notin \sum_{j=1}^{n} A(a_j - s_j) \right\} \tag{1.2}$$

and

$$\sigma_A^{\text{right}}(\mathbf{a}) = \left\{ \mathbf{s} \in \mathbb{C}^n : 1 \notin \sum_{j=1}^{n} (a_j - s_j)A \right\}. \tag{1.3}$$

These are compact subsets of $\mathbb{C}^n$, possibly empty, and subject to the "one-way spectral mapping theorem" ([2] Theorem 3.2):

$$\mathbf{f}(\sigma_A^{\text{left}}(\mathbf{a})) \subset \sigma_A^{\text{left}}(\mathbf{f}(\mathbf{a})), \quad \mathbf{f}(\sigma_A^{\text{right}}(\mathbf{a})) \subset \sigma_A^{\text{right}}(\mathbf{f}(\mathbf{a})). \tag{1.4}$$

Here $\mathbf{f} = (f_1, f_2, \ldots, f_m)$ is an m-tuple of "polynomials" in n variables on A; each f_k is a sum of products of scalar constants and co-ordinates $z_j : A^n \to A$. The "two-way spectral mapping theorem" ([2] Theorem 4.3) is the assertion that there is equality in (1.4) for commuting systems of elements $\boldsymbol{a} \in A^n$:

$$\mathbf{f}(\sigma_A^{\text{left}}(\mathbf{a})) = \sigma_A^{\text{left}}(\mathbf{f}(\mathbf{a})), \quad \mathbf{f}(\sigma_A^{\text{right}}(\mathbf{a})) = \sigma_A^{\text{right}}(\mathbf{f}(\mathbf{a})). \tag{1.5}$$

It follows that the sets (1.1)–(1.3) are nonempty for commuting systems of elements. In the case of a commutative algebra A these results follow from the Gelfand theory: we might also comment that the extension of our results to an infinite system of elements [4] actually gives a proof of Gelfand's theorem.

2. Again for an arbitrary system of elements $\mathbf{a} \in A^n$, recall the left and right *approximate point spectra* ([2] Definition 1.3)

$$\tau_A^{\text{left}}(\mathbf{a}) = \left\{\mathbf{s} \in \mathbb{C}^n : \inf_{\|b\| \geq 1} \sum_{j=1}^{n} \|(a_j - s_j)b\| = 0\right\} \tag{2.1}$$

and

$$\tau_A^{\text{right}}(\mathbf{a}) = \left\{\mathbf{s} \in \mathbb{C}^n : \inf_{\|b\| \geq 1} \sum_{j=1}^{n} \|b(a_j - s_j)\| = 0\right\}. \tag{2.2}$$

These are closed subsets of (1.2) and (1.3) respectively, subject to the analogue ([2] Theorem 3.2) of (1.4); for commuting systems $\mathbf{a} \in A^n$ the analogue of (1.5) has recently been obtained by Choi and Davis [1]. An essential component of our proof of (1.5) is the inclusion, for single elements $\boldsymbol{a} = a_1 \in A$ ([2] Lemma 4.1)

$$\partial\sigma_A^{\text{left}}(\mathbf{a}) \subset \tau_A^{\text{right}}(\mathbf{a}), \quad \partial\sigma_A^{\text{right}}(\mathbf{a}) \subset \tau_A^{\text{left}}(\mathbf{a}); \tag{2.3}$$

here ∂ denotes the topological boundary. Our purpose in this note is to raise the problem of an analogue of (2.3), valid for commuting systems of elements in non-commutative algebras: an application would be towards a converse for (1.5). Our candidate is latent in the work of Żelazko [5], [6]; but, as we shall see, modifications will be necessary.

3. For an arbitrary system of elements $\mathbf{a} \in A^n$ define

$$\partial_A^{\text{left}}(\mathbf{a}) = \left\{\mathbf{s} \in \mathbb{C}^n : \inf_{|b| \geq 1} \sum_{j=1}^{n} |(a_j - s_j)b| = 0\right\}; \tag{3.1}$$

here we are writing $|c|$ for the *spectral radius* of an element $c = c_1 \in A$, in contrast to the norm $\|c\|$ appearing in (2.1). Similarly

$$\partial_A^{\text{right}}(\mathbf{a}) = \left\{\mathbf{s} \in \mathbb{C}^n : \inf_{|b| \geq 1} \sum_{j=1}^{n} |b(a_j - s_j)| = 0\right\}. \tag{3.2}$$

In no particular order, the obvious questions about these sets would appear to be the following:

(3.3) In general, is the set $\partial_A^{\text{left}}(\mathbf{a})$ closed and bounded in $\mathbb{C}^n$?

(3.4) In general, is there inclusion $\mathbf{f}(\partial_A^{\text{left}}(\mathbf{a})) \subset \partial_A^{\text{left}}(\mathbf{f}(\mathbf{a}))$?

(3.5) In general, is there inclusion $\partial_A^{\text{left}}(\mathbf{a}) \subset \tau_A^{\text{left}}(\mathbf{a})$?

(3.6) Is there inclusion $\partial_B^{\text{left}}(\mathbf{a}) \subset \partial_A^{\text{left}}(\mathbf{a})$, where B is a closed subalgebra of A containing the elements a_j and the identity 1?

(3.7) Is there equality $\mathbf{f}(\partial_A^{\mathrm{left}}(\mathbf{a})) = \partial_A^{\mathrm{left}}(\mathbf{f}(\mathbf{a}))$, for commuting systems of elements $\mathbf{a} \in A^n$?

(3.8) Is there inclusion $\partial(\sigma_A^{\mathrm{right}}(\mathbf{a})) \subset \partial_A^{\mathrm{left}}(\mathbf{a})$, for single elements $\mathbf{a} = a_1 \in A$?

(3.9) Is there equality $\partial_A^{\mathrm{left}}(\mathbf{a}) = \{\phi(\mathbf{a}) : \phi \in (\partial\sigma)(A)\}$, for a commutative algebra A, where $(\partial\sigma)(A)$ is its Silov boundary?

If the algebra A is commutative then most of these inclusions follow from the corresponding arguments for the approximate point spectrum, noticing that the spectral radius is a sub-multiplicative seminorm. Żelazko's first paper [**5**] establishes the equality (3.9), while his second paper [**6**] is devoted to the inclusion (3.5).

For a general non-commutative algebra A the inclusion (3.6) is immediate, using the invariance of the spectral radius. We would like to demonstrate (3.8), and show the failure of (3.5), even for single elements.

4. The proof of (3.8) is an illustration of how we can sometimes work with the spectral radius in a non-commutative environment. If $s \in \mathbb{C}$ is in $\partial\sigma^{\mathrm{right}}(\boldsymbol{a})$ then of course s is in $\sigma^{\mathrm{right}}(\boldsymbol{a})$, and there are sequences (s_r) in $\mathbb{C}$ and (b_r) in A for which $s_r - s \to 0$ $(r \to \infty)$ and $(a - s_r)b_r = 1$ (all r). We claim that the spectral radius $|b_r| \geqslant 1/|s - s_r|$ (all r): for if not then

$$|1 - (a - s)b_r| = |(s - s_r)b_r| < 1,$$

giving $(a - s)b_r$ a two-sided inverse and excluding s from $\sigma^{\mathrm{right}}(\boldsymbol{a})$. Now $c = b_r/|b_r|$ gives $|c| = 1$ while

$$|(a - s)c| = |(a - s_r)c + (s_r - s)c| = \left| \frac{1}{|b_r|} + (s_r - s)c \right| \leqslant 2|s_r - s| \to 0. \quad \blacksquare$$

For the failure of (3.5) take A to be the algebra of all bounded operators on the sequence space l_2, and take $\boldsymbol{a} = a = a_1$ to be the shift

$$(x_1, x_2, x_3, \ldots) \to (0, x_1, x_2, \ldots).$$

It is familiar that

$$a^*a = 1 \neq aa^*;$$

in particular 0 is not in $\sigma^{\mathrm{left}}(a)$. Here a^* is the backward shift

$$(x_1, x_2, x_3, \ldots) \to (x_2, x_3, x_4, \ldots).$$

The element $b = 1 - aa^*$ is now a non-trivial projection; the spectral radius $|b| = 1$. Now observe

$$ba = a - aa^*a = 0,$$

and hence $(ab)^2 = a(ba)b = 0$. The spectral radius $|ab| = 0$, and so 0 is in $\partial^{\mathrm{left}}(a)$. ∎

This example is of course discouraging to this line of investigation. First aid can be applied by replacing the set (3.1) by

$$\partial'^{\,\mathrm{left}}_A(\boldsymbol{a}) = \partial^{\mathrm{left}}_A(\boldsymbol{a}) \cap \sigma^{\mathrm{left}}_A(\boldsymbol{a}): \tag{4.1}$$

then for the moment (3.5)′ is an open problem. We have no information about (3.4)′ or (3.7)′: the arguments for the approximate point spectrum are not available to us. We feel that some analogue of (2.3) does exist, and present this account in the hope that it may suggest the correct definition to someone.

References

1. Choi, M. D. and Davis C. The spectral mapping theorem for joint approximate point spectrum, *Bull. Amer. Math. Soc.* **80** (1974) 317–321.
2. Harte, R. E. Spectral mapping theorems, *Proc. Roy. Irish Acad.* **72**(A) (1972) 89–107.
3. Harte, R. E. The spectral mapping theorem in several variables, *Bull. Amer. Math. Soc.* **78** (1972) 871–875.
4. Harte, R. E. The spectral mapping theorem in many variables, *Seminar on Uniform Algebras, University of Aberdeen* 1973 pp. 59–63. (Aberdeen 1973).
5. Żelazko, W. A characterization of Silov boundary in function algebras, *Prace Mat.* **14** (1970) 59–64.
6. Żelazko, W. On a certain class of non-removable ideals in Banach algebras, *Studia Math.* **44** (1972) 87–92.

Orthogonal Measures on the State Space of a C^*-algebra

C. F. SKAU

Department of Mathematics, University of Oslo, Oslo, Norway

INTRODUCTION

In the classification of C^*-algebras a natural question is whether it is possible to write any concretely represented C^*-algebra as a direct integral of factorial C^*-algebras. This raises the more general question of decomposing a given representation of a C^*-algebra. The usual arguments show that it suffices to consider cyclic representations, i.e. the representations that arise by the Gelfand–Naimark–Segal construction from the states of the given C^*-algebra. It turns out that the question of decomposing the representation π_p of the C^*-algebra $\mathfrak{A}$ corresponding to a state p is equivalent to find certain measures on the state space of $\mathfrak{A}$ representing p. These measures, the so-called orthogonal measures, are in 1–1 correspondence with abelian von Neumann subalgebras of $\pi_p(\mathfrak{A})'$. Moreover, there is a close correspondence between the ordering of the measures in question as introduced by either Bishop–deLeeuw or Choquet, in fact, these two orderings coincide on the orthogonal measures, and the inclusion ordering of the corresponding abelian von Neumann subalgebras of $\pi_p(\mathfrak{A})'$. We shall investigate these and related matters in this paper.

1. PRELIMINARIES ON CHOQUET THEORY

We introduce some terminology and give a short resumé of results from convexity and Choquet theory that we shall need in the sequel. As a general reference we mention [1]. Let K by a compact convex set in a real locally convex Hausdorff space E. Let $C_{\mathbb{R}}(K)$ be the real-valued continuous functions on K in the supremum norm. We denote by $A(K)$ the linear subspace of

$C_{\mathbb{R}}(K)$ of *affine* functions on K; i.e. $a \in A(K)$ if $a \in C_{\mathbb{R}}(K)$ and

$$a(\lambda p + (1 - \lambda)q) = \lambda a(p) + (1 - \lambda)a(q)$$

for p, $q \in K$ and $\lambda \in [0, 1]$. $A(K)$ is closed in $C_{\mathbb{R}}(K)$ and the linear subspace $A(K; E)$ of $A(K)$ consisting of restrictions to K of continuous affine functions on E is dense in $A(K)$. Clearly $A(K; E) = E^* \oplus \mathbb{R}|_K$, where E^* is the topological dual of E. So $A(K)$ separates the point in K. If K lies in a closed hyperplane that does not meet 0 we have $A(K; E) = E^*|_K$.

We shall use the symbol $M_{\mathbb{R}}(K)$ to denote the vector space of (real) regular Borel measures on K. $M^+(K)$ will denote the non-negative measures in $M_{\mathbb{R}}(K)$ and $M_1^+(K)$ will denote the convex subset of $M^+(K)$ of normalized measures (probability measures). Recall that $M_{\mathbb{R}}(K)$ may be identified with the dual Banach space of $C_{\mathbb{R}}(K)$ and the (dual) norm is given by $\|\mu\| = |\mu|(K)$. By a well-known theorem $M_1^+(K)$ is a compact subset of $M_{\mathbb{R}}(K)$ in the $\sigma(M_{\mathbb{R}}(K), C_{\mathbb{R}}(K))$-topology, i.e. the w^*-topology.

Let $\mu \in M\ (K)$. Then $f \to \int_K f(s)\, d\mu s\ (f \in E^*)$ is a linear functional on E^*. A fundamental theorem in convexity theory says that there exists a unique $t_\mu \in E$ such that $f(t_\mu) = \int_K f(s) d\mu s$ for all $f \in E^*$. The vector t_μ is called the *resultant* for μ and we write $t_\mu = \int_K s d\mu$. The map $\mu \longrightarrow t_\mu$ is linear from $M_{\mathbb{R}}(K)$ to E. If $\mu \in M_1^+(K)$ we have $t_\mu \in K$. In this case we say t_μ is the *barycenter* of μ and we have $a(t_\mu) = \int_K a(s) d\mu s$ for all a in $A(K)$. If $p \in K$ and μ is a measure in $M_1^+(K)$ such that $\int_K s d\mu = p$ we say *μ represents p*. We write $M_p^+(K)$ for the measures in $M_1^+(K)$ that represent $p \in K$. $M_p^+(K)$ is a convex closed subset of $M_1^+(K)$, hence compact in the w^*-topology.

A question of primary importance in Choquet theory is to find measures in $M_p^+(K)$ "concentrated" on the extreme points $\partial_e K$ of K. If K is metrizable, which by Urysohn's Metrization Theorem is equivalent to K having a countable base, $\partial_e K$ is a G_δ-subset of K, hence a Borel set. If K is non-metrizable $\partial_e K$ may not be a Borel set. In 1956 Choquet [**4**] proved that if K is metrizable and $p \in K$ there exists a $\mu \in M_p^+(K)$ such that μ is *carried* by the extreme points of K, i.e. $\mu(\partial_e K) = 1$. In 1959 Bishop and deLeeuw [**2**] came up with an ingenious new proof of this fact which also gave some information in case K is non-metrizable. As their method of proof is of some interest in itself and will be referred to several times in this paper, we will describe it. Their basic idea is to introduce a (partial) ordering on measures in $M_p^+(K)$ for $p \in K$. Let μ, $\nu \in M_p^+(K)$. Then μ is less than ν in the Bishop–deLeeuw ordering, in symbols $\mu \ll \nu$, if $\int_K a(s)^2 d\mu s \leqslant \int_K a(s)^2 d\nu s$ for all $a \in A(K)$. This is a reflexive and transitive, but not an antisymmetric ordering. By a standard compactness and Zorn lemma argument we establish that there exists a maximal measure μ in $M_p^+(K)$ in the ordering $\ll$. It is then proved that the maximal measure μ is carried by $\partial_e K$ in case K is metrizable, and that in the non-metrizable case μ is *pseudo-carried* by $\partial_e K$, i.e. $\mu(C) = 0$ for every

Baire set C disjoint from $\partial_e K$. (The Baire sets are the σ-algebra generated by the compact G_δ-subsets of K. When K is metrizable the Borel sets and Baire sets coincide.)

In 1960 Choquet [**5**], inspired by the Bishop–deLeeuw approach, introduced another ordering on measures in $M_p^+(K)$ which he subsequently employed to give a new proof of the Bishop–deLeeuw result. Since we will make use of this ordering in the sequel we will describe it. Let $P(K)$ be the closed cone in $C_{\mathbb{R}}(K)$ consisting of convex functions on K; i.e. $f \in P(K)$ if $f \in C_{\mathbb{R}}(K)$ and

$$f(\lambda p + (1 - \lambda)q) \leqslant \lambda f(p) + (1 - \lambda)f(q)$$

for $p, q \in K$, $\lambda \in [0, 1]$. Since $P(K)$ is a sup-cone, i.e. $f, g \in P(K)$ implies $f \vee g \in P(K)$, it follows from Stone's Theorem that $P(K) - P(K)$ is dense in $C_{\mathbb{R}}(K)$. Let $\mu, \nu \in M_p^+(K)$ for some $p \in K$. Then μ is less than ν in the Choquet ordering, in symbols, $\mu < \nu$, if

$$\int_K f(s)d\mu s \leqslant \int_K f(s)d\nu s \quad \text{for all} \quad f \in P(K).$$

This is a reflexive, antisymmetric and transitive ordering. As a^2 is in $P(K)$ for a in $A(K)$ it follows immediately that for $\mu, \nu \in M_p^+(K)$ for some $p \in K$, $\mu < \nu$ implies $\mu \ll \nu$.

In 1963 Choquet and Meyer [**6**] presented a nice intrinsic characterization of measures in $M_p^+(K)$ maximal in the Choquet ordering $<$, which was due to Mokobodzki [**13**]. For K metrizable it says that $\mu \in M_p^+(K)$ is $<$-maximal if and only if μ is carried by $\partial_e K$, i.e. $\mu(\partial_e K) = 1$. In the same paper is an example of a non-metrizable K and a measure $\nu \in M_p^+(K)$, for some $p \in K$, such that ν is pseudo-carried by $\partial_e K$, but ν is not maximal in the ordering $<$. There is no intrinsic characterization of maximal measures in the ordering $\ll$. However, for K metrizable we do have that if $\mu \in M_p^+(K)$ $(p \in K)$ is $\ll$-maximal then μ is $<$-maximal.

We shall exhibit an example later on to show that the converse is not true.

We can always map a compact convex set by a homeomorphic affine map onto a *regularly embedded* compact convex set K in a locally convex Hausdorff space E. We say K is regularly embedded in E if K lies in a closed hyperplane that does not contain 0, E is the linear span of K and $A(K)$ is equal to the restriction to K of elements in E^* [**1**; p. 80].

If K is regularly embedded in E we say that K is a *Choquet simplex* if E is a vector lattice in the ordering defined by the cone $\tilde{K}$ generated by K. This is easily seen to be equivalent to $\tilde{K}$ being a lattice in the induced ordering. Also it can be shown that this is equivalent to that $\tilde{K}$ satisfies the Riesz decomposition property, i.e. if

$$u = \sum_{i=1}^{n} u_i = \sum_{j=1}^{m} v_j,$$

all u_i, v_j in $\tilde{K}$, then $\exists w_{ij} \in \tilde{K}$ such that

$$u_i = \sum_{j=1}^{m} w_{ij}, \qquad v_j = \sum_{i=1}^{n} w_{ij} \qquad (i = 1, \ldots, n; j = 1, \ldots, m).$$

In general, we say that a compact convex set is a Choquet simplex if its regular embedding is a Choquet simplex, and it is a simple observation that this is an unambiguous definition. In his 1956 paper Choquet [**4**] proved that if K is a metrizable compact convex set then K is a Choquet simplex if and only if there exists a *unique* maximal measure in $M_p^+(K)$ in the ordering $<$ for each p in K. In 1963 Choquet–Meyer [**6**] established this in general without metrizability.

Definition 1: Let K be a compact convex set and let $\mu \in M_p^+(K)$ for some p in K. We say μ is *simplicial* if μ is an extreme point in $M_p^+(K)$.

Since $M_p^+(K)$ is compact in the w^*-topology it follows by the Krein–Milman Theorem that there exist "many" simplicial measures in $M_p^+(K)$. In the finite-dimensional case, i.e. the hyperplane affinely spanned by K is finite-dimensional, it can be shown that a measure in $M_p^+(K)$ is simplicial if and only if it is supported by a finite number of affinely independent points $\{x_i\}$, $i = 1, \ldots, n$; i.e.

$$\sum_{i=1}^{n} \lambda_i x_i = 0 \qquad \text{and} \qquad \sum_{i=1}^{n} \lambda_i = 0$$

for $\lambda_i \in \mathbb{R}$ $(i = 1, \ldots, n)$ implies $\lambda_1 = \ldots = \lambda_n = 0$.

The following lemma gives a characterization of simplicial measures. A proof can be found in [**1**].

LEMMA 1. *Let K be a compact convex set and let $\mu \in M_p^+(K)$ for some $p \in K$. Then μ is simplicial if and only if $A(K)$ is dense in $L^1_{\mathbb{R}}(\mu)$.*

Remark: Note that $A(K)$ being dense in $L^1_{\mathbb{R}}(\mu)$ is equivalent to $A_{\mathbb{C}}(K)$, the complex-valued continuous affine functions on K, being dense in $L^1(\mu)$, the space of complex-valued integrable functions.

It can be shown that for each p in K there exists a simplicial $<$-maximal measure in $M_p^+(K)$ [**1**; Proposition I.6.12, p. 62]. This is a far-reaching generalization of a classical result by Carathéodory for finite-dimensional K.

The next lemma will be extremely useful to us later on. It was first proved by Cartier, Fell and Meyer [3] in 1964; a proof can also be found in [1].

LEMMA 2. *Let K be a compact convex set and let* μ, ν *be in* $M_p^+(K)$ *for some* $p \in K$. *Then* $\nu < \mu$ *if and only if for every decomposition* $\nu = \sum_{i=1}^{n} \nu_i$ *of* ν *into n positive components* ν_i *there exists a decomposition* $\mu = \sum_{i=1}^{n} \mu_i$ *of* μ *into n positive components* μ_i *such that resultant* $(\mu_i) =$ *resultant* (ν_i) *for* $i = 1, \ldots, n$.

2. DEFINITION AND BASIC PROPERTIES OF ORTHOGONAL MEASURES

Throughout the rest of this paper $\mathfrak{A}$ will be a C^*-algebra with a unit I. $\mathfrak{A}^s$ is the self-adjoint portion of $\mathfrak{A}$ and $\mathscr{S}$ will denote the state space of $\mathfrak{A}$. $\mathscr{S}$ is a subset of the continuous hermitian linear functionals on $\mathfrak{A}$ which we denote by $(\mathfrak{A}^*)^h$. Then $(\mathfrak{A}^*)^h$ may in a natural way be identified with the dual $(\mathfrak{A}^s)^*$ of $\mathfrak{A}^s$. Both $\mathfrak{A}^s$ and $(\mathfrak{A}^s)^*$ are real linear spaces, $\mathfrak{A}^s$ is a partially ordered vector space in the usual ordering of elements in $\mathfrak{A}^s$ and $(\mathfrak{A}^s)^*$ is a partially ordered vector space in the dual ordering of $\mathfrak{A}^s$; i.e. $p \leqslant q$ for $p, q \in (\mathfrak{A}^s)^*$ if $p(a) \leqslant q(a)$ for all $a \geqslant 0$, $a \in \mathfrak{A}^s$. $\mathscr{S}$ is a compact convex subset of $(\mathfrak{A}^s)^*$ in the $\sigma((\mathfrak{A}^s)^*, \mathfrak{A}^s)$-topology. We will henceforth assume that $\mathscr{S}$ is endowed with this topology. If $\mathfrak{A}$ is separable $\mathscr{S}$ is metrizable [17; p. 140].

For $a \in \mathfrak{A}$ we denote by $\hat{a}$ the continuous complex-valued affine function on $\mathscr{S}$ defined by $\hat{a}(p) = p(a)\,(p \in \mathscr{S})$. By the linear mapping $a \to \hat{a}$, $\mathfrak{A}^s$ is order-isomorphically and isometrically embedded into $C_{\mathbb{R}}(\mathscr{S})$. Since $\mathfrak{A}^s$ is complete it follows by [1; Corollary I.1.5, p. 4] that $A(\mathscr{S}) = \{\hat{a} \mid a \in \mathfrak{A}^s\}$. Hence we see that $\mathscr{S}$ is regularly embedded in $(\mathfrak{A}^s)^*$ as defined in Section 1. We also have $A_{\mathbb{C}}(\mathscr{S}) = \{\hat{a} \mid a \in \mathfrak{A}\}$, where we denote by $A_{\mathbb{C}}(\mathscr{S})$ the complex-valued continuous affine functions on $\mathscr{S}$.

Let p be a positive linear functional on $\mathfrak{A}$. We denote by π_p and H_p the representation and Hilbert space, respectively, engendered by p by the Gelfand–Naimark–Segal construction. For $b \in \mathfrak{A}$ let b_p denote the canonical image of b in H_p. Then we have $p(a) = (\pi_p(a)I_p, I_p)$ for each $a \in \mathfrak{A}$, and I_p is a cyclic vector for $\pi_p(\mathfrak{A})$ on H_p.

LEMMA 3. *Let* μ *be a regular non-negative Borel measure on the state space* $\mathscr{S}$ *of the* C^**-algebra* $\mathfrak{A}$. *Let* $p = \int_{\mathscr{S}} s\,d\mu$, *i.e.* p *is the positive linear functional on* $\mathfrak{A}$ *defined by*

$$p(a) = \int_{\mathscr{S}} \hat{a}(s)\,d\mu s \qquad (a \in \mathfrak{A})$$

*Then there exists a unique *-preserving, positive and norm-decreasing linear map $f \to K_f^\mu$ of $L^\infty(\mu)$ into $\pi_p(\mathfrak{A})'$ defined by the equation*

$$(K_f^\mu \pi_p(a)I_p, I_p) = \int_{\mathscr{S}} f(s)\hat{a}(s)d\mu s \qquad (a \in \mathfrak{A}).$$

The map $f \to K_f^\mu$ is continuous, where $L^\infty(\mu)$ is given the $\sigma(L^\infty(\mu), L^1(\mu))$-topology and $\pi_p(\mathfrak{A})'$ is given the weak-operator topology.

Proof. Uniqueness of the mapping is clear from the fact that I_p is cyclic for $\pi_p(\mathfrak{A})$. Now let $f \in L^\infty(\mu)$. To prove existence define a conjugate bilinear form B on $\pi_p(\mathfrak{A})I_p \times \pi_p(\mathfrak{A})I_p$ by

$$B(\pi_p(a)I_p, \pi_p(b)I_p) = \int_{\mathscr{S}} f(s)\widehat{b^*a}(s)d\mu s \qquad (a, b \in \mathfrak{A}).$$

B is well-defined since

$$\left|\int_{\mathscr{S}} f(s)\widehat{b^*a}(s)d\mu s\right| \leqslant \|f\|_\infty \int_{\mathscr{S}} |\widehat{b^*a}(s)|\, d\mu s$$

$$\leqslant \|f\|_\infty \int_{\mathscr{S}} (\widehat{b^*b}(s))^{1/2}(\widehat{a^*a}(s))^{1/2} d\mu s \leqslant \|f$$

$$\|f\|_\infty\left(\int_{\mathscr{S}} \widehat{a^*a}(s)d\mu s\right)^{1/2} \times \left(\int_{\mathscr{S}} \widehat{b^*b}(s)d\mu s\right)^{1/2}$$

$$= \|f\|_\infty(\pi_p(a^*a)I_p, I_p)^{1/2}(\pi_p(b^*b)I_p, I_p)^{1/2}$$

$$= \|f\|_\infty \|\pi_p(b)I_p\| \|\pi_p(a)I_p\|.$$

This also shows that B is bounded and can be extended to a bounded conjugate bilinear functional on $H_p \times H_p$ with bound $\leqslant \|f\|_\infty$. By Riesz' theorem there exists a bounded operator K_f^μ on H_p such that $\|K_f^\mu\| \leqslant \|f\|_\infty$ and

$$(K_f^\mu \pi_p(a)I_p, \pi_p(b)I_p) = \int_{\mathscr{S}} f(s)\widehat{b^*a}(s)d\mu s$$

for all a, b in $\mathfrak{A}$. Putting $b = I$ we get

$$(K_f^\mu \pi_p(a)I_p, I_p) = \int_{\mathscr{S}} f(s)\hat{a}(s)d\mu s \qquad (s \in \mathfrak{A}).$$

For a, b and c in $\mathfrak{A}$ we have

$$(K^{\mu}_{f}\pi_{p}(a)\pi_{p}(b)I_{p}, \pi_{p}(c)I_{p}) = \int_{\mathscr{S}} f(s)\widehat{c^{*}(ab)}(s)d\mu s = \int_{\mathscr{S}} f(s)\widehat{(a^{*}c)^{*}b}(s)d\mu s$$

$$= (K^{\mu}_{f}\pi_{p}(b)I_{p}, \pi_{p}(a^{*})\pi_{p}(c)I_{p}) = (\pi_{p}(a)K^{\mu}_{f}\pi_{p}(b)I_{p}, \pi_{p}(c)I_{p}).$$

Since I_p is cyclic for $\pi_p(\mathfrak{A})$ we get

$$K^{\mu}_{f}\pi_{p}(a) = \pi_{p}(a)K^{\mu}_{f}$$

for $a \in \mathfrak{A}$. Hence $K^{\mu}_{f} \in \pi_{p}(\mathfrak{A})'$. It is a simple observation to see that the map $f \to K^{\mu}_{f}$ is a *-preserving positive linear map. It remains to prove the continuity assertion. For $a, b \in \mathfrak{A}$ and $f \in L^{\infty}(\mu)$ we have

$$\int_{\mathscr{S}} f(s)\widehat{b^{*}a}(s)d\mu s = (K^{\mu}_{f}\pi_{p}(a)I_{p}, \pi_{p}(b)I_{p}).$$

This shows that $f \to (K^{\mu}_{f}\pi_{p}(a)I_{p}, \pi_{p}(b)I_{p})$ is a continuous linear functional on $L^{\infty}(\mu)$ with the $\sigma(L^{\infty}(\mu), L^{1}(\mu))$-topology. Since $\pi_{p}(\mathfrak{A})I_{p}$ is dense in H_p and $\| K^{\mu}_{f} \| \leqslant \| f \|_{\infty}$ it follows that $f \to (K^{\mu}_{f}x, y)$ is continuous on the unit ball of $L^{\infty}(\mu)$ in the $\sigma(L^{\infty}(\mu), L^{1}(\mu))$-topology for all x, y in H_p. By the Krein–Smŭlian Theorem we conclude that $f \to (K^{\mu}_{f}x, y)$ is continuous on $L^{\infty}(\mu)$ in the $\sigma(L^{\infty}(\mu), L^{1}(\mu))$-topology for all x, y in H_p.

This completes the proof.

We state a well-known result whose proof is analogous to the proof of Lemma 3, cf. [**8**; Proposition 2.5.1, p. 35].

LEMMA 4. *Let q and p be two positive linear functionals on $\mathfrak{A}$ such that $q \leqslant p$. Then there exists a unique V_q in $\pi_{p}(\mathfrak{A})'$ such that $0 \leqslant V_q \leqslant I$ and $q(a) = (V_{q}\pi_{p}(a)I_{p}, I_{p})$ for $a \in \mathfrak{A}$. Moreover, the mapping $q \to V_q$ is an order isomorphism between positive linear functionals $q \leqslant p$ and positive operators $V_q \leqslant I$ in $\pi_{p}(\mathfrak{A})'$.*

PROPOSITION 5. *Let $\mu \in M^{+}_{p}(\mathscr{S})$ for some $p \in \mathscr{S}$. Then μ is simplicial if and only if the mapping $f \to K^{\mu}_{f}(f \in L^{\infty}(\mu))$ is injective.*

Proof. Assume μ is simplicial and let $K^{\mu}_{f} = 0$ for some $f \in L^{\infty}(\mu)$. Then

$$0 = (K^{\mu}_{f}\pi_{p}(a)I_{p}, I_{p}) = \int_{\mathscr{S}} f(s)\hat{a}(s)d\mu s$$

for all $a \in \mathfrak{A}$. Since $A_{\mathbb{C}}(\mathscr{S}) = \{\hat{a} \mid a \in \mathfrak{A}\}$ and $L^{\infty}(\mu) \cong L^{1}(\mu)^{*}$ it follows from Lemma 1 that $f = 0$.

Conversely, assume $f \to K^{\mu}_{f}(f \in L^{\infty}(\mu))$ is injective. Let $g \in L^{\infty}(\mu)$ be such that $\int_{\mathscr{S}} g(s)\hat{a}(s)d\mu s = 0$ for all $a \in \mathfrak{A}$. Then

$$(K_g^\mu \pi_p(a)I_p, I_p) = \int_{\mathscr{S}} g(s)\hat{a}(s)d\mu s = 0$$

for all $a \in \mathfrak{A}$. This shows that $K_g^\mu = 0$ and hence $g = 0$. So $A_{\mathbb{C}}(\mathscr{S})$ is dense in $L^1(\mu)$ and by Lemma 1 we conclude that μ is simplicial.

Definition 2: Given positive linear functionals p and q on $\mathfrak{A}$, we say that p and q are *orthogonal*, in symbols, $p \perp q$, if for all positive linear functionals r on $\mathfrak{A}$, $r \leqslant p$ and $r \leqslant q$ imply that $r = 0$.

LEMMA 6. *Let s and q be two positive linear functionals on $\mathfrak{A}$ and let $p = s + q$. Let V_s and V_q be the positive operators $\leqslant I$ in $\pi_p(\mathfrak{A})'$ such that*

$$s(a) = (V_s\pi_p(a)I_p, I_p), \qquad q(a) = (V_q\pi_p(a)I_p, I_p)$$

for all $a \in \mathfrak{A}$. Then $s \perp q$ if and only if V_s (or V_q) is a projection.

Proof. By Lemma 4 we have that $s \perp q$ if and only if $0 \leqslant C \in \pi_p(\mathfrak{A})'$ and $C \leqslant V_s$, $C \leqslant V_q = I - V_s$ imply that $C = 0$. Now

$$0 \leqslant V_s(I - V_s) = V_s^{1/2}(I - V_s)V_s^{1/2} \leqslant V_s^{1/2}IV_s^{1/2} = V_s.$$

Likewise, $0 \leqslant V_s(I - V_s) \leqslant I - V_s$. Hence $s \perp q$ implies that $V_s(I - V_s) = 0$, i.e. V_s is a projection.

Conversely, if V_s is a projection we have $V_s(I - V_s) = 0$. If $0 \leqslant C \in \pi_p(\mathfrak{A})'$ and $C \leqslant V_s$, $C \leqslant V_q = I - V_s$ then for arbitrary x in H_p

$$\| C^{1/2}V_sx \|^2 = (CV_sx, V_sx) \leqslant ((I - V_s)V_sx, V_sx) = 0,$$

and similarly

$$\| C^{1/2}(I - V_s)x \|^2 = 0.$$

Hence $C^{1/2}V_s = C^{1/2}(I - V_s) = 0$ and so $C = 0$. We conclude that $s \perp q$.

Definition 3: Let μ be a regular non-negative Borel measure on $\mathscr{S}$ and let μ_S denote the restriction of μ to S for S a measurable set in $\mathscr{S}$, i.e. $\mu_S(T) = \mu(S \cap T)$ for T measurable in $\mathscr{S}$. If for all Borel sets S in $\mathscr{S}$ we have

$$\int_{\mathscr{S}} s\, d\mu_S \perp \int_{\mathscr{S}} s\, d\mu_{\mathscr{S}-S}$$

we say that μ is an *orthogonal measure* on $\mathscr{S}$.

THEOREM 7 (Tomita [22]). *Let μ be a non-negative regular Borel measure on $\mathscr{S}$. If μ is an orthogonal measure then the map $f \to K_f^\mu$ is a *-isomorphism of $L^\infty(\mu)$ into $\pi_p(\mathfrak{A})'$, where $p = \int_{\mathscr{S}} s\, d\mu$.*

*Conversely, if $f \to K_f^\mu(f \in L^\infty(\mu))$ is a *-homomorphism μ is orthogonal.*

Proof. By Lemma 3 we know that the map $f \to K_f^\mu (f \in L^\infty(\mu))$ is linear and *-preserving. Suppose that μ is orthogonal. If S is a measurable set in $\mathscr{S}$, S coincides with a Borel set S_0 almost everywhere and so $K_{\chi_S}^\mu = K_{\chi_{S_0}}^\mu$. ($\chi_S$ is the characteristic function of S, etc.) Now

$$\int_{\mathscr{S}} s\, d\mu_{S_0} + \int_{\mathscr{S}} s\, d\mu_{\mathscr{S}-S_0} = \int_{\mathscr{S}} s\, d\mu,$$

and by hypothesis

$$\int_{\mathscr{S}} s\, d\mu_{S_0} \perp \int_{\mathscr{S}} s\, d\mu_{\mathscr{S}-S_0}$$

and so by Lemma 6 $K_{\chi_{S_0}}^\mu$, hence $K_{\chi_S}^\mu$, is a projection.

If f and g are the characteristic functions of disjoint measurable sets, then $f \leqslant 1 - g$, hence $K_f^\mu \leqslant K_{1-g}^\mu = I - K_g^\mu$ and so $K_f^\mu K_g^\mu = 0$.

Now let f and g be the characteristic functions of arbitrary measurable sets in $\mathscr{S}$. Then as $f = fg + f(1-g)$ and $g = fg + (1-f)g$, we have

$$K_f^\mu K_g^\mu = (K_{fg}^\mu)^2 + K_{fg}^\mu K_{(1-f)g}^\mu + K_{f(1-g)}^\mu K_{fg}^\mu + K_{f(1-g)}^\mu K_{(1-f)g}^\mu = K_{fg}^\mu,$$

as K_{fg}^μ is a projection, fg and $(1-f)g$ are the characteristic functions of disjoint sets, etc. The relation $K_f^\mu K_g^\mu = K_{fg}^\mu$ extends to linear combinations of measurable characteristic functions. By Lemma 3 $\| K_f^\mu \| \leqslant \| f \|_\infty$ $(f \in L^\infty(\mu))$ and so by continuity it follows that $K_f^\mu K_g^\mu = K_{fg}^\mu$ for all f and g in $L^\infty(\mu)$.

We have

$$\| K_f^\mu I_p \|^2 = ((K_f^\mu)^* K_f^\mu I_p, I_p) = (K_{|f|^2}^\mu I_p, I_p) = \int_{\mathscr{S}} |f(s)|^2 d\mu s \ (f \in L^\infty(\mu)),$$

hence $\| f \|_\infty \neq 0$ implies $K_f^\mu \neq 0$.

This shows that the map is injective and so it is a *-isomorphism.

Conversely, suppose that $f \to K_f^\mu (f \in L^\infty(\mu))$ is a *-homomorphism. If S is a Borel set in $\mathscr{S}$,

$$K_{\chi_S}^\mu (I - K_{\chi_S}^\mu) = K_{\chi_S}^\mu K_{\chi_{\mathscr{S}-S}}^\mu = K_{\chi_S \chi_{\mathscr{S}-S}} = 0,$$

i.e. $K_{\chi_S}^\mu$ is a projection and so by Lemma 6 we have

$$\int_{\mathscr{S}} s\, d\mu_S \perp \int_{\mathscr{S}} s\, d\mu_{\mathscr{S}_S}.$$

Hence μ is orthogonal.

COROLLARY 1. *Let μ be an orthogonal probability measure on $\mathscr{S}$ with barycenter $p \in \mathscr{S}$. Then μ is simplicial in $M_p^+(\mathscr{S})$.*

Proof. By the theorem the map $f \to K_f^\mu (f \in L^\infty(\mu))$ is injective and so by Proposition 5 μ is simplicial.

COROLLARY 2. *Let $\mu \in M^+(\mathscr{S})$, with $p = \int_{\mathscr{S}} s\, d\mu$, be an orthogonal measure. Then $\mathscr{C}_\mu = \{K_f^\mu \mid f \in L^\infty(\mu)\}$ is an abelian von Neumann subalgebra of $\pi_p(\mathfrak{A})'$.*

Proof. By the theorem $\mathscr{C}_\mu$ is an abelian C^*-algebra contained in $\pi_p(\mathfrak{A})'$. As the map $f \to K_f^\mu$ is an isometry of $L^\infty(\mu)$ onto $\mathscr{C}_\mu$, being a *-isomorphism of C^*-algebras, we get from Lemma 3 that the unit ball of $\mathscr{C}_\mu$ is compact, hence closed, in the weak-operator topology. By Kaplansky's Density Theorem we conclude that $\mathscr{C}_\mu$ is a von Neumann algebra.

Observation. If μ is an orthogonal measure on $\mathscr{S}$ and $f \in L^\infty(\mu)$, $f \geqslant 0$, then it is easily shown that $dv = f\, d\mu$ is an orthogonal measure on $\mathscr{S}$.

Effros [9] gave an example to show that simplicial measures need not be orthogonal. Specifically, he showed that if $\mathfrak{A}$ is the C^*-algebra of all operators on a 2-dimensional Hilbert space H_2 and x_1, x_2 and x_3 are unit vectors in H_2, no two of which are linearly independent, then the probability measure on $\mathscr{S}$ concentrated on $\{\omega_{x_1}, \omega_{x_2}, \omega_{x_3}\}$ with $\mu(\omega_{x_i}) = \frac{1}{3}$ $(i = 1, 2, 3)$ is simplicial but not orthogonal. (Here ω_{x_i} denotes the state $a \to (ax_i, x_i)$ on $\mathfrak{A}$ $(i = 1, 2, 3)$.)

Godement [10] has proved that if p and q are positive linear functionals on $\mathfrak{A}$ then $p \perp q$ if and only if $\pi_{p+q} \cong \pi_p \oplus \pi_q$ under the linear isometry of H_{p+q} onto $H_p \oplus H_q$ which carries I_{p+q} to $I_p \oplus I_q$. This result has a natural generalization in our context which the next theorem will show. In fact, it turns out that for μ a positive regular Borel measure on $\mathscr{S}$, $\pi_{(\int \mathscr{S} s\, d\mu)}$ is naturally equivalent to $\int_{\mathscr{S}}^{\oplus} \pi_s d\mu s$ if and only if μ is orthogonal, in case $\mathfrak{A}$ is separable.

Assume $\mathfrak{A}$ is separable. Fix a norm-dense sequence $a_1, a_2, \ldots$ in $\mathfrak{A}$ with $a_1 = I$. Let μ be a positive regular Borel measure on $\mathscr{S}$ and let $p = \int_{\mathscr{S}} s\, d\mu$. Then $s \to H_s$ is a measurable field of separable Hilbert spaces over $\mathscr{S}$ with respect to the fundamental family of measurable vector fields $s \to (a_1)_s$, $s \to (a_2)_s, \ldots$; i.e. (i) the complex-valued functions $s \to ((a_i)_s, (a_j)_s)$ are measurable on $\mathscr{S}$ for all i and j, and (ii) for all $s \in \mathscr{S}$ the vectors

$$\{(a_i)_s \mid i = 1, 2, \ldots\}$$

span a dense set in H_s.

We form the Hilbert integral $H_\mu = \int_{\mathscr{S}}^{\oplus} H_s d\mu s$ and the direct integral representation $\pi_\mu = \int_{\mathscr{S}}^{\oplus} \pi_s d\mu s$ of the measurable field of representations $s \to \pi_s$ on $\mathscr{S}$. Then π_μ is a representation of $\mathfrak{A}$ on H_μ defined by $\pi_\mu(a) = \int_{\mathscr{S}}^{\oplus} \pi_s(a) d\mu s$ $(a \in \mathfrak{A})$; cf. [8, §8.1].

We denote by a_μ^i the vector in H_μ defined by the vector field $s \to (a_i)_s$ on $\mathscr{S}$. In particular, let $s \to I_s$ $(s \in \mathscr{S})$ be denoted by I_μ. Then we have

$$(\pi_\mu(a)I_\mu, I_\mu) = \int_{\mathscr{S}} (\pi_s(a)I_s, I_s) d\mu s = \int_{\mathscr{S}} \hat{a}(s) d\mu s = p(a) = (\pi_p(a)I_p, I_p)$$

for all $a \in \mathfrak{A}$.

The map $\pi_p(a)I_p \to \pi_\mu(a)I_\mu$ $(a \in \mathfrak{A})$ extends to a linear isometry J of H_p into H_μ with $J\pi_p(a) = \pi_\mu(a)J$ for $a \in \mathfrak{A}$ and $J(H_p) = \overline{\{\pi_\mu(a)I_\mu | a \in \mathfrak{A}\}}$.

Letting $E \in \pi_\mu(\mathfrak{A})'$ be the projection on $J(H_p)$ we have $J{:}\pi_p \cong (\pi_\mu)^E$, where $(\pi_\mu)^E$ is the subrepresentation of π_μ on $E(H_\mu)$. Hence $\pi_p \leqslant \pi_\mu$ and we have $J{:}\pi_p \cong \pi_\mu$ if and only if I_μ is cyclic for $\pi_\mu(\mathfrak{A})$.

The proof of the following theorem is due to Effros [**9**; Theorem 4].

THEOREM 8. *Let $\mathfrak{A}$ be separable and let μ be a positive regular Borel measure on $\mathscr{S}$ with $p = \int_{\mathscr{S}} s\, d\mu$.*

Let $\pi_\mu = \int_{\mathscr{S}}^{\oplus} \pi_s d\mu s$ be the direct integral representation of $\mathfrak{A}$ on $H_\mu = \int_{\mathscr{S}}^{\oplus} H_s d\mu s$ as described above. Then the following are equivalent:

(1) $J{:}\ _{\pi p} \cong \pi_\mu$.
(2) *I_μ is cyclic for $\pi_\mu(\mathfrak{A})$.*
(3) *μ is orthogonal.*

If one of these equivalent conditions are fulfilled π_μ is the direct integral decomposition of π_p with respect to the abelian von Neumann subalgebra $\mathscr{C}_\mu = \{K_f^\mu | f \in L^\infty(\mu)\}$ of $\pi_p(\mathfrak{A})'$.

Proof. We have already established (1) $\Leftrightarrow$ (2).

Suppose that I_μ is cyclic for $\pi_\mu(\mathfrak{A})$. We want to show that the mapping $f \to K_f^\mu$ of $L^\infty(\mu)$ into $\pi_p(\mathfrak{A})'$ is a *-isomorphism. By Theorem 7 we may then conclude that μ is orthogonal.

Let $f \to T_f$ be the natural isomorphism of $L^\infty(\mu)$ onto the diagonalizable operators on

$$H_\mu = \int_{\mathscr{S}}^{\oplus} H_s d\mu s,$$

i.e.

$$T_f = \int_{\mathscr{S}}^{\oplus} f(s) I_{H_s} d\mu s \qquad (f \in L^\infty(\mu)).$$

Since I_μ is cyclic J is a linear isometry of H_p onto H_μ. We claim that J carries K_f^μ into T_f for each $f \in L^\infty(\mu)$. In fact, for a, b in $\mathfrak{A}$ we have

$$\begin{aligned}
&((JK_f^\mu J^{-1})\pi_\mu(a)I_\mu, \pi_\mu(b)I_\mu) \\
&\quad = (K_f^\mu \pi_p(a)I_p, \pi_p(b)I_p) \\
&\quad = (K_f^\mu \pi_p(b^*a)I_p, I_p) = \int_{\mathscr{S}} f(s)\widehat{b^*a}(s) d\mu s \\
&\quad = \int_{\mathscr{S}} f(s)\,(\pi_s(b^*a)I_s, I_s) d\mu s = (T_f \pi_\mu(b^*a)I_\mu, I_\mu)
\end{aligned}$$

$$= (T_f \pi_\mu(a) I_\mu, \pi_\mu(b) I_\mu).$$

Since I_μ is cyclic for $\pi_\mu(\mathfrak{A})$ we have $T_f = JK_f^\mu J^{-1}$. Now $f \to J^{-1} T_f J = K_f^\mu$ ($f \in L^\infty(\mu)$) is a *-isomorphism and so μ is orthogonal. Note that the set of diagonalizable operators on H_μ is $J\mathscr{C}_\mu J^{-1}$ and so π_μ is the direct integral decomposition of π_p with respect to $\mathscr{C}_\mu \subset \pi_p(\mathfrak{A})'$ ([**8**; Théorème 8.3.2, p. 148]).

Conversely, assume that μ is orthogonal. Recall that a_μ^i is the vector in H_μ defined by the vector field $s \to (a_i)_s$ ($s \in \mathscr{S}$) for $i = 1, 2, \ldots$. The linear span of $\{T_f a_\mu^i | f \in L^\infty(\mu), i = 1, 2, \ldots\}$ is dense in $H_\mu = \int_{\mathscr{S}}^\oplus H_s d\mu s$ [**7**; Proposition 7, p. 150]. Thus to prove that I_μ is cyclic for $\pi_\mu(\mathfrak{A})$ it suffices to show that for any $\varepsilon > 0$, $f \in L^\infty(\mu)$ and natural number n there exists an $a \in \mathfrak{A}$ such that $\| \pi_\mu(a) I_\mu - T_f a_\mu^n \| < \varepsilon$. It is easily observed that a_μ^n equals $\pi_\mu(a_n) I_\mu$ and so we have for $a \in \mathfrak{A}$

$$\begin{aligned}
&\| \pi_\mu(a) I_\mu - T_f a_\mu^n \|^2 \\
&\quad = \| \pi_\mu(a) I_\mu - T_f \pi_\mu(a_n) I_\mu \|^2 \\
&\quad = (\pi_\mu(a) I_\mu, \pi_\mu(a) I_\mu) - (\pi_\mu(a) I_\mu, T_f \pi_\mu(a_n) I_\mu) - \\
&\qquad - (T_f \pi_\mu(a_n) I_\mu, \pi_\mu(a) I_\mu) + (T_f \pi_\mu(a_n) I_\mu, T_f \pi_\mu(a_n) I_\mu) \\
&\quad = (\pi_\mu(a^* a) I_\mu, I_\mu) - (T_{\bar{f}} \pi_\mu(a_n^* a) I_\mu, I_\mu) - (T_f \pi_\mu(a^* a_n) I_\mu, I_\mu) + \\
&\qquad + (T_{|f|^2} \pi_\mu(a_n^* a_n) I_\mu, I_\mu) \\
&\quad = \int_{\mathscr{S}} \widehat{a^* a}(s) d\mu s - \int_{\mathscr{S}} \bar{f}(s) \widehat{a_n^* a}(s) d\mu s - \int_{\mathscr{S}} f(s) \widehat{a^* a_n}(s) d\mu s + \\
&\qquad + \int_{\mathscr{S}} |f(s)|^2 \widehat{a_n^* a_n}(s) d\mu s \\
&\quad = (\pi_p(a^* a) I_p, I_p) - (K_{\bar{f}}^\mu \pi_p(a_n^* a) I_p, I_p) - (K_f^\mu \pi_p(a^* a_n) I_p, I_p) + \\
&\qquad + (K_{|f|^2}^\mu \pi_p(a_n^* a_n) I_p, I_p).
\end{aligned}$$

Since μ is orthogonal we have by Theorem 7 that

$$K_{|f|^2}^\mu = K_{\bar{f}}^\mu K_f^\mu = (K_f^\mu)^* K_f^\mu,$$

hence

$$\begin{aligned}
\| \pi_\mu(a) I_\mu - T_f a_\mu^n \|^2 &= (\pi_p(a) I_p, \pi_p)(a) I_p) - (\pi_p(a) I_p, K_f^\mu \pi_p(a_n) I_p) \\
&\quad - (K_f^\mu \pi_p(a_n) I_p, \pi_p(a) I_p) + (K_f^\mu \pi_p(a_n) I_p. K_f^\mu \pi_p(a_n) I_p) \\
&= \| \pi_p(a) I_p - K_f^\mu \pi_p(a_n) I_p \|.
\end{aligned}$$

Since I_p is cyclic for $\pi_p(\mathfrak{A})$ we make the latter small by picking an appropriate $a \in \mathfrak{A}$.

This completes the proof.

Corollary. *Let $\mathfrak{A}$ be separable and let μ be an orthogonal measure on $\mathscr{S}$. Then $A(\mathscr{S}) = \{\hat{a} \mid a \in \mathfrak{A}\}$ is dense in $L^2(\mu)$.*

Proof. It is sufficient to prove that given any $\varepsilon > 0$ and $f \in L^\infty(\mu)$ there exists an $a \in \mathfrak{A}$ such that

$$\int_{\mathscr{S}} |f(s) - \hat{a}(s)|^2 d\mu s < \varepsilon^2.$$

Let

$$T_f = \int_{\mathscr{S}}^{\oplus} f(s) I_{H_s} \, d\mu s.$$

By the theorem we can find $a \in \mathfrak{A}$ such that $\| \pi_\mu(a) I_\mu - T_f I_\mu \| < \varepsilon$. We get by the same computation as in the theorem,

$$\varepsilon^2 > \| \pi_\mu(a) I_\mu - T_f I_\mu \|^2 = \int_{\mathscr{S}} \widehat{a^* a}(s) d\mu s - \int_{\mathscr{S}} \bar{f}(s)\hat{a}(s) d\mu s$$

$$- \int_{\mathscr{S}} f(s)\widehat{a^*}(s) d\mu s + \int_{\mathscr{S}} |f(a)|^2 d\mu s \geqslant \int_{\mathscr{S}} \widehat{a^*}(s)\hat{a}(s) d\mu s - \int_{\mathscr{S}} \bar{f}(s)\hat{a}(s) d\mu s$$

$$- \int_{\mathscr{S}} f(s)\widehat{a^*(s)} d\mu s + \int_{\mathscr{S}} |f(s)|^2 d\mu s = \int_{\mathscr{S}} |f(s) - \hat{a}(s)|^2 d\mu s,$$

where we have used the Schwartz inequality to conclude that

$$\widehat{a^*}(s)\hat{a}(s) \leqslant \widehat{a^* a}(s).$$

Remark: Effros [**9**; Theorem 5] has shown that if μ is a non-negative regular Borel measure on $\mathscr{S}$ with $p = \int_{\mathscr{S}} s \, d\mu$ then π_p is *quasi-equivalent* to

$$\pi_\mu = \int_{\mathscr{S}}^{\oplus} \pi_s d\mu s,$$

i.e. every non-zero subrepresentation of $\pi_p(\pi_\mu)$ has a non-zero subrepresentation that is unitarily equivalent to a subrepresentation of $\pi_\mu(\pi_p)$.

3. ORTHOGONAL MEASURES AND ABELIAN SUBALGEBRAS

Let $\mathfrak{A}$ be a C^*-algebra, not necessarily separable, with a unit I and let $p \in \mathscr{S}$ be a state on $\mathfrak{A}$. We want to investigate the relationship between abelian von Neumann subalgebras of $\pi_p(\mathfrak{A})'$ and orthogonal measures in

$M_p^+(\mathscr{S})$, i.e. the orthogonal probability measures on $\mathscr{S}$ with barycenter p, which we will denote by $\Omega_p^+(\mathscr{S})$. By Corollary 2 of Theorem 7 we have that if $\mu \in \Omega_p^+(\mathscr{S})$ then $\mathscr{C}_\mu = \{K_f^\mu \mid f \in L^\infty(\mu)\}$ is an abelian von Neumann subalgebra of $\pi_p(\mathfrak{A})'$. We want to show that the mapping $\mu \to \mathscr{C}_\mu$ is 1–1 from $\Omega_p^+(\mathscr{S})$ and *onto* the family of all abelian von Neumann subalgebras of $\pi_p(\mathfrak{A})'$. Moreover, we shall show that the mapping is an order-isomorphism, where the subalgebras are ordered by inclusion and $\Omega_p^+(\mathscr{S})$ is ordered by either the Choquet ordering $<$, or the Bishop-deLeeuw ordering $\ll$, as described in Section 1. In fact, for measures in $\Omega_p^+(\mathscr{S})$ the two orderings coincide.

We shall need the following lemma which was communicated to us by E. G. Effros.

LEMMA 9. *Let K be a compact convex set in the locally convex Hausdorff space E. Let X be a compact Hausdorff space. Let $\theta: A_{\mathbb{C}}(K) \to C_{\mathbb{C}}(K)$ be a linear and positive map of the complex-valued continuous affine functions on K to the complex-valued continuous functions on X such that $\theta(1) = 1$. Then θ has a unique continuous extension $\bar{\theta}$ to a *-homomorphism of $C_{\mathbb{C}}(K)$ into $C_{\mathbb{C}}(X)$.*

Proof. Unicity is a consequence of the Stone–Weierstrass Theorem. For $x \in X$ the map $a \to \theta(a)(x)$ is a positive linear functional ψ_x on $A_{\mathbb{C}}(K)$ such that $\psi_x(1) = 1$. Then ψ_x is continuous with $\|\psi_x\| = 1$. By Hahn–Banach ψ_x has a continuous extension to a linear functional $\overline{\psi_x}$ on $C_{\mathbb{C}}(K)$ such that $\|\overline{\psi_x}\| = 1$. By a well-known result $\overline{\psi_x}$ is positive on $C_{\mathbb{C}}(K)$. Hence there exists by the Riesz Representation Theorem a probability measure μ_x on K such that $\overline{\psi_x}(f) = \int_K f(s)d\mu_x s$ for $f \in C_{\mathbb{C}}(K)$.

Let $\delta(x) \in K$ be the barycenter of μ_x. We want to show that $\delta: X \to K$ is continuous. Let $x_\alpha \to x$ in X. To show that $\delta(x_\alpha) \to \delta(x)$ in K it suffices to show, since K is compact, that every convergent subnet $\{\delta(x_\beta)\}$ of $\{\delta(x_\alpha)\}$ converges to $\delta(x)$. But if $\delta(x_\beta) \to y$, say, then $x_\beta \to x$ and

$$a(\delta(x_\beta)) = \int_K a(s)d\mu_{x_\beta}s = \overline{\psi_{x_\beta}}(a) = \psi_{x_\beta}(a) = \theta(a)(x_\beta) \to \theta(a)(x) = a(\delta(x)),$$

for $a \in A$ (K). Hence $a(y) = a(\delta(x))$ for all $a \in A_{\mathbb{C}}(K)$ and since the latter separates points of K we have $y = \delta(x)$. So δ is continuous. Define

$$\bar{\theta}: C_{\mathbb{C}}(K) \to C_{\mathbb{C}}(X)$$

by $\bar{\theta}(f)(x) = f(\delta(x))$ for $x \in X$ and $f \in C_{\mathbb{C}}(K)$. Clearly $\bar{\theta}$ is a *-homomorphism. For $a \in A_{\mathbb{C}}(K)$ we have $\bar{\theta}(a)(x) = a(\delta(x)) = \psi_x(a) = \theta(a)(x)\,(x \in X)$. Hence $\bar{\theta}$ is an extension of θ.

This concludes the proof.

Proposition 10. *Let $p \in \mathcal{S}$ and let $\mathcal{C}$ be an abelian von Neumann subalgebra of $\pi_p(\mathfrak{A})'$. Then there exists a $\mu \in \Omega_p^+(\mathcal{S})$ such that $\mathcal{C} = \mathcal{C}_\mu = \{K_f^\mu \mid f \in L^\infty(\mu)\}$.*

Proof. The vector $I_p \in H_p$, being cyclic for $\pi_p(\mathfrak{A})$, is separating for $\pi_p(\mathfrak{A})'$. By [7; p. 5, 7, 18] the projection e onto the subspace $[\mathcal{C}I_p]$ is in $\mathcal{C}'$ and the mapping $c \to ce$ is a *-isomorphism of $\mathcal{C}$ onto $\mathcal{C}e$. Since $\pi_p(\mathfrak{A}) \subset \mathcal{C}'$ we have $e\pi_p(\mathfrak{A})e \subset e\mathcal{C}'e = (\mathcal{C}e)'$ [7; Prop. 1, p. 16]. Besides $I_p \in e(H_p)$ is a cyclic vector for $\mathcal{C}e$ on $e(H_p)$ and so $\mathcal{C}e$ is a maximal abelian von Neumann algebra on $e(H_p)$, i.e. $(\mathcal{C}e)' = \mathcal{C}e$ [7; Corollaire 2, p. 90]. Moreover, I_p is cyclic for $\pi_p(\mathfrak{A})$ and so $e\pi_p(\mathfrak{A})e$ generates the abelian von Neumann algebra $\mathcal{C}e$ on $e(H_p)$. Let $C_{\mathbb{C}}(X)$ be the functional representation of $\mathcal{C}e$ by the *-isomorphism $\psi:\mathcal{C}e \to C_{\mathbb{C}}(X)$, where X is a compact Hausdorff space. The mapping $\hat{a} \to \psi(e\pi_p(a)e)$ is linear and positive from $A_{\mathbb{C}}(\mathcal{S})$ into $C_{\mathbb{C}}(X)$ and $1 \to 1$. By Lemma 9 it can be extended to a *-homomorphism of $C_{\mathbb{C}}(\mathcal{S})$ into $C_{\mathbb{C}}(X)$. Combining this with the *-isomorphism $\psi:\mathcal{C}e \to C_{\mathbb{C}}(X)\cdot$ and the *-isomorphism $c \to ce$ of $\mathcal{C}$ onto $\mathcal{C}e$ we get a *-homomorphism $\Gamma: C_{\mathbb{C}}(\mathcal{S}) \to \mathcal{C}$ of $C_{\mathbb{C}}(\mathcal{S})$ into $\mathcal{C}$ such that $\Gamma(\hat{a})e = e\pi_p(a)e$ $(a \in \mathfrak{A})$. Since $e\pi_p(\mathfrak{A})e$ generates the von Neumann algebra $\mathcal{C}e$ on $e(H_p)$ and $c \to ce$ is a *-isomorphism of $\mathcal{C}$ onto $\mathcal{C}e$ it follows that $\Gamma(C_{\mathbb{C}}(\mathcal{S}))$ generates the von Neumann algebra $\mathcal{C}$ on H_p [7: Corollaire 1, p. 54]. Let μ be the probability measure on $\mathcal{S}$ corresponding by the Riesz Representation Theorem to the bounded positive linear functional on $C_{\mathbb{C}}(\mathcal{S})$ defined by $f \to (\Gamma(f)I_p, I_p)$ $(f \in C_{\mathbb{C}}(\mathcal{S})$. For $f \in C_{\mathbb{C}}(\mathcal{S})$ and $a \in \mathfrak{A}$ we have

$$\int_{\mathcal{S}} f(s)\hat{a}(s)d\mu s = (\Gamma(f)\Gamma(\hat{a})I_p, I_p) = (\Gamma(f)\Gamma(\hat{a})eI_p, I_p)$$

$$= (\Gamma(f)e\pi_p(a)eI_p, I_p) = (\Gamma(f)\pi_p(a)eI_p, eI_p) = (\Gamma(f)\pi_p(a)I_p, I_p).$$

So $K_f^\mu = \Gamma(f)$ for $f \in C_{\mathbb{C}}(\mathcal{S})$ and by continuity the map $g \to K_g^\mu$ is a *-homomorphism of $L^\infty(\mu)$ into $\pi_p(\mathfrak{A})'$ (Lemma 3). By Theorem 7 μ is orthogonal and we have $\mathcal{C} = \{\Gamma(f) \mid f \in C_{\mathbb{C}}(\mathcal{S})\}'' = \{K_g^\mu \mid g \in L^\infty(\mu)\} = \mathcal{C}_\mu$. For $a \in \mathfrak{A}$ we have

$$\int_{\mathcal{S}} \hat{a}(s)d\mu s = (\Gamma(\hat{a})I_p, I_p) = (\Gamma(\hat{a})eI_p, I_p)$$

$$= (e\pi_p(a)eI_p, I_p) = (\pi_p(a)I_p, I_p) = p(a)$$

and so $\mu \in \Omega_p^+(\mathcal{S})$.

This completes the proof.

Observation. If $\mu \in \Omega_p^+(\mathcal{S})$ for some $p \in \mathcal{S}$ we have for $f \in L^\infty(\mu)$ and $a \in \mathfrak{A}$,

$$\int_{\mathcal{S}} f(s)\hat{a}(s)d\mu s = (K_f^\mu K_{\hat{a}}^\mu I_p, I_p) = (K_f^\mu \pi_p(a)I_p, I_p).$$

From this equation it is immediately verified that $K^\mu_{\hat{a}} e_\mu = e_\mu \pi_p(a) e_\mu$ for each $a \in \mathfrak{A}$, where $e_\mu \in \mathscr{C}'_\mu$ is the projection onto $[\mathscr{C}_\mu I_p]$, $\mathscr{C}_\mu$ being the abelian von Neumann subalgebra of $\pi_p(\mathfrak{A})'$ defined by $\{K^\mu_f \mid f \in L^\infty(\mu)\}$ (cf. Corollary 2 to Theorem 7). Furthermore, for $a_1, \ldots, a_n$ in $\mathfrak{A}$ we get

$$\int_{\mathscr{S}} \hat{a}_1(s) \ldots \hat{a}_n(s) d\mu s = (K^\mu_{\hat{a}_1} \ldots K^\mu_{\hat{a}_n} I_p, I_p) = (K^\mu_{\hat{a}_1} e_\mu \ldots K^\mu_{\hat{a}_n} e_\mu I_p, I_p)$$

$$= (e_\mu \pi_p(a_1) e_\mu \ldots e_\mu \pi_p(a_n) e_\mu I_p, I_p).$$

Thus if $\mu_1, \mu_2 \in \Omega_p^+(\mathscr{S})$ and $e_{\mu_1} = e_{\mu_2}$ it follows by the Stone–Weierstrass Theorem that $\mu_1 = \mu_2$.

THEOREM 11. *Let $\mathfrak{A}$ be a C^*-algebra and let $p \in \mathscr{S}$ be a state on $\mathfrak{A}$. For $\mu \in \Omega_p^+(\mathscr{S})$ let $\mathscr{C}_\mu = \{K^\mu_f \mid f \in L^\infty(\mu)\}$. Then the map $\mu \to \mathscr{C}_\mu$ is an order-isomorphism between $\Omega_p^+(\mathscr{S})$ in the Bishop–deLeeuw ordering $\preccurlyeq$ (i.e.*

$$\mu_1 \preccurlyeq \mu_2 \Leftrightarrow \int_{\mathscr{S}} \hat{a}(s)^2 d\mu_1 s \leqslant \int_{\mathscr{S}} \hat{a}(s)^2 d\mu_2 s$$

for all $a \in \mathfrak{A}^s$), and the family $\mathscr{F}$ of all abelian von Neumann subalgebras of $\pi_p(\mathfrak{A})'$, partially ordered by inclusion.

Proof. By Proposition 10 the map $\mu \to \mathscr{C}_\mu$ $(\mu \in \Omega_p^+(\mathscr{S}))$ is onto $\mathscr{F}$.

Now let $\mu \in \Omega_p^+(\mathscr{S})$ and let $a \in \mathfrak{A}$ with $a = a_1 + ia_2$ $(a_1, a_2 \in \mathfrak{A}^s)$. Let e_μ be the projection onto $[\mathscr{C}_\mu I_p]$. By the preceding observation we have

$$(*) \quad \int_{\mathscr{S}} (\hat{a}_1(s)^2 + \hat{a}_2(s)^2) d\mu s = \int_{\mathscr{S}} (\hat{a}_1(s) + i\hat{a}_2(s)(\hat{a}_1(s) - i\hat{a}_2(s)) d\mu s$$

$$= \int_{\mathscr{S}} \hat{a}(s)\widehat{a^*}(s) d\mu s = (e_\mu \pi_p(a^*) e_\mu e_\mu \pi_p(a) e_\mu I_p, I_p) = (e_\mu \pi_p(a) I_p, \pi_p(a) I_p)$$

Assume that $\mu_1, \mu_2 \in \Omega_p^+(\mathscr{S})$ and $\mu_1 \preccurlyeq \mu_2$. By (*) we have

$$(e_{\mu_1} \pi_p(a) I_p, \pi_p(a) I_p) \leqslant (e_{\mu_2} \pi_p(a) I_p, \pi_p(a) I_p)$$

for all $a \in \mathfrak{A}$. Hence $e_{\mu_1} \leqslant e_{\mu_2}$ since $[\pi_p(\mathfrak{A}) I_p] = H_p$. By [**21**] we conclude that $\mathscr{C}_{\mu_1} \subset \mathscr{C}_{\mu_2}$.

Conversely, assume $\mathscr{C}_{\mu_1} \subset \mathscr{C}_{\mu_2}$ for some $\mu_1, \mu_2 \in \Omega_p^+(\mathscr{S})$. Then $e_{\mu_1} \leqslant e_{\mu_2}$ and by (*) we have $\mu_1 \preccurlyeq \mu_2$. Finally, if $\mathscr{C}_{\mu_1} = \mathscr{C}_{\mu_2}$ for some $\mu_1, \mu_2 \in \Omega_p^+(\mathscr{S})$ we have $e_{\mu_1} = e_{\mu_2}$, and by the preceding observation we conclude that $\mu_1 = \mu_2$. So the map $\mu \to \mathscr{C}_\mu$ $(\mu \in \Omega_p^+(\mathscr{S}))$ is 1–1.

This completes the proof of the theorem.

Recall that the Choquet ordering $<$ on measures in $M_p^+(\mathscr{S})$ is defined as

$$\nu < \mu \Leftrightarrow \int_{\mathscr{S}} f(s) d\nu s \leqslant \int_{\mathscr{S}} f(s) d\mu s$$

for all convex functions f in $C_{\mathbb{R}}(\mathscr{S})$. The orderings $<$ and $\ll$ on measures in $M_p^+(\mathscr{S})$ are different as we shall show with an example later. However, for measures in $\Omega_p^+(\mathscr{S})$ ($\subset M_p^+(\mathscr{S})$) the two orderings coincide as the next theorem will show.

First we need the following lemma.

LEMMA 12. *Let μ and ν be two measures in $M_p^+(\mathscr{S})$ for some $p \in \mathscr{S}$. Then $\nu < \mu$ implies $\{K_f^\nu \mid f \in L^\infty(\nu)_1^+\} \subset \{K_g^\mu \mid g \in L^\infty(\mu)_1^+\}$. ($L^\infty(\cdot)_1^+$ denotes the positive unit ball of $L^\infty(\cdot)$.)*

If μ is simplicial then $\{K_f^\nu \mid f \in L^\infty(\nu)_1^+\} \subset \{K_g^\mu \mid g \in L^\infty(\mu)_1^+\}$ implies $\nu < \mu$.

Proof. Assume $\nu < \mu$ and let $f \in L^\infty(\nu)_1^+$. Let $d\nu_1 = f d\nu$. Then $0 \leqslant \nu_1 \leqslant \nu$ and $\nu = \nu_1 + \nu_2$ with $\nu_1, \nu_2 \geqslant 0$. By Lemma 2 we have a decomposition $\mu = \mu_1 + \mu_2$ ($\mu_1, \mu_2 \geqslant 0$) such that resultant (ν_i) = resultant (μ_i) ($i = 1, 2$). Hence we have

$$\int_{\mathscr{S}} \hat{a}(s) d\nu_1 s = \int_{\mathscr{S}} \hat{a}(s) d\mu_1 s$$

for all $a \in \mathfrak{A}$. Since $\mu_1 \leqslant \mu$ there is a $g \in L^\infty(\mu)_1^+$ by the Radon–Nikodym Theorem such that $d\mu_1 = g d\mu$. Then we have

$$\int_{\mathscr{S}} \hat{a}(s)\, g(s)\, d\mu s = \int_{\mathscr{S}} \hat{a}(s)\, d\mu_1 s = \int_{\mathscr{S}} \hat{a}(s)\, d\nu_1 s$$

$$= \int_{\mathscr{S}} \hat{a}(s)\, f(s)\, d\nu s = (K_f^\nu \pi_p(a) I_p, I_p)$$

for all $a \in \mathfrak{A}$. Hence $K_g^\mu = K_f^\nu$ and so we conclude that $\{K_f^\nu \mid f \in L^\infty(\nu)_1^+\} \subset \{K_g^\mu \mid g \in L^\infty(\mu)_1^+\}$.

Now assume μ is simplicial and $\{K_f^\nu \mid f \in L^\infty(\nu)_1^+\} \subset \{K_g^\mu \mid g \in L^\infty(\mu)_1^+\}$. Let $\nu = \sum_{i=1}^{n} \nu_i$ be a decomposition of ν into n positive components ν_i. Since $\nu_i \leqslant \nu$ there is by the Radon–Nikodym Theorem a $f_i \in L^\infty(\nu)_1^+$ such that $d\nu_i = f_i d\nu$ for each i. Clearly $\sum_{i=1}^{n} f_i = 1$. Let $g_i \in L^\infty(\mu)_1^+$ be such that $K_{fi}^\nu = K_{gi}^\mu$ and let $d\mu_i = g_i d\mu$ ($i = 1, \ldots, n$). Then we have for each i,

$$\int_{\mathscr{S}} \hat{a}(s)\, d\mu_i s = \int_{\mathscr{S}} \hat{a}(s)\, g_i(s)\, d\mu s = (K_{g_i}^\mu \pi_p(a) I_p, I_p) =$$

$$= (K_{f_i}^\nu \pi_p(a) I_p, I_p) = \int_{\mathscr{S}} \hat{a}(s)\, f_i(s)\, d\nu s = \int_{\mathscr{S}} \hat{a}(s)\, d\nu_i s \qquad (a \in \mathfrak{A}).$$

So resultant (μ_i) = resultant (ν_i) for $i = 1, \ldots, n$. Now

$$\sum_{i=1}^{n} K_{f_i}^{\nu} = K_{\sum_{i=1}^{n} f_i}^{\nu} = I_{H_p} = \sum_{i=1}^{n} K_{g_i}^{\mu} = K_{\sum_{i=1}^{n} g_i}^{\mu}$$

Since μ is simplicial we have by Proposition 5 that $\sum_{i=1}^{n} g_i = 1$. So $\mu = \sum_{i=1}^{n} \mu_i$ and by Lemma 2 we conclude that $\nu < \mu$.

Keeping the notation in Theorem 11 we can state the following theorem.

THEOREM 13. *Let μ and ν be two measures in $\Omega_p^+(\mathscr{S})$ for some $p \in \mathscr{S}$. Then the following are equivalent:*

(1) $\nu < \mu$
(2) $\nu \ll \mu$
(3) $\mathscr{C}_\nu \subset \mathscr{C}_\mu$.

Proof. The equivalence (2)⇔(3) was established in Theorem 11. Clearly (1)⇒(2) since $\hat{a}^2$ is a convex function for any $a \in \mathfrak{A}^s$. Assume (3). Then

$$\{K_f^{\nu} \mid f \in L^{\infty}(\nu)_1^+\} = (\mathscr{C}_\nu)_1^+ \subset (\mathscr{C}_\mu)_1^+ = \{K_g^{\mu} \mid g \in L^{\infty}(\mu)_1^+\}$$

(Theorem 7). By Lemma 12 we conclude that $\nu < \mu$ since μ is simplicial (Corollary 1 to Theorem 7). So (3)⇒(1), and the proof is complete.

Example: To illustrate that the two orderings $<$ and $\ll$ are different we present an example. Let K be the two-dimensional convex set shown in Figure 1, arising from the equilateral triangle ABC by moving the midpoints of each side outward by a small amount $\delta > 0$.

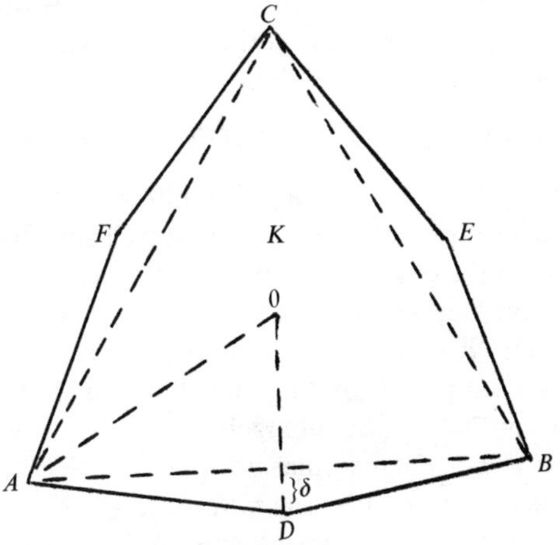

Figure 1.

Let μ be the probability measure on K concentrated on the extreme points $\{A, B, C\}$ with $\mu(A) = \mu(B) = \mu(C) = \frac{1}{3}$. Let ν be the probability measure on K concentrated on the extreme points $\{D, E, F\}$ with $\mu(D) = \mu(E) = \mu(F) = \frac{1}{3}$. Then μ and ν have the same barycenter 0. μ and ν are both maximal in the ordering $<$, being concentrated on extreme points. However, for δ sufficiently small we have $\nu \ll \mu$ and $\mu \not\ll \nu$, so that ν is not maximal in the ordering $\ll$.

[Indeed, it is a simple computation to establish that if a is a real-valued affine function on K then

$$\int_K a(s)^2 \, d\mu s = \tfrac{1}{2}\alpha^2 + a(0)^2,$$

where α is the maximum value of $|a - a(0)|$ on the circumference of the circle with center at 0 and radius equal to the distance r_{0A} between 0 and A. Likewise, we have

$$\int_K a(s)^2 \, d\nu s = \tfrac{1}{2}\beta^2 + a(0)^2,$$

where β is the maximum value of $|a - a(0)|$ on the circumference of the circle with center at 0 and radius equal to the distance r_{0D} between 0 and D. It is an easy observation that $\beta = (r_{0D}/r_{0A})\,\alpha$. By picking δ sufficiently small we get $\beta < \alpha$ and so

$$\int_K a(s)^2 \, d\nu s \leqslant \int_K a(s)^2 \, d\mu s$$

for all real-valued affine functions a on K, with strict inequality if a is not a constant.]

This example also exhibits two probability measures with the same barycenter that are comparable in the ordering $\ll$, but not in the ordering $<$.

4. MAXIMAL ORTHOGONAL MEASURES

As before $\mathfrak{A}$ is a C^*-algebra with a unit I and $\mathscr{S}$ is the state space of $\mathfrak{A}$. Let $p \in \mathscr{S}$. If $\mu \in \Omega_p^+(\mathscr{S})$ we may draw as a corollary from Theorem 13 that μ is maximal in $\Omega_p^+(\mathscr{S})$ in the ordering $<$ (or the ordering $\ll$) if and only if $\mathscr{C}_\mu$ is a maximal abelian von Neumann subalgebra of $\pi_p(\mathfrak{A})'$. However, more is true. It turns out that for $\mathfrak{A}$ separable, $\mu \in \Omega_p^+(\mathscr{S})$ is maximal in $M_p^+(\mathscr{S})$ in the ordering $<$, i.e. μ is supported by the pure states on $\mathfrak{A}$, if and only if $\mathscr{C}_\mu$ is a maximal abelian von Neumann subalgebra of $\pi_p(\mathfrak{A})'$. For $\mathfrak{A}$ non-separable we can show that if $\mathscr{C}_\mu$ is a maximal abelian von Neumann subalgebra of $\pi_p(\mathfrak{A})'$ then μ is pseudo-carried by the pure states on $\mathfrak{A}$.

For separable $\mathfrak{A}$ the result just referred to can be derived from Theorem 8, and the fact that the family of diagonalizable operators in a direct integral decomposition of a representation π of $\mathfrak{A}$ is maximal abelian in $\pi(\mathfrak{A})'$ if and only if the component representations in the direct integral are irreducible almost everywhere [**8**; Lemma 8.5.1, p. 153]. We shall, however, proceed from another angle and apply some Choquet theory to deduce the desired results. Along the way we draw as a corollary Sherman's Theorem which says that if the self-adjoint portion of $\mathfrak{A}$ is lattice-ordered in its usual ordering, then $\mathfrak{A}$ is commutative, cf. [**20**].

From now on we shall only use the Choquet ordering $<$ on measures in $M_p^+(\mathcal{S})$ for $p \in \mathcal{S}$.

LEMMA 14. *Let $p \in \mathcal{S}$. Then $\pi_p(\mathfrak{A})'$ is abelian if and only if there is a unique maximal measure in $M_p^+(\mathcal{S})$, namely the orthogonal measure corresponding to $\pi_p(\mathfrak{A})'$.*

Proof. Assume $\pi_p(\mathfrak{A})'$ is abelian and let $\mu \in \Omega_p^+(\mathcal{S})$ correspond to $\pi_p(\mathfrak{A})'$ (cf. Theorem 11). Let $\nu \in M_p^+(\mathcal{S})$. Then

$$\{K_f^\nu \mid f \in L^\infty(\nu)_1^+\} \subset (\pi_p(\mathfrak{A})')_1^+ = \{K_g^\mu \mid g \in L^\infty(\mu)_1^+\}$$

and so we have $\nu < \mu$ by Lemma 11. Thus μ is the unique maximal measure in $M_p^+(\mathcal{S})$.

Conversely, assume μ is the unique maximal measure in $M_p^+(\mathcal{S})$. Then μ is simplicial ([**1**; Proposition I.6.12]). Let $\nu \in \Omega_p^+(\mathcal{S})$. Then $\nu < \mu$ and so by Lemma 12, $(\mathcal{C}_\nu)_1^+ \subset \{K_f^\mu \mid f \in L^\infty(\mu)_1^+\}$, where $\mathcal{C}_\nu = \{K_g^\nu \mid g \in L^\infty(\nu)\}$ is the abelian von Neumann subalgebra of $\pi_p(\mathfrak{A})'$ corresponding to ν. Since this holds for all ν in $\Omega_p^+(\mathcal{S})$ we conclude by Theorem 11 that $(\pi_p(\mathfrak{A})')_1^+ \subset \{K_f^\mu \mid f \in L^\infty(\mu)_1^+\}$. Clearly $\{K_f^\mu \mid f \in L^\infty(\mu)_1^+\} \subset (\pi_p(\mathfrak{A})')_1^+$ (Lemma 3) and so $(\mu_p(\mathfrak{A})')_1^+ = \{K_f^\mu \mid f \in L^\infty(\mu)_1^+\}$. Since μ is simplicial the map $f \to K_f^\mu\ (f \in L^\infty(\mu))$ is injective (Proposition 5). Hence extreme points in $L^\infty(\mu)_1^+$ are mapped onto extreme points in $(\pi_p(\mathfrak{A})')_1^+$ by this mapping. Let B be a Borel subset of $\mathcal{S}$. Then χ_B, the characteristic function of B, is an extreme point of $L^\infty(\mu)_1^+$ and so $K_{\chi_B}^\mu$ is an extreme point of $(\pi_p(\mathfrak{A})')_1^+$. Now the extreme points of the positive unit ball of a C^*-algebra are the projections [**17**; Proposition I.6.2, p. 12]. So $K_{\chi_B}^\mu$ is a projection. It follows from the proof of Theorem 7 that μ is orthogonal. Then $\mathcal{C}_\mu = \{K_f^\mu \mid f \in L^\infty(\mu)\} = \pi_p(\mathfrak{A})'$ is abelian and the proof is complete.

Definition 4: A representation π of $\mathfrak{A}$ is said to be *multiplicity-free* if $\pi(\mathfrak{A})'$ is abelian.

LEMMA 15. *Let $\mathfrak{A}$ be a C^*-algebra such that each cyclic representation of $\mathfrak{A}$ is multiplicity-free. Then $\mathfrak{A}$ is commutative.*

Proof. Assume $\mathfrak{A}$ is not commutative. Then there exist b, c in $\mathfrak{A}$ such that

$bc \neq cb$. Since the states on $\mathfrak{A}$ separate points in $\mathfrak{A}$ there exists by the Krein–Milman Theorem a pure state p on $\mathfrak{A}$ such that $p(bc) \neq p(cb)$. Now $p(a) = (\pi_p(a) I_p, I_p)$ for all $a \in \mathfrak{A}$. The Hilbert space H_p must be of larger dimension that one since otherwise $\pi_p(bc) = \pi_p(b)\,\pi_p(c) = \pi_p(c)\,\pi_p(b) = \pi_p(bc)$, which implies $p(bc) = (\pi_p(bc) I_p, I_p) = (\pi_p(cb) I_p, I_p) = p(cb)$, contrary to assumption.

Let x_1 and x_2 be two linearly independent unit vectors in H_p and let $\pi = \pi_p \oplus \pi_p$. Then π is a cyclic representation of $\mathfrak{A}$ on $H_p \oplus H_p$ with $x_1 \oplus x_2$ a cyclic vector for $\pi(\mathfrak{A})$.

In fact, by Segal's Theorem π_p is an irreducible representation of $\mathfrak{A}$ [**18**]. Let $y_1 \oplus y_2 \in H_p \oplus H_p$. By Kadison's Transitivity Theorem [**12**] there exists an $a \in \mathfrak{A}$ such that $\pi_p(a)\,x_1 = y_1$ and $\pi_p(a)\,x_2 = y_2$. Then we have

$$\pi(a)\,(x_1 \oplus x_2) = \pi_p(a)\,x_1 \oplus \pi_p(a)\,x_2 = y_1 \oplus y_2,$$

and so $x_1 \oplus x_2$ is cyclic for $\pi(\mathfrak{A})$.

However, π is not multiplicity-free. In fact, let S be the operator on $H_p \oplus H_p$ defined by $S(z_1 \oplus z_2) = 0 \oplus z_1$ for z_1 and z_2 in H_p. It is easy to verify that $S \in \pi(\mathfrak{A})'$. On the other hand, let $P \in \pi(\mathfrak{A})'$ be the projection onto $H_p \oplus 0$. Then $PS \neq SP$ and so π is not multiplicity-free, contradicting the hypothesis.

Hence we conclude that $\mathfrak{A}$ is commutative.

Remark: Lemma 15 is also true for C^*-algebras without unit. In fact, by adding a unit the proof is reduced to the case when the C^*-algebra has a unit.

Assume now that $\mathfrak{A}^s$, the self-adjoint part of the C^*-algebra $\mathfrak{A}$, is lattice-ordered in its usual ordering. Then the continuous hermitian linear functionals on $\mathfrak{A}$ is a lattice in the dual ordering of $\mathfrak{A}^s$. In fact, we prove that a hermitian functional $h \in \mathfrak{A}^*$ has a Jordan (minimal) decomposition $h = h^+ - h^-$ $(h^+, h^- \geqslant 0)$ in just the same way as we do for a continuous linear functional on $C_{\mathbb{R}}(X)$ (X a compact Hausdorff space). Specifically,

$$h^+(a) = \sup\{h(b) | 0 \leqslant b \leqslant a\}, \qquad h^-(a) = \sup\{-h(c) | 0 \leqslant c \leqslant a\}$$

for $a \geqslant 0$ in $\mathfrak{A}$. So we have that the state space $\mathscr{S}$ of $\mathfrak{A}$ is a Choquet simplex when $\mathfrak{A}^s_n$ is lattice-ordered. Hence by Choquet's Theorem there is a unique maximal measure in $M_p^+(\mathscr{S})$ for each $p \in \mathscr{S}$ (see Section 1). We then have the following theorem as a consequence of the two preceding lemmas.

THEOREM 16 (Sherman [**20**]). *Let $\mathfrak{A}$ be a C^*-algebra such that the self-adjoint part of $\mathfrak{A}$ is lattice-ordered. Then $\mathfrak{A}$ is commutative.*

LEMMA 17. *Let $\tau: \mathfrak{A} \to \mathfrak{B}$ be a *-homomorphism of the C^*-algebra $\mathfrak{A}$ into the C^*-algebra $\mathfrak{B}$, both with unit, such that τ maps the unit of $\mathfrak{A}$ to the unit of $\mathfrak{B}$. Let $\tau^*: \mathfrak{B}^* \to \mathfrak{A}^*$ be the dual transformation of τ defined by $\tau^*(p)(a) = p(\tau(a))$; $p \in \mathfrak{B}^*$, $a \in \mathfrak{A}$. Then τ^* is a linear map continuous in the weak *-topologies and τ^**

maps the state space of $\mathfrak{B}$ *into the state space of* $\mathfrak{A}$. *If* τ *is a map* onto $\mathfrak{B}$ *then the pure states on* $\mathfrak{B}$ *are mapped into the pure states on* $\mathfrak{A}$ *by* τ^*.

Proof. We need only prove that τ^* maps pure states on $\mathfrak{B}$ into pure states on $\mathfrak{A}$ when τ is a map onto $\mathfrak{B}$ since the rest is well known.

Let p be a pure state on $\mathfrak{B}$. We have

$$(\pi_{\tau^*(p)}(a)I_{\tau^*(p)}, I_{\tau^*(p)}) = \tau^*(p)(a) = p(\tau(a)) = (\pi_p(\tau(a))\, I_p, I_p)$$

for $a \in \mathfrak{A}$. Now I_p is a cyclic vector for $\pi_p(\tau(\mathfrak{A}))$ since τ is onto $\mathfrak{B}$, and so $\pi_{\tau^*(p)} \cong \pi_p \circ \tau$. By Segal's Theorem π_p is an irreducible representation of $\mathfrak{B}$, and so we get that $\pi_{\tau^*(p)}$ is an irreducible representation of $\mathfrak{A}$ since τ is onto $\mathfrak{B}$. Again by Segal's Theorem we conclude that $\tau^*(p)$ is a pure state on $\mathfrak{A}$.

In Lemma 14 we proved that if $\pi_p(\mathfrak{A})'$ is abelian then the corresponding orthogonal measure is maximal in $M_p^+(\mathscr{S})$. The next theorem is a generalization of this. In fact, the proof is accomplished by first reducing to the multiplicity-free case and then applying Lemma 14.

THEOREM 18. *Let* $\mathfrak{A}$ *be a* C^**-algebra with state space* $\mathscr{S}$ *and let* $p \in \mathscr{S}$. *Let* $\mathscr{C}$ *be a maximal abelian von Neumann subalgebra of* $\pi_p(\mathfrak{A})'$. *Let* μ *be the orthogonal measure in* $M_p^+(\mathscr{S})$ *corresponding to* $\mathscr{C}$, *i.e.* $\mathscr{C} = \mathscr{C}_\mu = \{K_f^\mu \mid f \in L^\infty(\mu)\}$. *Then* μ *is pseudo-carried by the pure states* $\partial_e\mathscr{S}$ *on* $\mathfrak{A}$. *In particular, if* $\mathfrak{A}$ *is separable* $\partial_e\mathscr{S}$ *is a Baire set and so* $\mu(\partial_e\mathscr{S}) = 1$. *So* μ *is a maximal measure in* $M_p^+(\mathscr{S})$ *if* $\mathfrak{A}$ *is separable.*

Proof. Let $\mathfrak{B}$ be the C^*-algebra of operators on H_p generated by $\pi_p(\mathfrak{A})$ and $\mathscr{C}$, and let p_1 be the state on $\mathfrak{B}$ defined by $p_1(b) = (bI_p, I_p)$ for $b \in \mathfrak{B}$. We have $\mathfrak{B}' = \mathscr{C}$ since $\mathscr{C}$ is maximal abelian in $\pi_p(\mathfrak{A})'$. Since I_p is cyclic for $\mathfrak{B}$ we have that the representation of $\mathfrak{B}$ engendered by the state p_1 on H_{p_1} is unitarily equivalent, by the unitary extension of the mapping $\pi_{p_1}(b)\, I_{p_1} \to bI_p (b \in \mathfrak{B})$, with the identity representation of $\mathfrak{B}$ on H_p. So we may identify the two. Let $\mathscr{S}_1$ be the state space of $\mathfrak{B}$ and let μ_1 be the orthogonal measure in $M_{p_1}^+(\mathscr{S}_1)$ corresponding to $\mathscr{C} = \mathfrak{B}'$. By Lemma 14 we have that μ_1 is a maximal measure in $M_{p_1}^+(\mathscr{S}_1)$ and so is pseudo-carried by $\partial_e\mathscr{S}_1$ (see Section 1).

Let τ: $\mathfrak{A} \to \mathfrak{B}$ be the map defined by $\tau(a) = \pi_p(a)$ for $a \in \mathfrak{A}$. Then τ is a *-homomorphism of $\mathfrak{A}$ into $\mathfrak{B}$ and by Lemma 17 the dual map τ^*: $\mathfrak{B}^* \to \mathfrak{A}^*$ is a weak *-continuous linear map that maps $\mathscr{S}_1$ into $\mathscr{S}$. Clearly $\tau = i \circ \pi_p$, where π_p: $\mathfrak{A} \to \pi_p(\mathfrak{A})$ and i: $\pi_p(\mathfrak{A}) \to \mathfrak{B}$ is the injection map. Hence $\tau^* = \pi_p^* \circ i^*$. Let $\mu = \tau^*\mu_1$ be the probability measure on $\mathscr{S}$ defined by $\mu(B) = \mu_1(\tau^{*-1}(B))$ for B a Borel subset of $\mathscr{S}$. Since $\tau^{*-1}(D)$ is a compact G_δ-set in $\mathscr{S}_1$ if D is a compact G_δ-set in $\mathscr{S}$ it follows easily that $\tau^{*-1}(E)$ is a Baire set in $\mathscr{S}_1$ if E is a Baire set in $\mathscr{S}$. So to show that μ is pseudo-carried by $\partial_e\mathscr{S}$ it is sufficient to show that τ^* maps pure states on $\mathscr{S}_1$ into pure states on $\mathscr{S}$. Since π_p: $\mathfrak{A} \to \pi_p(\mathfrak{A})$ is an onto *-homomorphism it follows by Lemma 17 that π_p^* maps pure states on $\pi_p(\mathfrak{A})$ into pure states on $\mathfrak{A}$. So it is sufficient to establish that i^* maps pure states on $\mathfrak{B}$ into pure states on $\pi_p(\mathfrak{A})$. Now $i^*: \mathfrak{B}^* \to \pi_p(\mathfrak{A})^*$ is equal to the restriction map, i.e. $i^*(q) = q|\pi_p(\mathfrak{A})$ for $q \in \mathfrak{B}^*$.

Assume that q is a pure state on $\mathfrak{B}$. Then we have $q(b) = (\pi_q(b)\,I_q, I_q)$ for $b \in \mathfrak{B}$ and $\pi_q(\mathfrak{B})$ acts irreducibly on H_q. Clearly the linear span of $\{ac | a \in \pi_p(\mathfrak{A}), c \in \mathscr{C}\}$ is norm-dense in $\mathfrak{B}$. Since $\mathscr{C}$ is the center of $\mathfrak{B}$ it follows that $\pi_q(\mathscr{C})$ is in the center of $\pi_q(\mathfrak{B})$, hence must be scalar multiples of the identity operator on H_q. Combining this we get that $\pi_q(\mathfrak{A})$ acts irreducibly on H_q. Since I_q is a cyclic vector for $\pi_q(\mathfrak{A})$ on H_q and $i^*(q)(a) = q(a) = (\pi_q(a)\,I_q, I_q)$ for $a \in \pi_p(\mathfrak{A})$ it follows by Segal's Theorem that $i^*(q)$ is a pure state on $\pi_p(\mathfrak{A})$. We conclude that $\mu = \tau^*\mu_1$ is pseudo-carried by the pure states on $\mathfrak{A}$.

It remains to show that μ is the orthogonal measure in $M_p^+(\mathscr{S})$ corresponding to $\mathscr{C}$. Let e be the projection in H_p onto the closed subspace $[\mathscr{C}I_p]$. Then we have for $a_1, \ldots, a_n$ in $\mathfrak{A}$,

$$\int_{\mathscr{S}} \hat{a}_1(s) \ldots \hat{a}_n(s)\, d\mu s = \int_{\mathscr{S}_1} \hat{a}_1(\tau^*(s_1)) \ldots \hat{a}_n(\tau^*(s_1))\, d\mu_1 s_1$$

$$= \int_{\mathscr{S}} \widehat{\pi_p(a_1)}(s_1) \ldots \widehat{\pi_p(a_n)}(s_1) d\mu_1 s_1 = (e\pi_p(a_1)\, e \ldots e\pi_p(a_n)\, eI_p, I_p),$$

and so μ is the orthogonal measure in $M_p^+(\mathscr{S})$ corresponding to $\mathscr{C}$ by the observation preceding Theorem 11.

This completes the proof of the theorem.

Remark: For $\mathfrak{A}$ non-separable it is an open question whether the measure μ in the theorem is actually *maximal* in $M_p^+(\mathscr{S})$. As the example mentioned in Section 1 shows it is not sufficient for that to happen to know that μ is pseudo-carried by $\partial_e\mathscr{S}$.

5. CENTRAL MEASURE

Definition 5: Let $\mathfrak{A}$ be a C^*-algebra with unit and let $\mathscr{S}$ be the state space of $\mathfrak{A}$. Let $p \in \mathscr{S}$ and let $\mathscr{Z}$ be the center of $\pi_p(\mathfrak{A})'$. The unique orthogonal measure in $M_p^+(\mathscr{S})$ corresponding to $\mathscr{Z}$ is called the *central measure of p*. The orthogonal measures in $M_p^+(\mathscr{S})$ corresponding to von Neumann subalgebras of $\mathscr{Z}$ are called *subcentral* measures.

The central measure of $p \in \mathscr{S}$ is of particular importance since by Theorem 8 it is directly related to the central decomposition of the representation π_p engendered by p when $\mathfrak{A}$ is separable. The central decomposition of π_p, i.e. the direct integral decomposition of π_p with respect to the center $\mathscr{Z} \subset \pi_p(\mathfrak{A})'$, is a direct integral of factor representations that are pairwise disjoint almost everywhere [**8**; Lemma 8.4.1, p. 148]. This fact is reflected in what we are going to show, without resort to decomposition theory, namely, that the central measure of $p \in \mathscr{S}$ is carried by the *factorial states* $\mathfrak{F}$, i.e. $s \in \mathfrak{F}$ if and only if π_s is a factor representation. (It can be shown that $\mathfrak{F}$ is a Baire set when $\mathfrak{A}$ is separable [**17**; Corollary 3.4.5, p. 142].) The non-separable analogue of this result is that the central measure is pseudo-carried by $\mathfrak{F}$.

Wils [25] has obtained a nice geometric characterization of the central measure of $p \in \mathscr{S}$ when $\mathfrak{A}$ is separable, which we are going to prove. Specifically, it says that the central measure in $M_p^+(\mathscr{S})$ is the largest measure in the Choquet ordering $<$ that are dominated by all $<$-maximal measures in $M_p^+(\mathscr{S})$.

First we shall single out in the next proposition the subcentral measures from the orthogonal measures in $M_p^+(\mathscr{S})$ by means of a condition of *disjointness* which is stronger than orthogonality.

Recall that two representations π_1 and π_2 of $\mathfrak{A}$ are said to be *disjoint*, in symbols, $\pi_1 \between \pi_2$, if they have no unitarily equivalent non-zero subrepresentations.

Definition 6: Let p and q be two positive linear functionals on $\mathfrak{A}$. Then p and q are *disjoint*, in symbols, $p \between q$, if the representation π_p engendered by p is disjoint from the representation π_q engendered by q.

LEMMA 19. *Let p and q be two positive linear functionals on $\mathfrak{A}$ such that $p \between q$. Then $p \perp q$.*

Proof. Assume $p \not\perp q$. Then there exists a non-zero positive linear functional r on $\mathfrak{A}$ such that $r \leqslant p, r \leqslant q$. By Lemma 4 there are vectors x and y in H_p and H_q, respectively, such that $r(a) = (\pi_p(a)\,x, x) = (\pi_q(a)\,y, y)$ for all $a \in \mathfrak{A}$. Let π_1 be the subrepresentation of π_p on $[\pi_p(\mathfrak{A})x]$ and let π_2 be the subrepresentation of π_q on $[\pi_q(\mathfrak{A})y]$. Then $0 \neq \pi_r \cong \pi_1$ and $0 \neq \pi_r \cong \pi_2$, and we conclude that $p \not\between q$. Hence $p \between q$ implies $p \perp q$.

PROPOSITION 20. *Let $p \in \mathscr{S}$ be a state on $\mathfrak{A}$. Assume μ is an orthogonal measure in $M_p^+(\mathscr{S})$. Then $\mathscr{C}_\mu (= \{K_f^\mu \mid f \in L^\infty(\mu)\})$, the abelian von Neumann subalgebra of $\pi_p(\mathfrak{A})'$ corresponding to μ, is contained in the center $\mathscr{Z}$ of $\pi_p(\mathfrak{A})'$, i.e. μ is subcentral, if and only if $\int_S s\, d\mu \between \int_{\mathscr{S}-S} s\, d\mu$ for each Borel subset S of $\mathscr{S}$.*

Proof. Assume $\mathscr{C}_\mu \subset \mathscr{Z}$ and let S be a Borel subset of $\mathscr{S}$. Let $r = \int_S s\, d\mu$ and $t = \int_{\mathscr{S}-S} s\, d\mu$. Then we have

$$r(a) = \int_{\mathscr{S}} \hat{a}(s)\, \chi_S(s)\, d\mu s = (Q\pi_p(a)\, I_p, I_p) = (\pi_p(a)\, QI_p, QI_p),$$

and

$$t(a) = \int_{\mathscr{S}} \hat{a}(s)\,\chi_{\mathscr{S}-S}(s)\,d\mu s = ((I-Q)\,\pi_p(a)\,I_p, I_p) = (\pi_p(a)(I-Q)I_p, (I-Q)I_p)$$

for $a \in \mathfrak{A}$, where we have denoted by Q the projection $K^{\mu}_{\chi_S}$ in $\mathscr{Z}$. The representations π_r and π_t of $\mathfrak{A}$ engendered by r and t, respectively, are unitarily equivalent to the subrepresentations of π_p on the subspaces $[\pi_p(\mathfrak{A})\,QI_p] = Q(H_p)$ and $[\pi_p(\mathfrak{A})(I-Q)\,I_p] = (I-Q)(H_p)$, respectively. If $r \not\diamond t$ there must be a $0 \neq V \in \pi_p(\mathfrak{A})'$ such that V is a partial isometry with initial space $E \leqslant Q$ and final space $F \leqslant I - Q$. But this is impossible since Q is in the center of $\pi_p(\mathfrak{A})'$. So $r \diamond t$.

Conversely, assume that for each Borel subset S of $\mathscr{S}$ we have $\int_S s\,d\mu \diamond \int_{\mathscr{S}-S} s\,d\mu$. Let S_0 be a Borel subset of $\mathscr{S}$ and set $Q = K^{\mu}_{\chi_{S_0}}$. Then Q is a projection in $\pi_p(\mathfrak{A})'$ and letting $r = \int_S s\,d\pi$, $t = \int_{\mathscr{S}-S} s\,d\mu$ we have

$$r(a) = \int_{\mathscr{S}} \hat{a}(s)\,\chi_S(s)\,d\mu s = (Q\pi_p(a)\,I_p, I_p) = (\pi_p(a)\,QI_p, QI_p),$$

$$t(a) = \int_{\mathscr{S}} \hat{a}(s)\,\chi_{\mathscr{S}-S}(s)\,d\mu s = ((I-Q)\,\pi_p(a)\,I_p, I_p) = (\pi_p(a)(I-Q)I_p, (I-Q)I_p)$$

for $a \in \mathfrak{A}$. So π_r and π_t are unitarily equivalent to the subrepresentations of π_p on the subspaces $[\pi_p(\mathfrak{A})\,QI_p] = Q(H_p)$ and $[\pi_p(\mathfrak{A})(I-Q)\,I_p] = (I-Q)(H_p)$, respectively.

Since by assumption $\pi_r \diamond \pi_t$ we have that Q is a central projection in $\pi_p(\mathfrak{A})'$ [**8**; Proposition 5.2.1, p. 101]. So $K^{\mu}_{\chi_{S_0}} = Q$ is in $\mathscr{Z}$. Since linear combinations of $\{\chi_S \mid S \text{ Borel subset of } \mathscr{S}\}$ are norm-dense in $L^{\infty}(\mu)$ we conclude that $\mathscr{C}_{\mu} = \{K^{\mu}_f \mid f \in L^{\infty}(\mu)\}$ is contained in $\mathscr{Z}$.

This completes the proof.

In proving the following theorem we use the same device employed in the proof of Theorem 18, namely by first reducing to the multiplicity-free case.

THEOREM 21 (Sakai [**16**]). *Let $\mathscr{Z}$ be the center of $\pi_p(\mathfrak{A})'$, where $p \in \mathscr{S}$ is a state on $\mathfrak{A}$. Let μ be the central measure of p. Then μ is pseudo-carried by the factorial states $\mathfrak{F}$ of $\mathfrak{A}$. In particular, if $\mathfrak{A}$ is separable $\mathfrak{F}$ is a Baire set and so $\mu(\mathscr{S}\backslash\mathfrak{F}) = 0$.*

Proof. Let $\mathfrak{B}$ be the C^*-algebra of operators on H_p generated by $\pi_p(\mathfrak{A})$ and $\pi_p(\mathfrak{A})'$. Then

$$\mathfrak{B}' = \{\pi_p(\mathfrak{A}) \cup \pi_p(\mathfrak{A})'\}' = \pi_p(\mathfrak{A})' \cap \pi_p(\mathfrak{A})'' = \mathscr{Z}.$$

If $\mathscr{S}_1$ is the state space of $\mathfrak{B}$ and $p_1 \in \mathscr{S}_1$ is defined by $p_1(b) = (bI_p, I_p)$ for

$b \in \mathfrak{B}$, then the orthogonal measure $\mu_1 \in M^+_{p1}(\mathscr{S}_1)$ corresponding to $\mathscr{Z}$ is maximal by Lemma 14. In the same way as in the proof of Theorem 18 we transfer the measure μ_1 to a measure μ on $\mathscr{S}$, which we show to be the orthogonal measure in $M^+_p(\mathscr{S})$ corresponding to $\mathscr{Z}$; in other words, μ is the central measure of p. Specifically, $\mu = \tau^*\mu_1$ where $\tau^*: \mathfrak{B}^* \to \mathfrak{A}^*$ is the dual mapping to the *-homomorphism $\tau: \mathfrak{A} \to \mathfrak{B}$ defined by $\tau(a) = \pi_p(a)$ for $a \in \mathfrak{A}$. It remains to show that $\tau^*: \mathfrak{B}^* \to \mathfrak{A}^*$ maps pure states on $\mathfrak{B}$ into factorial states on $\mathfrak{A}$.

Let $q \in \mathscr{S}_1$ be a pure state on $\mathfrak{B}$. Then $\tau^*(q)$ is the state on $\mathfrak{A}$ defined by $\tau^*(q)(a) = q(\pi_p(a))$, $a \in \mathfrak{A}$. We want to show that $\tau^*(q)$ is a factorial state on $\mathfrak{A}$. Let π_q be the representation of $\mathfrak{B}$ engendered by q, and let $\pi_{\tau^*(q)}$ be the representation of $\mathfrak{A}$ engendered by $\tau^*(q)$. Then we have

$$(\pi_{\tau^*(q)}(a)\, I_{\tau^*(q)}, I_{\tau^*(q)}) = \tau^*(q)(a) = q(\pi_p(a)) = (\pi_q(\pi_p(a))\, I_q, I_q)$$

for $a \in \mathfrak{A}$. Hence

$$\pi_{\tau^*(q)} \cong \pi_q \circ \pi_p \big|_{e(H_q)},$$

where e is the projection onto the subspace $[\pi_q(\pi_p(\mathfrak{A}))\, I_q]$ of H_q. Clearly $e \in \pi_q(\pi_p(\mathfrak{A}))'$.

We want to show that the strong-operator closure of $\{\pi_q(\pi_p(\mathfrak{A}))\}e$ on $e(H_q)$ is a factor. For that it is sufficient to show that the strong-operator closure of $\pi_q(\pi_p(\mathfrak{A}))$ on H_q is a factor. Assume

$$c \in \pi_q(\pi_p(\mathfrak{A}))' \cap \pi_q(\pi_p(\mathfrak{A}))''.$$

Now

$$\begin{aligned}\pi_q(\pi_p(\mathfrak{A}))' \cap \pi_q(\pi_p(\mathfrak{A}))'' &= \{\pi_q(\pi_p(\mathfrak{A}))' \cup \pi_q(\pi_p(\mathfrak{A}))''\}' \\ &\subset \{\pi_q(\pi_p(\mathfrak{A})') \cup \pi_q(\pi_p(\mathfrak{A}))\}' = \pi_q(\mathfrak{B})',\end{aligned}$$

since obviously

$$\pi_q(\pi_p(\mathfrak{A})') \subset \pi_q(\pi_p(\mathfrak{A}))'.$$

Since q is a pure state on $\mathfrak{B}$ we have that $\pi_q(\mathfrak{B})'$ are the scalars on H_q. Hence c is a scalar, and we conclude that $\pi_q(\pi_p(\mathfrak{A}))''$ is a factor.

This completes the proof.

Definition 7: Let K be a compact convex set. A probability measure μ on K is said to be a *simple* measure if μ is supported by a finite number of points.

We denote by ε_p the point measure at $p \in K$. Then μ is a simple measure on K if $\mu = \sum_{i=1}^{n} \lambda_i \varepsilon_{p_i}$ for some $p_i \in K$, $\lambda_i \geqslant 0$, $\sum_{i=1}^{n} \lambda_i = 1$. In proving theorems about measures in $M_p^+(K)$ for $p \in K$, it is an often used device first to study simple measures in $M_p^+(K)$ and then pass to a general element in $M_p^+(K)$ thanks to the following lemma whose proof can be found in [**1**; Proposition I.2.3, p. 13].

LEMMA 22. *Let K be a compact convex set in a locally convex Hausdorff space. Then every measure $\mu \in M_1^+(K)$ can be approximated in the w*-topology by simple measures on K with the same barycenter as μ and dominated by μ in the Choquet ordering $<$.*

For $\mathcal{S}$ the state space of a C^*-algebra and μ an orthogonal measure in $M_p^+(\mathcal{S})$ for $p \in \mathcal{S}$, the same proof as for Lemma 22 works to establish that μ can be approximated in the w^*-topology by simple orthogonal measures in $M_p^+(\mathcal{S})$ dominated by μ [**1**; Lemma II.8.17, p. 182]. We shall, however, give another proof of this fact.

LEMMA 23. *Let μ be an orthogonal measure in $M_p^+(\mathcal{S})$ for some $p \in \mathcal{S}$. Then μ can be approximated in the w*-topology by simple orthogonal measures in $M_p^+(\mathcal{S})$ that are dominated by μ in the Choquet ordering $<$.*

Proof. Let $\mathcal{C}_\mu$ be the abelian von Neumann subalgebra of $\pi_p(\mathfrak{A})'$ corresponding to μ. Let e be the projection onto the subspace $[\mathcal{C}_\mu I_p]$ of H_p. Let $\mathfrak{J}$ be the collection of all finite-dimensional von Neumann subalgebras of $\mathcal{C}_\mu$ partially ordered by inclusion. For $\mathcal{C} \in \mathfrak{J}$ let $e_\mathcal{C}$ be the projection onto the subspace $[\mathcal{C} I_p]$. Then $\{e_\mathcal{C}\}_{\mathcal{C}\in\mathfrak{J}}$ is an increasing net of projections dominated by e, and it is easy to see that e is the least upper bound of $\{e_\mathcal{C}\}_{\mathcal{C}\in\mathfrak{J}}$. Then $e_\mathcal{C} \to e$ in the strong-operator topology [**7**; p. 321]. For $\mathcal{C} \in \mathfrak{J}$ let $\mu_\mathcal{C}$ be the orthogonal measure in $M_p^+(\mathcal{S})$ corresponding to $\mathcal{C}$. Then clearly $\mu_\mathcal{C}$ is a simple measure and also $\mu_\mathcal{C} < \mu$ by Theorem 13. By the observation preceding Theorem 11 we have for $a_1, \ldots, a_n$ in $\mathfrak{A}$,

$$\int_{\mathcal{S}} \hat{a}_1(s) \ldots \hat{a}_n(s)\, d\mu_\mathcal{C} s = (e_\mathcal{C}\pi_p(a_1)\, e_\mathcal{C} \ldots e_\mathcal{C}\pi_p(a_n)\, e_\mathcal{C} I_p, I_p) \to$$

$$\to (e\pi_p(a_1)\, e \ldots e\pi_p(a_n)\, eI_p, I_p) = \int_{\mathcal{S}} \hat{a}_1(s) \ldots \hat{a}_n(s)\, d\mu s.$$

By the Stone–Weierstrass Theorem we conclude that $\mu_\mathcal{C} \to \mu$ in the w^*-topology.

Remark: Note that if μ is a subcentral measure in the above lemma we can approximate μ by simple subcentral measures. In fact, if ν is an orthogonal measure in $M_p^+(\mathcal{S})$ then $\nu < \mu$ implies that ν is subcentral by Theorem 13.

The next lemma is crucial in proving that a subcentral measure in $M_p^+(\mathcal{S})$ is majorized by all $<$-maximal measures in $M_p^+(\mathcal{S})$.

LEMMA 24. *Let* $\mu = \sum_{i=1}^{n} \lambda_i \varepsilon_{p_i}$, $\nu = \sum_{j=1}^{m} \delta_j \varepsilon_{q_j}$ *be two simple measures in* $M_p^+(\mathscr{S})$ *for* $p \in \mathscr{S}$, μ *being subcentral. Then* μ *and* ν *have an upper bound* $\theta \in M_p^+(\mathscr{S})$ *in the Choquet ordering* $<$.

Proof. We assume that the p_i's are distinct and the λ_i's are $\neq 0$. We make the same assumption for the q_j's and δ_j's.

Let $\mathscr{C}_\mu$ be the abelian von Neumann subalgebra of $\pi_p(\mathfrak{A})'$ corresponding to μ. Then $\mathscr{C}_\mu$ is contained in the center $\mathscr{Z}$ of $\pi_p(\mathfrak{A})'$. Set $C_i = K^\mu_{\chi p_i}$ $(i = 1, \ldots, n)$ and $D_j = K^\nu_{\chi_{q_j}}$ $(j = 1, \ldots, m)$. Then $C_i \in \mathscr{Z}$ for all $i = 1, \ldots, n$ and $\sum_{i=1}^{n} C_i = \sum_{j=1}^{m} D_j = I$. We have

$$\lambda_i p_i(a) = \int_{\mathscr{S}} \chi_{p_i}(s)\, \hat{a}(s)\, d\mu s = (C_i \pi_p(a) I_p, I_p)$$

for $a \in \mathfrak{A}$ and $i = 1, \ldots, n$. Similarly,

$$\delta_j q_j(a) = (D_j \pi_p(a) I_p, I_p)$$

for $a \in \mathfrak{A}$ and $j = 1, \ldots, m$. Define the positive linear functional r'_{ij} on $\mathfrak{A}$ by

$$r'_{ij}(a) = (C_i D_j \pi_p(a) I_p, I_p),$$

for $a \in \mathfrak{A}$ and $i = 1, \ldots, n$; $j = 1, \ldots, m$. Note that r'_{ij} is positive since $D_j \geqslant 0$ and $C_i \geqslant 0$, $C_i \in \mathscr{Z}$. For each i and j such that $r'_{ij} \neq 0$ let r_{ij} be the state on $\mathfrak{A}$ defined by $r_{ij} = r'_{ij}/\gamma_{ij}$, where $\gamma_{ij} = \|r'_{ij}\|$. For $r'_{ij} = 0$, $\gamma_{ij} = 0$ and let r_{ij} be any state on $\mathfrak{A}$. Let

$$\theta = \sum_{i,j} \gamma_{ij} \varepsilon_{r_{ij}} = \sum_{i=1}^{n} \sum_{j=1}^{m} \gamma_{ij} \varepsilon_{r_{ij}}.$$

Then it is a simple observation that θ is a simple measure in $M_p^+(\mathscr{S})$. We also have that $\lambda_i p_i = \sum_{j=1}^{m} \gamma_{ij} r_{ij}$ $(i = 1, \ldots, n)$ and $\delta_j q_j = \sum_{i=1}^{n} \gamma_{ij} r_{ij}$ $(j = 1, \ldots, m)$. By a simple adaptation of Lemma 2 we conclude that $\mu < \theta$, $\nu < \theta$.

Remark: We can actually prove that the simple measure $\theta \in M_p^+(\mathscr{S})$ occurring in the above proof is the *least* upper bound of μ and ν in $M_p^+(\mathscr{S})$, cf. [**26**; Proposition 3.20].

LEMMA 25. *Let* $\mu, \nu \in M_p^+(\mathscr{S})$ *for some* $p \in \mathscr{S}$, *and assume* μ *is subcentral. Then* μ *and* ν *have an upper bound in* $M_p^+(\mathscr{S})$ *in the Choquet ordering* $<$.

Proof. By Lemma 22 and Lemma 23 there are nets of simple measures $\{\mu_\alpha\}$ and

$\{\nu_\beta\}$ in $M_p^+(\mathscr{S})$, the μ_α's being subcentral, such that $\mu_\alpha \to \mu$ and $\nu_\beta \to \nu$ in the w^*-topology and $\mu_\alpha < \mu$ for all α, $\nu_\beta < \nu$ for all β. For all α and β let $\theta_{\alpha\beta} \in M_p^+(\mathscr{S})$ be an upper bound for μ_α and ν_β (Lemma 24). Since $M_p^+(\mathscr{S})$ is w^*-compact the net $\{\theta_{\alpha\beta}\}$, the indices being directed by the product ordering, has a cluster point θ in $M_p^+(\mathscr{S})$. Let $f \in C_{\mathbb{R}}(\mathscr{S})$ be a convex function. Then

$$\int_{\mathscr{S}} f(s)\, d\mu_\alpha s \leqslant \int_{\mathscr{S}} f(s)\, d\theta_{\alpha\beta} s, \qquad \int_{\mathscr{S}} f(s)\, d\nu_\beta s \leqslant \int_{\mathscr{S}} f(s)\, d\theta_{\alpha\beta} s$$

for all α and β. Since

$$\int_{\mathscr{S}} f(s)\, d\mu_\alpha s \to \int_{\mathscr{S}} f(s)\, d\mu s, \qquad \int_{\mathscr{S}} f(s)\, d\nu_\beta s \to \int_{\mathscr{S}} f(s)\, d\nu s$$

and θ is a cluster point for $\{\theta_{\alpha\beta}\}$ it follows that

$$\int_{\mathscr{S}} f(s)\, d\mu s \leqslant \int_{\mathscr{S}} f(s)\, d\theta s \quad \text{and} \quad \int_{\mathscr{S}} f(s)\, d\nu s \leqslant \int_{\mathscr{S}} f(s)\, d\theta s.$$

So we conclude that $\mu < \theta$ and $\nu < \theta$. Hence θ is an upper bound for μ and ν.

Remark: With μ and ν as in the above lemma, we can prove the stronger result that μ and ν have a *least* upper bound in $M_p^+(\mathscr{S})$, cf. [**26**; Proposition 3.20].

THEOREM 26 (Wils [**25**]): *Let μ be a subcentral measure in $M_p^+(\mathscr{S})$ for some $p \in \mathscr{S}$. Then μ is dominated by all $<$-maximal measures in $M_p^+(\mathscr{S})$.*

Proof. Let $\nu \in M_p^+(\mathscr{S})$ be a $<$-maximal measure. Then the upper bound of μ and ν in $M_p^+(\mathscr{S})$, which exists by Lemma 25, must be ν. Hence $\mu < \nu$.

THEOREM 27 (Wils [**25**]): *Let $\mathfrak{A}$ be a separable C^*-algebra and let $p \in \mathscr{S}$. Then the central measure μ of p is the largest measure in $M_p^+(\mathscr{S})$ in the Choquet ordering $<$ that is dominated by all $<$-maximal measures in $M_p^+(\mathscr{S})$.*

Proof. By Theorem 26 we have that μ is dominated by all $<$-maximal measures in $M_p^+(\mathscr{S})$. By Theorem 18 the orthogonal measures in $M_p^+(\mathscr{S})$ corresponding to maximal abelian von Neumann subalgebras of $\pi_p(\mathfrak{A})'$ are $<$-maximal. Assume that $\nu \in M_p^+(\mathscr{S})$ is dominated by all $<$-maximal measures in $M_p^+(\mathscr{S})$. Let $\mathscr{C}$ be a maximal abelian von Neumann subalgebra of $\pi_p(\mathfrak{A})'$ corresponding to a $<$-maximal orthogonal measure $\tau \in M_p^+(\mathscr{S})$. Then $\nu < \tau$ and by Lemma 12 we have

$$\{K_f^\nu \mid f \in L^\infty(\nu)_1^+\} \subset \{K_g^\tau \mid g \in L^\infty(\tau)_1^+\} = (\mathscr{C})_1^+.$$

Since this holds for each maximal abelian von Neumann subalgebra $\mathscr{C}$ of $\pi_p(\mathfrak{A})'$ we get

$$\{K_f^\nu \mid f \in L^\infty(\nu)_1^+\} \subset (\mathscr{Z})_1^+ = \{K_g^\mu \mid g \in L^\infty(\mu)_1^+\},$$

and so $\nu < \mu$ by Lemma 12. ($\mathscr{Z}$ is the center of $\pi_p(\mathfrak{A})'$ and is equal to the intersection of all maximal abelian von Neumann subalgebras of $\pi_p(\mathfrak{A})'$.)

This concludes the proof.

6. HISTORICAL NOTES

The first one to consider measures on the state space of a C^*-algebra in connection with decomposition theory was Segal in 1951 [**19**]. For p a state on the separable C^*-algebra $\mathfrak{A}$ and $\mathscr{C}$ an abelian von Neumann subalgebra of $\pi_p(\mathfrak{A})'$, he proved the existence of a positive measure on the state space of $\mathfrak{A}$ which gives rise in the way we have described in Theorem 8 to the direct integral decomposition of π_p relative to $\mathscr{C}$. In 1956 Tomita [**22**] introduced the concept of orthogonal measure (cf. Definition 3) and he established Theorem 7. He discussed the correspondence between orthogonal measures and decomposition theory as presented in Theorem 8. This was further explored in his 1959 paper [**23**] where the proved the 1-1 correspondence between orthogonal measures in $M_p^+(\mathscr{S})$ and abelian von Neumann subalgebras of $\pi_p(\mathfrak{A})'$.

Effros [**9**] in his 1961 thesis introduced the concept of a simplicial measure and subsequently showed that an orthogonal measure is simplicial. In 1965 Sakai [**16**] defined the central measure $\mu \in M_p^+(\mathscr{S})$ of $p \in \mathscr{S}$ by the formula

$$p(za) = \int_{\mathscr{S}} \Phi(z)(s)\, \hat{a}(s)\, d\mu s,$$

where $a \in \mathfrak{A}$ and z is in $\mathscr{Z}$, the center of the enveloping von Neumann algebra $\mathfrak{A}^{**}$ of $\mathfrak{A}$, and Φ is a weak*-continuous homomorphism of $\mathscr{Z}$ onto $L^\infty(\mu)$. (Recall that any state on $\mathfrak{A}$ has a unique weak*-continuous extension to a state on $\mathfrak{A}^{**}$.) It is a simple matter to show that Sakai's definition of central measure coincides with ours. By applying von Neumann decomposition theory for separable C^*-algebras, Sakai proved for a separable C^*-algebra $\mathfrak{A}$ the existence and uniqueness of the central measure, that the set of factorial states $\mathfrak{F}$ is measurable for all regular Borel measures on $\mathscr{S}$ and that the central measure is carried by $\mathfrak{F}$.

In 1968 Wils [**24**] extended Sakai's result to the non-separable case. His proof was quite new in that he invoked Choquet theory in the form of the Bishop-deLeeuw theorem, which states that a maximal measure in the ordering $\ll$ is pseudo-carried by the extreme points. In 1969 Wils [**25**] announced Theorem 27. He also extended the notion of a central measure to an arbitrary compact convex set in a locally convex Hausdorff space. For a comprehensive exposition of that part of the theory we refer to [**26**] and [**1**].

Ruelle [**15**] in 1970 established the (1)⇔(3) part of Theorem 13.

An aspect of the theory that has not been treated in this paper is the theory of G-invariant states, ergodic states and the ergodic decomposition. We refer to [**14**], [**15**], [**17**] and [**11**] for information and results on this subject.

References

1. Alfsen, E. M., *Compact Convex Sets and Boundary Integrals*. Springer-Verlag, Berlin, 1971.
2. Bishop, E. and deLeeuw, K., The representation of linear functionals by measures on sets of extreme points. *Ann. Inst. Fourier* (Grenoble), **9** (1959), 305–331.
3. Cartier, P., Fell, J. M. G. and Meyer, P. A., Comparison des mesures portées par un ensemble convexe compact. *Bull. Soc. Math. France,* **92** (1964), 435–445.
4. Choquet, G., Existence et unicité des représentations intégrales au moyen des points extrémaux dans les cônes convexes. *Seminaire Bourbaki,* 139, Dec. 1956, 15 pp.
5. Choquet, G., Le théorème de représentation intégrale dans les ensemble convexes compacts. *Ann. Inst. Fourier* (Grenoble), **10** (1960), 333–344.
6. Choquet, G. and Meyer, P. A., Existence et unicité des représentation intégrales dans les convexes compacts quelconque. *Ann. Inst. Fourier* (Grenoble), **13** (1963), 139–154.
7. Dixmier, J., *Les algèbres d'opérateurs dans l'espace Hilbertien.* Gauthier-Villars, Paris, 2nd edition, 1969.
8. Dixmier, J., *Les C*-algèbres et leurs représentations.* Gauthier-Villars, Paris, 2nd edition, 1969.
9. Effros, E. G., *On the Representations of C*-algebras.* Thesis, Harvard University, Cambridge, Mass., 1961.
10. Godement, R., Les fonctions de type positif et la théorie des groupes. *Trans. Amer. Math. Soc.* **63** (1948), 1–84.
11. Guichardet, A. and Kastler, D., Désintégration des états quasi-invariants des C^*-algèbres. *J. Math. Pures et Appl.,* **49** (1970), 349–380.
12. Kadison, R. V., Irreducible operator algebras. *Proc. Nat. Acad. Sci.* (U.S.A.) **43** (1957), 273–276.
13. Mokobodzki, G., Balayage défini par un cône convexe de fonctions numériques sur un espace compact. *C.R. Acad. Sci. Paris,* **254** (1962), 803–805.
14. Phelps, R. R., *Lectures on Choquet's Theorem.* Math. Studies, Van Nostrand, Princeton, 1966.
15. Ruelle, D., Integral representation of states on a C^*-algebra. *J. of Funct. Analysis,* **6** (1970) 116–151.
16. Sakai, S., On the central decomposition for positive functionals on C^*-algebras. *Trans. Amer. Math. Soc.* **118** (1965), 406–419.
17. Sakai, S., *C*-algebras and W*-algebras*. Springer-Verlag, Berlin, 1971.
18. Segal, I. E., Irreducible representations of operator algebras. *Bull. Amer. Math. Soc.* **53** (1947), 73–88
19. Segal, I. E., Decomposition of operator algebras I. *Memoirs Amer. Math. Soc.* **9** (1951), 1–67.

20. Sherman, S., Order in operator algebras. *Amer. J. Math.* **73** (1951), 227–232.
21. Skau, C. F., *Commutative Projections and Abelian Subalgebras,* Thesis, University of Pennsylvania, Philadelphia, Pa., 1973.
22. Tomita, M., Harmonic analysis on locally compact groups. *Math. J. Okayama Univ.* **5** (1956), 133–193.
23. Tomita, M., Spectral theory of operator algebras I. *Math. J. Okayama Univ.* **9** (1959), 63–98.
24. Wils, W., Désintégration centrale des formes positives sur les C^*-algèbres. *C.R. Acad. Sci. Paris,* **267** (1968), 810–812.
25. Wils, W., Désintégration centrale dans une partie convexe compacte d'un espace localement convexe. *C.R. Acad. Sci. Paris,* **269** (1969), 702–704.
26. Wils, W., The ideal center of partially ordered vector spaces. *Acta Math.,* **127** (1971), 41–77.

INDEX

Abelian semigroup, 149
Absolutely p-convex set, 220
Action, irreducible, 107
Adams, J. F., 174, 185
Adjoint operator, 102
Admissible contour, 70
  map, 98
Affine function, 273
A-holomorphic function, 207
Akkar, M., 207, 249
Alfsen, E. M., 272, 275, 276, 291, 298, 302
Algebra, *b*-, 199
  Banach, 63
  C^*-, 102, 263, 272
  continuous inverse, 215
  group, 177
  locally multiplicatively convex, 200
  measure, 176
  normed, 63
  operator, 102
  pseudo-Banach, 244
  twisted group, 264
  uniform, 265
  von Neumann, 108
Algebra boundedness, 199
Allan, G. R., 200, 244, 249
Allan, G. R., H. G. Dales and J. P. McClure, 205, 206, 244, 250
Allan boundedness, 206, 209, 212, 244, 247
Almost periodic function, 19
  map, 26
Amenable Banach algebra, 95
  group, 94
A-module, 172
Analytic cocycle, 31, 58
  function (on a group), 3
  measure, 13
Analyticity, 28
Aoki, T., 198
Approximate identity, 80
  point spectrum, 269
  unit, 80
Archimedean dual, 17
Arens, R., 17, 36, 57, 60, 118, 143, 145, 185, 245, 256, 257
Arens, R. and E. A. Michael, 200
Arens-Calderón trick, 193, 208, 216
Arens-Royden theorem, 118, 119, 143, 175, 181
Arens theorem, 145, 245
Arveson, W., 19, 55, 60
Asplund, E. (with I. Namioka), 91, 100
Atiyah, M. F., 119, 120, 163, 173, 174, 185
Automorphism, 87
Auxiliary norm, 91

Bachelis, G. F., 255
Bagchi, S. C., 57, 60
b-algebra, 199
Banach, S., 92, 99
Banach algebra, 63
  amenable, 95
  non-commutative, 74
  real, 63
  unital, 64, 139
Banach $\mathfrak{A}$-module, left, 75
Barycenter, 273
Barnes, B. A., 255
Base point, 122
Bass, H., 173, 175, 185
Besicovitch function, 20, 38
  norm, 20
Beurling, A., 4, 5, 60
Beurling's first theorem, 5
Beurling's main theorem, 9, 13
Beurling's second theorem, 5
b-ideal, 223
Bishop, E. and K. deLeeuw, 273, 302
Blaschke cocycle, 44, 54, 55, 56, 58

Bochner, S., 17, 60
Bohr, H., 20
Bollobas, B., 256, 257
Borel flow, 57
  version, 31, 38, 46, 47
Bornifying operator, 207
Bott periodicity theorem, 163, 169
Bound (of operator), 102
Boundary, Shilov, 261, 268
Bounded structure, 198
    equicontinuous, 202
    standard, 198
  subset (of vector space), 197, 198
Boundedness, 198
  algebra, 199
  convex, 198
  equibornifying, 207
  equiregular, 211
  idempotent, 200
  Mackey, 198
  pseudo-convex, 220
  separated, 198
  vector space, 198
Browder, A., 265, 267
b-space, 199
b-subspace, 223
deLeeuw, K. and I. Glicksberg, 17, 18,
C^*-algebra, 102, 263, 272
Cancellative semigroup, 149
Carlson, C. G. R., 38, 46, 57, 58, 59, 60
Carried (measure), 273
Cartier, P., J. M. G. Fell and P. A. Meyer, 276, 302
Casson, A., 174
Category, 121
Cauchy sequence, 198
Cayley transform, 112
Čech cochain, 132
  cohomology, 131
  cohomology group, 134
Central measure, 294
Chern character, 174
Choi, M. D. and C. Davis, 269, 271
Choquet, G., 273, 274, 275, 292, 302
Choquet, G. and P. A. Meyer, 274, 275, 302
Choquet simplex, 274, 275
  theory, 272, 273
Clopen set, 110
Cnop, I., 248, 250, 258, 259
Coboundary, 23, 49, 58, 133
  $\bar{\partial}$-, 243
Cochain, Čech, 132
Cocycle, 23, 49, 58, 133
  analytic, 31, 58
  Blaschke, 44, 56, 58
  continuous, 50
  $\bar{\partial}$-, 243
  singular, 44, 58
  trivial, 23, 49, 58
Cocycles, structure of, 37
Coefficient group, 125
Cofunctor, 121
  continuous, 122
Cohen, P. J., 179
Cohen's idempotent theorem, 179
Cohomologous, 23
Cohomology, 84
  axioms for, 120
  Čech, 131
  $\bar{\partial}$-, 229, 243
  Helemskiĭ's, 96
  relative, 98
Cohomology group, Čech, 134
  Hochschild, 89
Cohomology theory, 123, 126
  reduced generalised, 124
Commutant, 111
Commutation relation (Weyl), 22
Commutative diagram, 121
Compact pairs, 125
Comparison Theorem, 114
  theory of projections, 114
Completant bounded set, 199
  structure, 199
Completely-additive state, 116
Cone, 103, 129
Conjugate-analytic function, 45
Continuous cocycle, 50
  cofunctor, 124
  index, 180
  inverse algebra, 215
Contour, admissible, 70
Contractible (pointed space), 127
Contravariant functor, 121
Convex boundedness, 198
Convolution, 176
  operator, 180
Craw, I. G., 260, 265
Cross-section, 18

Cyclic representation, 103
vector, 103

Dales, H. G. (with G. R. Allan and J. P. McClure), 205, 206
Davie, A. M., 267
Davis, C., (with M. D. Choi), 269, 271
Deformation retract, 128
de Leeuw, K. and I. Glicksberg, 17, 18, 58, 60
deLeeuw, K. (with E. Bishop), 273, 302
$\bar{\partial}$-coboundary, 243
$\bar{\partial}$-cocycle, 243
$\bar{\partial}$-cohomology, 229, 243
Derivation, 87
Diagram, commutative, 121
Dimension axiom, 125
Direct integral, 113,
sum, 142
Disconnected, extremely, 110
Discrete index, 180
Disjoint functionals, 295
representations, 295
Dixmier, J., 87, 99, 101, 278, 281, 283, 286, 291, 295, 296, 298, 302
Double Commutant Theorem, 111
Douglas, R. G. and J. L. Taylor, 173, 180, 181, 185
Dual, archimedean, 17
Dual Banach module, 91
Dunford, N., 189
Dyer, E., 120, 131, 174, 184, 185

Effros, E. G., 282, 284, 285, 301, 302
Eigenfunction, 59
Eilenberg, S. and N. Steenrod, 131, 134, 174, 185
Eilenberg–Steenrod axioms, 126
Embedding, regular, 274
Engendered representation, 104
Equibornifying boundedness, 207
set, 207
Equicontinuous bounded structure, 202
Equiregular boundedness, 211, 212
set, 211
Equivalent differential operators, 43
extensions, 86
projections, 113
Ergodic theory, 59
Etter, D. O., Jr., 218, 250
Etter integrable, 218, 219
Exact sequence, 97, 124
Extension (of Banach algebra), 85
trivial, 146
Extensions, equivalent, 86
Extremely disconnected, 110

Factor, 109
of type I_n, 112
of type II, II_1, II_∞, III, 114
Factorial state, 295, 301
Faithful representation, 103
Fantappiè, L., 189
Fell, J. M. G. (with P. Cartier and P. A. Meyer), 276, 302
Ferrier, J.-P., 248, 249, 250, 258, 259
Figure 1, 289
Final projection, 113
Finite projection, 114
Fixed point, 90
Fixed point theorem, Ryll–Nardzewski, 91
Flow, Borel, 57
weakly mixing, 59
Foiaş, C. (with B. Sz.-Nagy), 57, 61
Forelli, F., 17, 50, 57, 58, 59, 60
Forster, O., 119, 154, 174, 175, 178, 186
Free module, 173
Function, affine, 273
A-holomorphic, 207
almost periodic, 19
analytic (on a group), 3
inner, 6
Lipschitz, 238
Lorch analytic, 217
outer, 6, 12
pointwise analytic, 30
weakly analytic, 30
Function algebra, 65
Functional calculus, holomorphic, 189
one-variable, 70
Functions, disjoint, 295
orthogonal, 279
Functor, 121
contravariant, 121
Furstenburg, H., 39, 60, 61

Gamelin, T. W., 2, 10, 12, 33, 39, 47, 49, 50, 58, 61, 120, 141, 144, 181, 186
Gamelin's representation theorem, 48

Gauge, 199
Gelfand, I. M., 189
Gelfand and Neumark, theorem of, 102
Gelfand topology, 68
  transform, 68
Generalised F. and M. Riesz theorem, 16, 17, 58
Gleason, A. M., 260, 265, 267
Glicksberg, I., 17, 58, 61
Glicksberg, I. (with K. deLeeuw), 17, 18, 58, 60
GNS construction, 103, 262, 272, 276
Godement, R., 281, 302
Graham, C. C., 260
Gramsch, B., 218, 250
Grauert, H., 145, 154
Greenleaf, F. P., 95, 99
Group, amenable, 94
  locally compact abelian, 176
Group algebra, 177
  twisted, 264
Guichardet, A. and D. Kastler, 302
Gunning, R. C. and H. Rossi, 190, 191, 250

Haar measure, 176
Halmos, P. R., 5, 19, 61
Harte, R. E., 261, 268, 269, 271
Heaviside calculus, 221
Helemskiĭ, A. Ja., 96, 99
Helson, H., 1, 5, 24, 39, 42, 46, 56, 57, 61
Helson, H. and J.-P. Kahane, 23, 61
Helson, H. and D. Lowdenslager, 14, 35, 61
Hochschild cohomology groups, 89
Hoffman, K., 2, 29, 44, 61
Hofmann, K. H., 178, 185, 186
Hogbe-Nlend, H., 200, 207, 245, 250
Holomorphic functional calculus, 189
Hull, 73
Hull-kernel topology, 74
Husemoller, D., 120, 186

Idempotent, 154
  boundedness, 200
Identity, adjunction of, 64
  approximate, 80
  resolution of, 110
Identity element (of Banach algebra), 64
Implementation, 116
Index, 181
  continuous, 180
  discrete, 180
Inequality, Jensen's, 9, 27
Infinite projection, 114
Initial projection, 113
Inner derivation, 87
  function, 6
Integral, direct, 113
Intersection (of projections), 109
Invariant mean, 94
  subspace, 4, 12, 58
Inverse (of element in Banach algebra), 65
Irreducible action, 107
  left Banach 𝔄-module, 75
  representation, 107
Isometry, partial, 113
Isomorphism, 121

Jackson, Judy, vii
Jensen's inequality, 9, 27
Johnson, B. E., 63, 84, 96, 99, 100, 188, 189, 191, 193, 246
Joint spectrum, 225, 268
Jordan decomposition (of functionals), 292
  multiplication, 215

Kadison, R. V., 101, 292, 302
Kadison's Transitivity Theorem, 292
Kahane, J.-P. (with H. Helson), 23, 61
Kaplansky Density Theorem, 111, 281
Karoubi, M., 186
Kastler, D. (with A. Guichardet), 302
Kato, T., 25, 61, 262
Katznelson, Y., 3, 61
Katznelson, Y. and W. Rudin, 255
Kernel, 73
  left, 104
Kolmogoroff and Krein, theorem of, 14
König, H., 57
Köthe, G., 198, 251
Krein, see Kolmogoroff and Krein, theorem of,
Krein-Smŭlian Theorem, 278
K-theory, 169
  for *L*(*G*), 184
  for *M*(*G*), 183
Künneth formula, 185

Left approximate point spectrum, 261, 269
  Banach 𝔄-module, 75
  kernel (of a state), 104
  spectrum, 261, 268
Length (of vector), 102
Lewittes, J., 265, 267
Linear loop, 164
Lipschitz function, 238
Locally bounded, 198
  compact abelian group, 176
    space, 122
  multiplicatively convex algebra, 200
  p-convex, 203
  pseudo-convex, 203
Logarithm, 142
Long Exact Sequence of Homology Theorem, 97
Loop, 164
  linear, 164
  polynomial, 166
Lorch, E. R., 189, 217, 250
Lorch analytic function, 217
Lowdenslager, D., 20, 38
Lowdenslager, D., (with H. Helson), 14, 35, 61
Ludvik, Peter, vii, 187, 188
Lumer, G., 57

Mackey, G. W., 57, 61
Mackey boundedness, 198
Mackey's imprimitivity theorem, 57
Mandrekar, V. and M. Nadkarni, 17, 58, 61
Mandrekar, V., M. Nadkarni and D. Patil, 57, 61
Map, proper, 122
  admissible, 98
Maximal ideal, 67
    space, 68
  orthogonal measure, 290
Mazur, S. and W. Orlicz, 218, 250
McClure, J. P. (with G. R. Allan and H. G. Dales), 205, 206
McIntosh, Alan, 262
Mean, 94
Measure, analytic, 13
  central, 294
  Haar, 176
  orthogonal, 263, 272, 276, 279
Measure—*continued*
  simple, 297
  simplicial, 275
  subcentral, 294
Measure algebra, 176
Meyer, P. A. (with P. Cartier and J. M. G. Fell), 276, 302
Meyer, P. A. (with G. Choquet), 274, 275, 302
Michael, E. A., 200
Michael, E. A. (with R. Arens), 200
Milaszewicz, J. P., 53
Minimal projection, 112
Mitjagin, B., S. Rolewicz and W. Żelazko, 216, 251
Modular left ideal, 76
  maximal ideal, 69
Module, 172
  free, 173
  projective, 96, 173
  relatively projective, 98
Module homomorphism, 97
Mokobodzki, G., 274, 302
Moore, C. C., 41
Moran, W., 19
Morphism (of bounded structures), 198
Muhly, P. S., 3, 32, 57, 58, 61
Multiplication operator, 106
Multiplicative linear functional, 67
Multiplicity-free representation, 291
Multiplier, 182

Nadkarni, M. 57, 61
Nadkarni, M. (with V. Mandrekar), 17, 58, 61
Nadkarni, M. (with V. Mandrekar and D. Patil), 57, 61
Namioka, I. and E. Asplund, 91, 100
Negative (group element), 3
Nerve, 132
n-fold reduced suspension, 123
Nöel, G., 223, 242, 250
Non-commutative Banach algebra, 74
Norm (in Hilbert space), 102
  (of operator), 102
  auxiliary, 91
  tensor, 87
  uniqueness of, 78
Norm topology, 102
Normal state, 116

Normed algebra, 63
Novadvorskii, M. E., 119, 154, 172, 186

Operator, 102
  bornifying, 207
  convolution, 180
  multiplication, 106
  positive, 103
  Wiener-Hopf, 180
Operator algebra, 102
Order structure (of C^*-algebra), 103
Ordered (group), 3
Orlicz, W. (with S. Mazur), 218, 250
Orthogonal functionals, 279
  measure, 263, 272, 276, 279
Outer function, 6, 12

Parameterized family of maps, 127
Parrott, S., 39, 61
Parry, W., 60, 61
Partial isometry, 113
Partition of unity, 140
Patil, D. (with V. Mandrekar and M. Nadkarni), 57, 61
Periodicity theorem (Bott), 163, 169
Phelps, R. R., 302
Picard group, 175
  of $M(G)$, 185
Pietsch, A., 222, 251
p-norm, 203
Pointed compact space, 122
Pointwise analytic function, 30
Polyhedron, polynomial, 192
Polynomial loop, 166
  polyhedron, 192
Polynomially convex hull, 192
Positive (group element), 3
  operator, 103
Product, smash, 123
Projection, final, 113
  finite, 114
  infinite, 114
  initial, 113
  minimal, 112
  range, 109
Projections, equivalent, 113
Projective module, 96, 173
  resolution, 97
Proper map, 122
Przeworska-Rolewicz, D., and S. Rolewicz, 218, 250
p-semi-norm, 203
Pseudo-Banach algebra, 244
Pseudo-carried (measure), 273
Pseudo-convex boundedness, 220
p-simplex, 132
Pure state, 105

q-algebra, 224
q-space, 223
Quasi-equivalence (of representations), 284

Radical of Banach algebra, 78
  of commutative Banach algebra, 69
Radius, spectral, 193
Range projection, 109
Rapidly decreasing sequence, 203
  set, 204
Real Banach algebra, 63
Reduced generalised cohomology theory, 124
  suspension, 123
Reduction of Dimension Theorem, 89, 90, 93, 95
Refinement homomorphism, 133
Reflexive (Banach module), 91
Regular Banach algebra, 74
  element of b-algebra, 206
    of Banach algebra, 65
  embedding, 274
Relative cohomology, 98
  p-cochains, 132
Relatively projective module, 98
Representation by measure, 273
  cyclic, 103
  faithful, 103
  irreducible, 103
  multiplicity-free, 291
  of Banach algebra, 75
  of C^*-algebra, 103
  of state, 105
Representations, disjoint, 295
Resolution, projective, 97
  of the identity, 110
Resultant, 273
Retract, deformation, 128
Riesz, F. and M. theorem, 13, 17
  generalised, 16, 17, 58

Right approximate point spectrum, 269
  spectrum, 268
Rolewicz, S., 198
Rolewicz, S. (with B. Mitjagin and W. Żelazko), 216, 251
Rolewicz, S. (with D. Przeworska-Rolewicz), 218, 250
Rossi, H (with R. C. Gunning), 190, 191, 250
Royden, H. L., 118, 143, 186
Rudin, W., 2, 3, 17, 61, 176, 177, 179, 186
Rudin, W. (with Y. Katznelson), 255
Ruelle, D., 302
Runge's theorem, 192, 246, 249
Ryll-Nardzewski fixed point theorem, 91

Sakai, S., 101, 276, 291, 295, 296, 301, 302
Schatten, R., 222, 251
Sebastião e Silva, J., 248, 250
Segal, I. E., 103, 292, 301, 302
Segal's Theorem, 292, 293, 294
Self-adjoint operator, 102
Semigroup, 149
  abelian, 149
  cancellative, 149
Semi-simple Banach algebra, 78
  commutative Banach algebra, 70
Separated boundedness, 198
Separation Theorem, 108
Sequence, exact, 97, 124
Set, absolutely p-convex, 220
  equiregular, 211
Sherman, S., 291, 292, 303
Shilov, G. E., 118, 141, 175, 186, 256, 257
Shilov boundary, 261, 268
Shilov idempotent theorem, 118, 141, 175
Silov: see Shilov
Simple measure, 297
Simplex, Choquet, 274, 275
Simplicial measure, 275
Simply invariant subspace, 21, 58
Simpson's integration formula, 236
Singular cocycle, 44, 58
  (element of Banach algebra), 65
Skau, C. F., 263, 272, 287, 303
Smash product, 123
Smith, Harvey, 264
Smŭlian: see Krein–Smŭlian Theorem
Space, b-, 199
Spanier, E. H., 131, 134, 186
Spectral decomposition, 140
  mapping theorem, 268
  radius, 193
  resolution, 20
  Theorem, 110
Spectrum, approximate point, 269
  joint, 225, 268
  left, 261, 268
  left approximate point, 261, 269
  of element in Banach algebra, 65, 105
  of n-tuple, 234, 237
  right, 268
  right approximate point, 261, 269
Srinivasan, T. P., 5, 12, 57, 61
Star-invariant subspace, 56
State, 103
  completely additive, 116
  factorial, 295, 301
  normal, 116
  pure, 105
  vector, 105
State space, 263, 272, 276
Steenrod, N. (with S. Eilenberg), 131, 134, 174, 185
Stone's theorem, 22, 274
Strict version, 47
Strong-operator topology, 107
Structure, bounded, 198
Submodule, 173
Subcentral measure, 294
Subspace, invariant, 4, 12, 58
  simply invariant, 21, 58
  star-invariant, 56
  Wiener, 21, 58
Support (of $p + 1$-tuple of integers), 133
  (of state), 117
Suspension, reduced 123
Synthesis, problem of, 73
Szegö, G., 6, 61
Szegö's theorem, 6, 35
Sz.-Nagy, B., 57, 61
Sz.-Nagy, B., and C. Foias, 57, 61

Taylor, J. L., 118, 119, 144, 175, 177, 178, 179, 180, 181, 183, 186, 245, 249, 250
Taylor, J. L. (with R. G. Douglas), 180, 181, 185
Tensor norm, 87
Tits, J. and L. Waelbroeck, 248, 250

Tomita, M., 279, 301, 303
Topology, Gelfand, 68
  norm, 102
  strong-operator, 107
  weak-operator, 107
Transform, Gelfand, 68
Translation invariant mean, 94
Trivial cocycle, 23, 49, 58
  extension, 146
Turpin, P., 215, 216, 250
Turpin, P., and L. Waelbroeck, 219, 221, 250
Twisted group algebra, 264
Type I_n factor, 112
Type II, II_1, II_∞ factor, 114
Type III factor, 114

Uniform algebra, 65, 265
Uniformly convex, 92
Union (of projections), 109
Uniqueness of norm, 78
Unit, approximate, 80
Unital Banach algebra, 64, 139
  homomorphism, 139
Unitary Implementation Theorem, 116
Universal group (of semigroup), 149
Urysohn's theorem, 239
Uss, P., 207, 250

Vector, cyclic, 103
Vector state, 105
Version, 5
  Borel, 31, 38, 46, 47
Version—*continued*
  strict, 47
von Neumann algebra, 108

Waelbroeck, L., 187, 201, 209, 216, 218, 219, 239, 244, 245, 247, 248, 251, 258, 259
Waelbroeck, L. (with J. Tits), 248, 250
Waelbroeck, L. (with P. Turpin), 219, 221, 250
Wagner, J., 53, 62
Wang, J.-K., 57
Weak exactness, 148, 158
Weak-operator topology, 107
Weakly analytic function, 30
  mixing flow, 59
Weight at infinity, 44
Wermer, J., 53, 62
Weyl commutation relation, 22
Whitney, H., 215, 251
Wiener, N., 21
Wiener–Hopf operator, 180
Wiener subspace, 21, 58
Wilkie, H. C., vii
Wils, W., 295, 299, 300, 301, 302, 303
Wood, R., 186

Yale, K., 39, 62

Żelazko, W., 269, 270, 271
Żelazko, W. (with B. Mitjagin and S. Rolewicz), 216, 251